Face Method

Ping-Qi Pan

Face Method

Toward the New Generation of Linear Programming

Springer

Ping-Qi Pan
School of Mathematics
Southeast University
Nanjing, Jiangsu, China

ISBN 978-3-031-93593-0 ISBN 978-3-031-93594-7 (eBook)
https://doi.org/10.1007/978-3-031-93594-7

This Springer imprint is published by the registered company Springer Nature Switzerland AG
The registered company address is: Gewerbestrasse 11, 6330 Cham, Switzerland

Dedicated to parents, Chao Pan and Yinyun Yan and to wife Minghua Jiang, son Yunpeng, daughter-in-law Kong Wen, and granddaughter Heidi

Preface

Linear programming (LP) emerged when young American mathematician George B. Dantzig invented the simplex method in 1947, at the end of World War II. The goal is to optimize a multivariable linear function subject to linear equality/inequality constraints. Due to the extreme universality and fundamentality of the model, LP and the simplex method developed very rapidly. They spawned Optimization and OR/MS, such as nonlinear programming, integer programming, combinatorial Optimization, and the like.

In the past, several contributions related to LP have been awarded Nobel Prizes in Economics. Russian-American economist Wassily Leontief won the Nobel Prize in Economics in 1973 for his input-output model. Kantorovich and Koopmans shared the Nobel Prize in Economics in 1975 for their contributions to resource allocation theory based on LP. Later, K. Arrow, P. Samuelson, H. Simon, and L. Herwricz et al. won the Nobel Prize in Economics when they paid close attention to LP.

Since its birth, the simplex method has dominated the area of LP for nearly 80 years. Its applications have brought about enormous economic and social benefits. Today, the simplex method is considered one of the most well-known and widely used mathematical tool. Nevertheless, the simplex method has exponential time computational complexity [15], while a "nice" one is expected to be polynimial time. The method even cannot terminate forever because of cycling [1, 13]. Researchers have suggested cycling-free pivot rules. Unfortunately, they are far inferior to the Dantzig Rule in practice (e.g., [4, 20]).

The behavior of the simplex method stems from the following philosophy:

> Moving from vertex to vertex, in the underlying polyhedron, until achieving an optimal vertex.

Moreover, the simplex method even stays at a *degenerate* vertex for *too long* before exiting it. In addition, its stability has long been an overlooked problem: the selected pivot may be too close to zero, yielding an unusable basis. Occasionally, the simplex method performs unsatisfactorily in solving some highly degenerate and challenging LP problems.

1984 is the time node when Karmarkar published his interior-point algorithm, which goes across the interior of the feasible region [14]. It was not only of polynomial time complexity but also appeared fast in practice. It drew immediate attention from researchers worldwide, giving rise to an upsurge in the interior-point method. For about 20 years, researchers published a lot of papers in the subject. Some implementations were so efficient that many scholars considered the interior-point method the winner against the simplex method for solving large-scale and sparse LP problems.

Unlike the simplex method, however, the interior-point method can only *approach* an optimal solution on the boundary. Moreover, it cannot be "warmly" started; hence, it is not applicable for solving integer LP problems, which form the primary domain of LP applications. For this sake, the domination of the simplex method is still unshakable (Bixby [3]). Therefore, the upsurge yielded nothing or little compared with the vast amount of effort—just a wrong road! The author pointed out at the end of the article published then [21]:

> It is unwise to focus on *nonlinear* approaches for solving *linear* problems.

Today is an era of big data. The trend of globalization is fast and irresistible, and even hyper-scale LP models constantly spring up, calling for more powerful LP solvers ever.

> Where is the way out?

The preceding question has circled since we were involved in LP in the 1980s. We proposed and explored pivot rules, deficient-basis and projection methods, and so on, but failed to make fundamental progress. We summarized our findings in the monograph entitled Linear Programming Computation (in Chinese) [33], and the English monograph of the same title [34] (for a review, see Shi et al. [38]). We gradually realize that the problem with the simplex method is its rigid limitation of iteration to vertices. In such a setting, the difficulty of degeneracy is impossible to solve. We must find some suitable approach differing from the simplex and the interior-point methods. Given the optimal face covering the optimal vertices, extending the iteration to the face is natural, direct, and logical. Supported by auspicious numerical results, prototypes of the face method based on Cholesky factorization debuted in our monograph in 2012 and the monograph in 2014. Then, prototypes based on LU factorization came out in the second edition of Linear Programming Computation in 2023 [36]. At this point, the following passway is clear:

> Going from face to face in the underlying polyhedron until achieving an optimal face and solution.

After many years, it is time to integrate new developments of the face method with existing results for systematic, in-depth improvements and revision. Consequently, we gather all these findings to form this first monograph on the face method, putting forward the theme to the academic world.

First, we review the state of the art of the simplex method in Chap. 1, which is the source of the face method. In the following chapters, we present several versions of the face method, including their generalizations for solving variable-bounded LP problems. Although the design of face algorithms aims at real-world and large-scale problems, we offer small examples to help the reader understand the detailed solution process.

The contents of this book are as follows. In Chaps. 2 and 3, we present the face method and its generalization, respectively. Chapters 4 and 5 present the dual face method and its generalization. Chapter 6 first introduces the reduced face tableau and then derives the reduced face method. This method is of particular interest because of its potential application for handling integer LP problems. Chapter 7 is devoted to the dual reduced face method and its generalization. Chapters 8 and 9 present the Cholesky-factorization-based face and dual face methods.

Nanjing, China Ping-Qi Pan

Acknowledgments

The author profoundly cherishes the memory of Professor Xu-Chu He, the former mentor at Nanjing University aroused interest in Operations Research and Optimization. The author honors the father of linear programming and the simplex method, Professor George B. Dantzig, for his encouragement of the author's early research, which forms the origin of the face method. The author thanks Professor Michael Saunders of Stanford University for his valuable comments and suggestions. He selflessly offered the MINOS 5.51 package, from which the author's work benefited greatly. The author would like to thank Professor R. Tyrrell Rockafellar of the University of Washington, Professor Thomas F. Coleman of Cornell University, Professor Weide Zheng of Nanjing University, and Professor Yuan-ming Wang of Southeast University for their kind assistance. Thanks also go to the following colleagues: Professor James V. Burke of the University of Washington, Professor Leena Suhl of Paderborn University, Professor Uwe Suhl of Freie University, Professor Achim Koberstein of Europa-Universitat Viadrina Frankfurt (Oder), Professor Li-Zhi Liao of Hong Kong Baptist University, Professor Li-Qun Qi of Hong Kong Polytechnic University, Professor Xing Luh of National Zhengchi University, Professor Ren-Zhi Yao of National Sun Yat-sen University, Dr. Pablo Guerrero-Garcia of the University of Malaga, Dr. Yang-Yang Shi of Meta Platforms Inc., Professor Lei-Hong Zhang of Soochow University, Professor Wen-Xing Zhu of Fuzhou University, and Dr. Jia-Dong Wang of Lehigh University.

Competing Interests The author has no competing interests to declare that are relevant to the content of this manuscript.

Mountain View, CA, USA
March, 2025

Ping-Qi Pan

Contents

Notation

In this book, uppercase English letters denote matrices, lowercase letters denote vectors, and lowercase Greek letters denote reals. Unless indicated otherwise, all vectors are columns.

LP:	Linear programming.
$\mathcal{R}^n$:	Euclidean space of dimension n.
e_i:	Unit vector with the ith component 1.
e:	Vector of all ones.
I:	Unit matrix.
A:	Coefficient matrix of the equality constraint or the index set of columns of A.
m:	Number of rows of A.
n:	Number of columns of A.
b:	Right-hand side of the equality constraint.
c:	Cost vector, i.e., coefficient vector of the objective function.
B:	Basis matrix or the set of basic indices.
N:	Nonbasic columns of A or the set of nonbasic indices.
a_j:	The jth column of matrix A.
$a_{i,j}$:	The entry of the ith row and jth column of matrix A.
$\square^{\mathrm{T}}$:	Transpose of matrix $\square$.
$\square_J$:	Submatrix consisting of columns of matrix $\square$ indexed by set J.
v_j:	jth component of vector v.
v_I:	Subvector consisting of components of v indexed by row index set I.
$\|v\|$:	Euclidean norm of vector v.
X:	Diagonal matrix with components of x as diagonals.
$\nabla f(x)$:	Gradient of function $f(x)$.
$\Pi \subset \Gamma$:	Π is subset of Γ.
$\Pi \cup \Gamma$:	Union of Π and Γ.
$\Pi \cap \Gamma$:	Intersection of Π and Γ.
$\Pi \backslash \Gamma$:	Complementary set, i.e., $\{\tau \mid \tau \in \Pi, \tau \notin \Gamma\}$.

$\emptyset$:	Empty set.
$\|\tau\|$:	Absolute value of real τ or cardinality of set τ.
$Sign\ \tau$:	Sign of real τ.
range $\square$:	Column space of matrix $\square$.
null $\square$:	Null space of matrix $\square$.

There are the following two FORTRAN simplex codes involved:

MINOS: The sophisticated smooth optimization package developed by Bruce A. Murtagh and Michael A. Saunders at the Department of Management Science and Engineering of Stanford University [18]. Based on the simplex method, the LP option of the sparsity-exploiting code served as the benchmark and platform for our empirical evaluation of LP solvers.

RSA: Our working code without exploiting sparsity. It is based on the revised double-phase simplex method, using the Harris two-pass row rule.

Chapter 1
Simplex Method: The State of the Art

In this chapter, we review the simplex method for solving LP problems, which is the source of the face method. Indeed, it affords us many inspirations and lessons that merit attention.

We focus on the core of the simplex method even if there is rich content concerning. It comprises two parts: the first part includes the simplex algorithm, dual simplex algorithm, duality theorems, and optimality condition, and the second part involves our contributions supported by solid computational experiments, such as the double-phase simplex and dual simplex algorithms, the powerful N-M rule, the simplex interior-point algorithm, and the introduction of the optimal face, which is not only of significance in the simplex context but also serves as the source and foundation of the face method. In particular, the introduced "duality gaps" associated with indices bring about a basis of the pivot rules for the face method. We will give special attention to fundamental ideas and geometrical interpretations behind algorithms.

Most results in this chapter will be stated without proof. For details, refer to Pan [36].

1.1 Introduction

The LP problems are mathematically expressed as minimization/maximization of a linear function with multiple variables over a set of linear equalities/inequalities. Various such problems can be converted to the *standard* LP problem.

Let us consider the standard minimization problem below:

P.-Q. Pan, *Face Method*, https://doi.org/10.1007/978-3-031-93594-7_1

$$
\begin{aligned}
\min\ f = {} & c_1x_1 + c_2x_2 + \cdots + c_nx_n \\
\text{s.t.} \quad & a_{11}x_1 + a_{12}x_2 + \cdots + a_{1n}x_n = b_1 \\
& a_{21}x_1 + a_{22}x_2 + \cdots + a_{2n}x_n = b_2 \\
& \qquad \vdots \\
& a_{m1}x_1 + a_{m2}x_1 + \cdots + a_{mn}x_n = b_m \\
& x_j \geq 0, \ j = 1, \ldots, n.
\end{aligned} \tag{1.1}
$$

Using the following vector and matrix symbols:

$$A = (a_{ij}), \quad x = [x_1, \ldots, x_n]^{\mathrm{T}}, \quad c = [c_1, \ldots, c_n]^{\mathrm{T}}, \quad b = [b_1, \ldots, b_m]^{\mathrm{T}},$$

we put the standard problem (1.1) into the following concise form:

$$
\begin{aligned}
& \min\ f = c^{\mathrm{T}}x \\
& \text{s.t.}\ \ Ax = b, \qquad x \geq 0,
\end{aligned} \tag{1.2}
$$

where $A \in \mathcal{R}^{m\times n}$, $c \in \mathcal{R}^n$, $b \in \mathcal{R}^m$, $m < n$. It is assumed that Rank $(A) = m$, except indicated otherwise.

The $f = c^{\mathrm{T}}x$ is called *objective (function)*, and c is *cost*. The system $Ax = b$ is *equality constraint*, $x \geq 0$ *nonnegative constraint*.

A solution fulfilling all the constraints is a *feasible solution*, and the set of feasible solutions is the *feasible region*. A feasible solution with the minimum objective value over the feasible region is *optimal solution*. The minimization problem is said to be (lower) unbounded if its objective value in the feasible region can be arbitrarily small. If no feasible solution exists, the problem is said to be *infeasible*.

George B. Dantzig created the LP model and the simplex method in 1947, marking the birth of LP. It is the cornerstone of operations research, decision science, management science, and the like.

1.2 Canonical Form of Linear System

The linear system $Ax = b$ forms the main part of the constraints. This section shows how to convert it to its useful *canonical form*.

Linear systems are equivalent if they have the same solution set. There are the following two types of (equivalent) transformations.

Proposition 1.2.1 *A system, resulting from multiplying an equation of it by any nonzero, is equivalent to the original.*

Proposition 1.2.2 *A system, resulting from adding a multiple of any equation to another, is equivalent to the original.*

The preceding operations are called *elementary (row) transformations*, helpful in introducing zero entries into the system. In particular, a series of such transformations, termed *Gauss-Jordan elimination*, converts the constraint system $Ax = b$ of the problem (1.2) to a so-called canonical form, which exhibits the solution set or detects nonexistence of it.

The canonical form is closely related to the following concept, which plays a central role in the simplex methodology.

Definition 1.2.1 A nonsingular $m \times m$ submatrix of A is a basis.

Let (B, N) be a partition of A with the basis B such that

$$B = \{j_1, \ldots, j_m\}, \qquad N = A \backslash B = \{j_{m+1}, \ldots, j_n\},$$

where N is termed *nonbasis*

Instead of the $Ax = b$ itself, it is beneficial to perform the Gauss-Jordan elimination with the following tableau.

x_B x_N	rhs
B N	b

Assume that $a_{1\,j_1}$ is nonzero, which is called *pivot*. Multiply by $1/a_{1\,j_1}$ the first row to turn the pivot to 1; then add $-a_{i\,j_1}$ times of the first row to the ith row to turn the first entry of the ith row to 0 for $i = 2, \ldots, m$, and so on. Without loss of generality, assume that the m pivots were in the natural order $(1, j_1), (2, j_2), \ldots, (m, j_m)$, and zeros were introduced such that the basis B becomes the unit matrix, as shown in Tableau 1.2.1.

Canonical Tableau 1.2.1

x_{j_1}	x_{j_2}	$\cdots$	x_{j_m}	$x_{j_{m+1}}$	$\cdots$	x_{j_n}	rhs
1				$\bar{a}_{1\,j_{m+1}}$	$\cdots$	$\bar{a}_{1\,j_n}$	$\bar{b}_1$
	1			$\bar{a}_{2\,j_{m+1}}$	$\cdots$	$\bar{a}_{2\,j_n}$	$\bar{b}_2$
		$\ddots$		$\vdots$	$\vdots$	$\vdots$	$\vdots$
			1	$\bar{a}_{m\,j_{m+1}}$	$\cdots$	$\bar{a}_{m\,j_n}$	$\bar{b}_m$

Tableau 1.2.1 represents the following system:

$$\begin{cases} x_B = \bar{b} - \bar{N} x_N \\ x_N = x_N \end{cases}$$

which exhibits the set of solutions to $Ax = b$ if the nonbasics x_N on the right-hand side are viewed as independent variables.

In particular, the solution, yielding from setting nonbasics to zero, is termed *basic solution*. Thus, the canonical tableau renders the basic solution below:

$$\bar{x}_B = \bar{b}, \qquad \bar{x}_N = 0.$$

If, further, $\bar{b} \geq 0$, $\bar{x}$ is termed *feasible basic solution.*

Starting from a feasible basic solution, the simplex method generates a series of such solutions with decreasing objective value until achieving an optimal basic solution.

1.3 Geometric Survey of Optimality

In this section, we explore the optimality issue geometrically, such as the heuristic characterization of an optimal solution, vertex and optimal vertex, face and optimal face, and the like. We give some insight into existing LP solvers, including the simplex, interior-point, and face methods.

The feasible region is the set of solutions satisfying the constraints, i.e.,

$$P = \{x \in \mathcal{R}^n \mid Ax = b,\ x \geq 0\}. \tag{1.3}$$

As the intersection of a finite number of closed half-spaces, the P is a polyhedral convex set in $\mathcal{R}^n$. If bounded, it is a *polyhedron*, a compact set. It could degenerate into a segment or point or even an empty set.

There are only the following three cases arising:

(a) P is empty: there is no feasible solution to the problem.
(b) P is lower unbounded: the objective value can be arbitrarily small over it.
(c) An optimal solution exists on the boundary of P.

Let us bring up the following maximization problem with 2-dimensions:

$$\begin{array}{llr} \max\ f = 2x + 5y & & \\ \text{s.t.} & -2x - 3y \geq & -12 \\ & -4x - \ y \geq & -5 \\ & -y \geq & -3 \\ & x,\ y \geq & 0 \end{array} \tag{1.4}$$

In Fig. 1.1, the area of polygon OABCD represents the feasible region P. Our task is to determine a point over the area such that the objective value reaches the highest value.

In the figure, the equation $2x + 5y = 0$ of the isoline of the objection function corresponds to the straight line OE, going through the origin, all points corresponding to the same objective value 0. The line's slope, the tangent of the

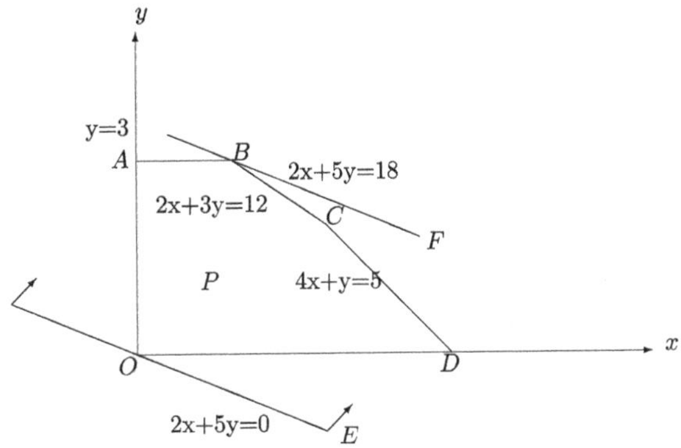

Fig. 1.1 Graphic solution to problem (1.4)

angle between it and the x axis, is $-2/5 = -0.4$. Therefore, the corresponding isoline parallel shifts to the upper-right side as the objective value increases from 0. The points on the intersection of the line and the area $OABCD$ are feasible solutions. We push parallel shifting to make the corresponding objective value as great as possible while maintaining the intersection nonempty. The farthest shifting line is BF though the vertex B. Thus, the B is the optimal convex, corresponding to

$$\bar{x} = 1.5, \qquad \bar{y} = 3$$

with optimal value $f = 18$, which is the solution to the system below:

$$2x + 3y = 12, \qquad y = 3.$$

Therefore, the key is to find those inequalities binding at the optimal vertex! With this respect, a careful observation leads to the following proposition (Pan [20]):

Proposition 1.3.1 (Heuristic Characteristic of Optimal Solution 1) *Gradients of the constraints binding at optimal solution tend to form the maximum angles with the negative objective gradient.*

As indicated, the preceding is *heuristic*. It is valuable to offer a clue toward optimality.

Formally, a point $x \in P$ is a *vertex* if it is not an interior point of any segment in P.

Lemma 1.3.1 *A point $\bar{x} \in P$ is a vertex if and only if it is a feasible basic solution.*

We state the so-called fundamental theorem below:

Theorem 1.3.2 *If there is an optimal solution, there is an optimal basic solution (vertex).*

The preceding theorem serves as the source of the simplex method, which goes from vertex to vertex to attain the optimal vertex. The canonical form of the system is a suitable algebraical tool for this purpose. Unfortunately, there is no one-to-one correspondence between vertex and basis. The so-called degeneracy would occur whenever more than $n-m$ components of the feasible basic solution are zero; hence, the current vertex corresponds to multiple bases. In this case, the basis change of the canonical form would not lead to any vertex change. Consequently, the solution process could pause at a degenerate vertex for too long before exiting it. Scholars even constructed some cycling instances that stay at a degenerate vertex forever!

To overcome this difficulty, several finite pivot rules were created to control the basis change more smartly but they performed unsatisfactorily (Bland [4], Pan [20]). In addition, although simplex variants were suggested from time to time, they failed to get out of trouble completely (for details, the reader is referred to Pan [36]).

The problem seems to be the strategy of the simplex method searching within vertices. In fact, there is no reason to limit iterate to vertex only, let alone there exist infinitely many nonvertex optimal solutions, in general.

The interior-point method culminated in the 1980s generates a series of interior points, which bypasses degenerate vertices. Since optimality is only attained on the boundary of the feasible region, the method can only approach optimality. Moreover, it cannot be "warmly" rebooted and hence not suitable for dealing with integer or mixed integer problems.

We will expand the category of iterate to the so-called face:

Definition 1.3.1 A set is called a *face* if whenever it includes an interior point of any segment of the feasible region P, it includes the segment.

The vertex of P can be viewed as a *zero*-dimensional face and the $(n-m)$-dimensional feasible region P itself is the maximum face. The edge of P is *one*-dimensional face, and the intersection of P and any coordinate plane is a *two*-dimensional face, in the *three*-dimensional space.

In general, the face can be defined algebraically.

Assume that $A = (B, N)$ is a partition, and $0 \leq k \leq n - m$ is an integer. Then, it is straightforward to show that the following subset of P is $(n-k)$-dimensional face:

$$P(B) = \{x \in \mathcal{R}^n \mid Bx_B = b, \;\; x_N = 0\}, \tag{1.5}$$

where $B \in \mathcal{R}^{m\times(n-k)}$ is termed *face matrix*, $N \in \mathcal{R}^{m\times k}$ is *nonface matrix*, and rank $B = m$.

The face $P(B)$ plays a central role in the face method. If, in particular, $k = n-m$, then the B is a typical basis used in the simplex method.

It will be clear later that there is an optimal face if an optimal solution exists. Instead of vertices, the face method moves from face to face until reaching an optimal face together with an optimal solution unless lower unboundedness is detected. Therefore, it would bypass the degeneracy pitfall and open up new prospects.

1.4 Simplex Algorithm: Tableau Form

In this section, we formulate the simplex algorithm in tableau form.

Assume B is basis, and N is nonbasis. Recall that the canonical form of the system $Ax = b$ helps determine basic solutions. We have to consider the objective function to solve the LP problem (1.2).

So, we insert the coefficients of the objective equation

$$c_B^T x_B + c_N^T x_N - f = 0$$

to the bottom of the canonical tableau 1.2.1 and do the according f column. Then, eliminating the entry c_B^T in the bottom line by elementary transformations using the diagonals of the unit matrix forms the following so-called simplex tableau:

Simplex Tableau 1.4.1

x_{j_1}	x_{j_2}	$\cdots$	x_{j_m}	$x_{j_{m+1}}$	$\cdots$	x_{j_n}	f	rhs
1				$\bar{a}_{1\ j_{m+1}}$	$\cdots$	$\bar{a}_{1\ j_n}$		$\bar{b}_1$
	1			$\bar{a}_{2\ j_{m+1}}$	$\cdots$	$\bar{a}_{2\ j_n}$		$\bar{b}_2$
		$\ddots$		$\vdots$	$\vdots$	$\vdots$		$\vdots$
			1	$\bar{a}_{m\ j_{m+1}}$	$\cdots$	$\bar{a}_{m\ j_n}$		$\bar{b}_m$
				$\bar{z}_{j_{m+1}}$	$\cdots$	$\bar{z}_{j_n}$	-1	$-\bar{f}$

where the bottom line is termed *objective row* and $\bar{z}_N$ termed *reduced costs/gradient*. Note that the ith basic variable x_{j_i} is at the ith row of the tableau $(i = 1, \ldots, m)$.

Thereafter, we will not distinguish between the simplex tableau 1.4.1 and its vector/matrix form below:

x_B	x_N	f	rhs
I	$\bar{N}$		$\bar{b}$
	$\bar{z}_N^T$	-1	$-\bar{f}$

(1.6)

The simplex tableau is a nice tool to illustrate the simplex method, as it gives the following basic solution:

$$\bar{x}_B = \bar{b}, \qquad \bar{x}_N = 0. \tag{1.7}$$

From the bottom row of the simplex tableau, it is evident that the solution $\bar{x}$ corresponds to the objective value $\bar{f}$.

Assume that $\bar{b} \geq 0$, and hence the $\bar{x}$ is a feasible basic solution.

We make the following claims without proof.

Lemma 1.4.1 *If the reduced costs are nonnegative, the feasible simplex tableau is optimal, giving an optimal basic solution.*

Now, assume that for some $q \in N$ it holds that

$$\bar{z}_q < 0. \tag{1.8}$$

Introduce the following row index set:

$$I = \{i = 1, \ldots, m \mid \bar{a}_{i,q} > 0\}. \tag{1.9}$$

Lemma 1.4.2 *Assume $\bar{z}_q < 0$. If the set I is empty, the LP problem is lower unbounded.*

Proof Defining the n-dimensional vector Δx such that

$$\Delta x_B = \bar{a}_q, \quad \Delta x_q = -1; \quad \Delta x_j = 0, \quad j \in N,\ j \neq q. \tag{1.10}$$

Consider the following update formula:

$$\hat{x} = \bar{x} - \alpha \Delta x, \tag{1.11}$$

where α is termed *stepsize*. The new solution $\hat{x}$ satisfies the equality constraint for any α. If the set I is empty, then $\hat{x}$ is nonnegative and feasible for all $\alpha \geq 0$.

From the bottom row of the simplex tableau, it is clear that the corresponding new objective value is

$$\hat{f} = \bar{f} + \alpha \bar{z}_q,$$

which along with (1.8) leads to

$$\hat{f} \leq \bar{f}.$$

Thus, the objective value decreases strictly for any given $\alpha > 0$. Consequently, we have

$$\hat{f} \longrightarrow -\infty \quad \text{as} \quad \alpha \longrightarrow \infty.$$

Therefore, the problem is lower unbounded in the feasible region. □

The $-\Delta x$ is termed *descent search direction*.

If the set I is nonempty, then the stepsize α is limited to maintain the nonnegativity of $\hat{x}$. The maximum possible α and the associated row index is determined by the following rule.

Rule 1.4.1 (Row Pivot Rule) Determine a row index p and stepsize α such that

$$\alpha = \bar{b}_p/\bar{a}_{pq} = \min_{j \in I} \bar{b}_i/\bar{a}_{i\,q} \geq 0. \tag{1.12}$$

Taking $\bar{a}_{pq}$ as pivot, we transform the simplex tableau (1.6) by multiplying the pth row by $1/\bar{a}_{pq}$, then add $-\bar{a}_{i\,q}$ times of pth row to the ith row ($i = 1, \ldots, m,\ i \neq p$), and $-\bar{z}_q$ times of the row to the bottom row. The iteration finishes with (B, N) adjusted by exchanging j_p and q.

It is obvious that the resulting simplex tableau corresponds to the new feasible basic solution $\hat{x}$. However, if α vanishes, the basic solution remains unchanged even if the basis changes.

Definition 1.4.1 If $\bar{b}_i = 0$ holds for some $j \in I$, the feasible basic solution is degenerate.

If the current solution $\bar{x}$ degenerates, no objective decrease is made.

Since any index $j \in N$ corresponding to negative $\bar{z}_j$ is eligible, in general, there are multiple choices for entering index q.

Originally, Dantzig used the following rule to select the pivot column [5].

Rule 1.4.2 (Column Rule) Select a column index q such that

$$q \in \arg\min_{j \in N} \bar{z}_j. \tag{1.13}$$

So, the rule selects the column with the most negative reduced cost as the pivot column. Obviously, for a unit increment of the nonbasic variable x_q, the rule leads to the maximum decrease in the objective value.

Thus, the simplex method generates a series of equivalent simplex tableaus to achieve optimality, as is formulated as follows (Dantzig [5, 6]).

Algorithm 1.4.1 (Simplex Algorithm: Tableau Form)
Initial: a feasible simplex tableau of form (1.6). This algorithm solves the standard LP problem (1.2).

1. *Determine the column index* $q \in \arg\min_{j \in N} \bar{z}_j$.
2. *Stop if* $\bar{z}_q \geq 0$.
3. *Stop if* $I = \{i = 1, \ldots, m \mid \bar{a}_{i\,q} > 0\} = \emptyset$.
4. *Determine a pivot row index* $p \in \arg\min_{i \in I} \bar{b}_i/\bar{a}_q$.
5. *Convert* $\bar{a}_{pq}$ *to 1, and eliminate the other nonzeros in the column by elementary transformations.*
6. *Go to Step 1.*

We make the following claim without proof.

Theorem 1.4.3 *Under the nondegeneracy assumption, the simplex algorithm terminates either at:*

(i) Step 2, generating an optimal basic solution.
(ii) Step 3 detects the lower unboundedness of the problem.

The preceding simplex algorithm is described at a scalar level. The simplex tableau used is computed as a whole in each iteration. However, only a part of it is applied, that is, the reduced costs $\bar{z}_N$, the right-hand side $\bar{b}$, and the pivot column $\bar{a}_q$, including the pivot $\bar{a}_{pq}$ in particular. Therefore, a more compact form of the algorithm is needed for practical use. We will revise it in the next section.

1.5 Simplex Algorithm

In this section, we derive another more compact and practicable version of the simplex method, *revised simplex algorithm*. We first introduce the so-called revised simplex tableau formed with the original data and then revise the simplex algorithm based on the equivalence of the two tableaus.

To this end, premultiplying the system $Ax = b$ by B^{-1} with transposition of terms, we have the following system:

$$x_B + B^{-1}Nx_N = B^{-1}b,$$

substituting which to the objective equation

$$c_B^{\mathrm{T}}x_B + c_N^{\mathrm{T}}x_N - f = 0$$

gives

$$(c_N^{\mathrm{T}} - c_B^{\mathrm{T}}B^{-1}N)x_N - f = -c_B^{\mathrm{T}}B^{-1}b.$$

The following tableau can represent the preceding two equations:

x_B	x_N	f	rhs
I	$B^{-1}N$		$B^{-1}b$
	$c_N^{\mathrm{T}} - c_B^{\mathrm{T}}B^{-1}N$	-1	$-c_B^{\mathrm{T}}B^{-1}b$

(1.14)

which is associated with the basic solution

$$\bar{x}_B = B^{-1}b, \qquad \bar{x}_N = 0.$$

The validity of the following claim is apparent.

Proposition 1.5.1 *Any simplex tableau and revised simplex tableau associated with the same basis are equivalent.*

The preceding means that the corresponding entries of the two tableaus are equal.

Based on the equivalence, we may only compute the necessary items from the original data provided the inverse B^{-1} is available, as is shown as follows:

(a) The $\bar{x}_B$ equals the right-hand side:

$$\bar{x}_B = B^{-1}b. \tag{1.15}$$

We will update it in the solution process.

(b) The so-called simplex multiplier $\bar{y}$ and reduced costs $\bar{z}_N$:

$$\bar{y} = B^{-T}c_B, \qquad \bar{z}_N = c_N - N^T\bar{y}, \tag{1.16}$$

where calculating of $\bar{z}_N$ is referred to as *pricing*.

(c) The pivot column:

$$\bar{a}_q = B^{-1}a_q. \tag{1.17}$$

Let us derive the update of the inverse B^{-1}.

Assume that column index q and row index p are already determined. Replacing the pth column of B with nonbasic column a_q gives the new basis below:

$$\widehat{B} = (a_{j_1}, \dots, a_{j_{p-1}}, a_q, a_{j_{p+1}}, \dots, a_{j_m}). \tag{1.18}$$

Then, compute $\widehat{B}^{-1}$ as follows.

Taking $\bar{a}_{pq}$ as the pivot, the associated elementary transformations amount to premultiplying the first m rows of the tableau by the following $m \times m$ elementary matrix:

$$E_p = \begin{bmatrix} 1 & & -\bar{a}_{1q}/\bar{a}_{pq} & & \\ & \ddots & \vdots & & \\ & & -\bar{a}_{p-1,q}/\bar{a}_{pq} & & \\ & & 1/\bar{a}_{pq} & & \\ & & -\bar{a}_{p+1,q}/\bar{a}_{pq} & & \\ & & \vdots & \ddots & \\ & & -\bar{a}_{mq}/\bar{a}_{pq} & & 1 \end{bmatrix} \begin{matrix} \\ \\ \\ p \\ \\ \\ \\ \end{matrix} \tag{1.19}$$

$$p$$

which may be obtained by executing the same elementary transformations on the unit matrix. It is the same as the unit matrix except for the pth column determined by $\bar{a}_q$ only.

Then, combining (1.18) and (1.19) gives

$$\begin{aligned} E_pB^{-1}\widehat{B} &= E_p(B^{-1}a_{j_1}, \dots, B^{-1}a_{j_{p-1}}, B^{-1}a_q, B^{-1}a_{j_{p+1}}, \dots, B^{-1}a_{j_m}) \\ &= E_p(e_1, \dots, e_{p-1}, \bar{a}_q, e_{p+1}, \dots, e_m) \end{aligned}$$

$$= (E_pe_1, \ldots, E_pe_{p-1}, E_p\bar{a}_q, E_pe_{p+1}, \ldots, E_pe_m) = I,$$

from which the recursion formula follows, i.e.,

$$\widehat{B}^{-1} = E_pB^{-1}. \tag{1.20}$$

Based on the preceding discussions, it is straightforward to transform Algorithm 1.4.1 to the following revision (Dantzig and Orchard-Hays [7]):

Algorithm 1.5.1 (Simplex Algorithm I)
Initial: (B, N), $B^{-1}, \bar{x}_B \geq 0$ *and* $\bar{f}$. *This algorithm solves the standard LP problem (1.2).*

1. *Compute* $\bar{y}$, $\bar{z}_N$ *by (1.16).*
2. *Determine column index* $q \in \arg\min_{j\in N} \bar{z}_j$.
3. *Stop if* $\bar{z}_q \geq 0$. *(optimality achieved)*
4. *Compute pivot column* $\bar{a}_q = B^{-1}a_q$ *(1.17).*
5. *Stop if* $I = \{i = 1, \ldots, m \mid \bar{a}_{i\,q} > 0\}$ *is empty. (lower unbounded problem)*
6. *Determine* α *and row index p such that* $\alpha = \bar{x}_{j_p}/\bar{a}_{p\,q} = \min_{i\in I} \bar{x}_{j_i}/\bar{a}_{i\,q}$.
7. *If* $\alpha \neq 0$, *update* $\bar{x}_B = \bar{x}_B - \alpha\bar{a}_q,\ \bar{x}_q = \alpha,\ \bar{f} = \bar{f} + \alpha\bar{z}_q$.
8. *Update* B^{-1} *by (1.20).*
9. *Adjust* (B, N) *by exchanging* j_p *and* q.
10. *Go to Step 1.*

It is possible to update the reduced costs to get another version of the simplex algorithm. For this purpose, compute the pivot row w_N by

$$v = B^{-T}e_p, \qquad w_N = N^{\mathrm{T}}v, \tag{1.21}$$

Algorithm 1.5.2 (Simplex Algorithm II)
Initial: (B, N), B^{-1}, $\bar{x}_B \geq 0$, $\bar{z}_N$ *and* $\bar{f}$. *This algorithm solves the standard LP problem (1.2).*

1. *Determine column index* $q \in \arg\min_{j\in N} \bar{z}_j$.
2. *Stop if* $\bar{z}_q \geq 0$. *(optimality achieved)*
3. *Compute* $\bar{a}_q = B^{-1}a_q$ *(1.17).*
4. *Stop if* $I = \{i = 1, \ldots, m \mid \bar{a}_{i\,q} > 0\}$ *is empty. (lower unbounded problem)*
5. *Determine* α *and row index p such that* $\alpha = \bar{x}_{j_p}/\bar{a}_{p\,q} = \min_{i\in I} \bar{x}_{j_i}/\bar{a}_{i\,q}$.
6. *If* $\alpha \neq 0$, *update* $\bar{x}_B = \bar{x}_B - \alpha\bar{a}_q,\ \bar{x}_q = \alpha,\ \bar{f} = \bar{f} + \alpha\bar{z}_q$.
7. *Compute pivot row* w_N *by (1.21).*
8. *Update by:* $\bar{z}_N = \bar{z}_N + \beta w_N,\ \bar{z}_{j_p} = \beta$, *where* $\beta = -\bar{z}_q/\bar{a}_{p\,q}$.
9. *Update* B^{-1} *by (1.20).*
10. *Update* (B, N) *by exchanging* j_p *and* q.
11. *Go to Step 1.*

A variety of variants of the simplex algorithm exist.

On the one hand, the rule for selection of the column index q has infinitely many choices since any index $q \in N$ is eligible if $\bar{z}_q < 0$.

On the other hand, the algorithm's structure is improvable.

Firstly, although the inverse B^{-1} is used in the algorithm explicitly, it is not suitable for practice since the inverse can be very dense even if the basis itself is sparse (Orchard-Hays [19]).

Formulating the revised simplex algorithm by updating LU factors of the basis instead of its inverse is not difficult. If the LU factorization $B = LU$ is available, the computation of (1.17) can be replaced by solving the following two triangular systems successively:

$$Lu = a_q, \qquad U\bar{a}_q = u.$$

Moreover, the computation of (1.16) can be replaced by solving the two triangular systems below:

$$U^T v = c_B, \qquad L^T \bar{y} = v.$$

Because of its suitability for handling large-scale sparse LP problems, LU factorization is widely applied in modern simplex codes, like the optimization package MINOS, developed by Professors Murtagh and Saunders of Stanford University [18].

1.6 Double-Phase Simplex Algorithm

The simplex algorithms, stated in the previous section, must start from a feasible basic solution. Since no such a solution is available in general, a so-called phase I procedure is required to get it started. Therefore, one of the simplex algorithms should be taken as the phase II.

Several phase I procedures are recommended (Chapter 14, Pan [36]). In this section, we pick the simplest one using the *most-obtuse-angle column rule*, in tableau form and revised form.

Let (1.6) be the current simplex tableau. At first, execute the following rule.

Rule 1.6.1 (Row Rule) Select the row index p such that

$$p \in \arg\min\{\bar{b}_i \mid i = 1, \ldots, m\}. \tag{1.22}$$

If $\bar{b}_p \geq 0$, then the basic solution is feasible.

In the other case, execute the following column rule, which has a geometric interpretation showing its plausibility (Section 14.4, Pan [36]),

Rule 1.6.2 (Most-Obtuse-Angle Column Rule) Determine the column index

$$q \in \arg\min_{j \in N} \bar{a}_{pj}. \tag{1.23}$$

In case of a tie, select that corresponding to the minimum reduced cost.

It is straightforward to show that the problem is infeasible if $\bar{a}_{pq} \geq 0$.

Once (p, q) is determined with $\bar{a}_{pq} < 0$, we perform the preliminary transformation to update the simplex tableau and complete the iteration. The process is repeated until the right-hand side becomes nonnegative or the infeasibility of the problem is detected.

Associated steps are included in the following procedure.

Subalgorithm 1.6.1 (Tableau Phase I: Most-Obtuse-Angle Column Rule)
Initial: simplex tableau. This algorithm seeks a feasible basic solution.

1. *Select row index* $p \in \arg\min\{\bar{b}_i \mid i = 1, \ldots, m\}$.
2. *Return if* $\bar{b}_p \geq 0$.
3. *Determine column index q by Rule 1.6.2.*
4. *Return if* $\bar{a}_{pq} \geq 0$.
5. *Perform elementary transformations to convert* $\bar{a}_{pq}$ *to* 1, *and eliminate all other nonzeros of the column.*
6. *Go to Step 1.*

Theorem 1.6.1 *Assume termination of Subalgorithm 1.6.1. It terminates either at:*

(i) Step 2, obtaining a basic feasible solution
(ii) Step 4, detecting infeasibility of the problem

The termination of the preceding subalgorithm is not ensured. In fact, Guerrero-Garcia and Santos-Palomo [11] provided an artificial cycling instance for it. Nevertheless, it can still be expected that cycling hardly happens in practice.

Let us turn to the revision of Subalgorithm 1.6.1.

Let $A = (B, N)$ be a partition, and let factorization $B = LU$ be available. The LU factors are updated in the solution process (see Sections 4.4 and 4.5, Pan [36]).

The following quantities are to be computed from the original data.

(i) The nonbasic part of the pth row v_N^T.

After the row index p is determined, solve the following two triangular systems successively:

$$U^T w = e_p, \qquad L^T h = w. \tag{1.24}$$

Then, the needed part of pth row is computed by

$$v_N = N^{\mathrm{T}} h. \tag{1.25}$$

(ii) The qth column $\bar{a}_q$.

Once the pivot column index q is available, solve the following two triangular systems to obtain the qth column:

$$Lu = a_q, \qquad U\bar{a}_q = u \tag{1.26}$$

(iii) Updating $\bar{x}_B$ once both $\bar{a}_q$ and v_N are available.
Using $\alpha = \bar{x}_{j_p}/v_q$, update

$$\bar{x}_B = \bar{x}_B - \alpha\bar{a}_q, \qquad \bar{x}_q = \alpha. \tag{1.27}$$

Initially, solve the following two triangular systems for $\bar{x}_B$:

$$Lu = b, \qquad U\bar{x}_B = u. \tag{1.28}$$

Note that $\bar{x}_N$ is always 0 in the solution process.

Then, it is straightforward to obtain the following revised form.

Subalgorithm 1.6.2 (Phase I: Most-Obtuse-Angle Column Rule)
Initial: (B, N), $B = LU$, $\bar{x}_B$. *This algorithm seeks a feasible basic solution.*

1. *Select pivot row index* $p \in \arg\min\{\bar{x}_{j_i} \mid i = 1, \ldots, m\}$.
2. *Return if* $\bar{x}_{j_p} \geq 0$ *(feasibility achieved).*
3. *Solve two triangular systems (1.24), and compute* v_N *by (1.25).*
4. *Determine column index* $q \in \arg\min_{j\in N} v_j$.
5. *Return if* $v_q \geq 0$ *(infeasible problem).*
6. *Solve two triangular systems (1.26) for* $\bar{a}_q$.
7. *Update* $\bar{x}_B$ *and* $\bar{x}_q$ *by (1.27).*
8. *Update* $B = LU$.
9. *Adjust* (B, N) *by exchanging* j_p *and* q.
10. *Go to Step 1.*

Note: Taking the row rule in Step 1 is only for composition. In practice, it should be much better to use some "dual N-M row rule" (see the last paragraph of Sect. 1.12.3).

Any of the simplex algorithms formulated in the previous section supported by a phase I procedure is called *double-phase simplex algorithm.*

Example 1.6.1 Solve the following standard problem by the double-phase simplex algorithm:

$$A = \begin{bmatrix} 1\,0\,0\,0 & 4 & -6 & 0 & 5 \\ 0\,1\,0\,0 & 2/3 & -11 & 8 & -2 \\ 0\,0\,1\,0 & -6 & 3 & -1/2 & 7 \\ 0\,0\,0\,1 & -2 & 0 & 5 & -4 \end{bmatrix} \qquad b = \begin{bmatrix} -1/2 \\ -12 \\ 10 \\ -11/3 \end{bmatrix}$$

$$c = [0, 0, 0, 0, 2, 7, 8, -6]^T$$

Inserting the objective coefficients as the bottom line, we have the initial feasible simplex tableau below (f column is omitted):

x_1	x_2	x_3	x_4	x_5	x_6	x_7	x_8	rhs
1				4	−6		5	−1/2
	1			2/3	−11	8	−2	−12
		1		−6	3	−1/2	7	10
			1	−2		5	−4	−11/3
				2	7	8	−6	

Phase-1: Call Subalgorithm 1.6.1

Iteration 1:

1. $\min\,[-1/2, -12, 10, -11/3] = -12,\ \ p = 2,\ j_p = 2.$
2. $\bar{b}_2 = -12 < 0.$
3. $\min\,[2/3, -11, 8, -2] = -11, q = 6.$
4. $\bar{a}_{2,6} = -11 < 0.$
5. Multiply row 2 by $-1/11$, then add $6, -3, -7$ times of row 2 to rows 1,3,5:

x_1	x_2	x_3	x_4	x_5	x_6	x_7	x_8	rhs
1	−6/11			40/11		−48/11	67/11	133/22
	−1/11			−2/33	1	−8/11	2/11	12/11
	3/11	1		−64/11		37/22	71/11	74/11
			1	−2		5	−4	−11/3
	7/11			80/33		144/11	−80/11	−84/11

Iteration 2:

1. $\min\,[133/22, 12/11, 74/11, -11/3] = -11/3,\ \ p = 4,\ j_p = 4.$
2. $\bar{b}_4 = -11/3 < 0.$
3. $\min\,[0, -2, 5, -4] = -4, q = 8.$
4. $\bar{a}_{2,6} = -11 < 0.$
5. Multiply row 4 by -4, then add $-67/11, -2/11, -71/11, 80/11$ times of row 4 to rows 1,2,3,5:

x_1	x_2	x_3	x_4	x_5	x_6	x_7	x_8	rhs
1	−6/11		67/44	13/22		13/4		61/132
	−1/11		1/22	−5/33	1	−1/2		61/66
	3/11	1	71/44	−199/22		39/4		107/132
			−1/4	1/2		−5/4	1	11/12
	7/11		−20/11	200/33		4		−32/33

The right-hand side is nonnegative, feasibility is achieved.

Phase II: Call Algorithm 1.4.1.

Iteration 3:

1. $\min[7/11, -20/11, 200/33, 4] = -20/11,\ q = 4$.
2. $\bar{z}_4 < 0$.
3. $I = \{1, 2, 3\} \neq \emptyset$.
4. $\min[(61/132)/(67/44), (61/66)/(1/22), (107/132)/(71/44)] = 61/201,\ p = 1$.
5. Multiply row 1 by $1/(67/44)$, then add $-1/22, -71/44, 1/4, 20/11$ times of rows 1 to rows 2,3,4,5:

x_1	x_2	x_3	x_4	x_5	x_6	x_7	x_8	rhs
44/67	−24/67		1	26/67		143/67		61/201
−2/67	−5/67			−34/201	1	−40/67		61/67
−71/67	57/67	1		−648/67		845/134		43/134
11/67	−6/67			40/67		−48/67	1	133/134
80/67	−1/67			1360/201		528/67		−28/67

Iteration 4:

1. $\min[80/67, -1/67, 1360/201, 528/67] = -1/67,\ q = 2$.
2. $\bar{z}_2 < 0$.
3. $I = \{3\} \neq \emptyset$.
4. $\min[(43/134)/(57/67)],\ p = 3$.
5. Multiply row 1 by $1/(57/67)$, then add $24/67, 5/67, 6/67, 1/67$ times of row 3 to rows 1,2,4,5:

x_1	x_2	x_3	x_4	x_5	x_6	x_7	x_8	rhs
4/19		8/19	1	−70/19		91/19		25/57
−7/57		5/57		−58/57	1	−5/114		107/114
−71/57	1	67/57		−216/19		845/114		43/114
1/19		2/19		−8/19		−1/19	1	39/38
67/57		1/57		376/57		911/114		−47/114

Now, the reduced costs are all nonnegative; hence, optimality is achieved. The optimal basic solution and objective value are:

$$\bar{x} = (0, 43/114, 0, 25/57, 0, 107/114, 0, 39/38)^{\mathrm{T}}, \qquad \bar{f} = 47/114.$$

1.7 Duality and Duality Theorem

Duality is an essential part of LP theory. This section introduces the dual problem of an LP problem and duality theorems, which reveals the profound relationship between the two problems.

Using the data of the standard LP problem (1.2), we can construct another LP problem as follows:

$$\begin{aligned} &\max\ g = b^{\mathrm{T}} y \\ &\text{s.t. } A^{\mathrm{T}} y + z = c, \qquad z \geq 0. \end{aligned} \tag{1.29}$$

which is called *dual problem* of the standard problem. Accordingly, the latter is called *primal problem*.

The set

$$D = \{(y, z) \in \mathcal{R}^m \times \mathcal{R}^n \mid A^{\mathrm{T}} y + z = c,\ z \geq 0\}$$

is defined as *dual feasible region*. Any solution in D is a *dual feasible solution*.

For a given basis B, setting $z_B = 0$ in $B^{\mathrm{T}} y + z_B = c_B$ gives

$$\bar{y} = B^{-T} c_B, \qquad \bar{z}_B = 0, \qquad \bar{z}_N = c_N - N^{\mathrm{T}} \bar{y}, \tag{1.30}$$

which is called *dual basic solution*. If $\bar{z}_N \geq 0$, the $(\bar{y}, \bar{z})$ is a *dual feasible basic solution*. Clearly, $(\bar{y} = 0,\ \bar{z} = c)$ is a dual feasible solution if $c \geq 0$.

Notably, the dual solutions play special roles in the primal simplex context: the $\bar{y}$ is just the simplex multiplier, and $\bar{z}_N$ is the reduced costs.

The following claim is verified.

Theorem 1.7.1 (Symmetry) *The dual problem of the dual problem is the primal problem.*

Lemma 1.7.2 *If x and (y, z) are primal and dual solutions, it holds that*

$$c^T x - b^T y = x^T z. \tag{1.31}$$

Proof The condition of the lemma means that

$$Ax = b, \qquad A^T y + z = c. \tag{1.32}$$

Premultiplying by x^T the second equation of (1.32) gives

$$x^T A^T y + x^T z = x^T c. \tag{1.33}$$

On the other hand, the first equation of (1.32) is equivalent to

$$x^T A^T = b^T.$$

Substituting the preceding into (1.33) leads to

$$b^T y + x^T z = c^T x,$$

which is equivalent to (1.31). Therefore, the claim is valid. □

Theorem 1.7.3 (Weak Duality) *If x and (y, z) are feasible primal and dual solutions, it holds that*

$$c^T x \geq b^T y. \tag{1.34}$$

Proof The condition of the theorem implies the satisfaction of (1.31). Since they are feasible, and hence the x and z are nonnegative, it holds that

$$x^T z \geq 0,$$

from which and (1.31), the inequality (1.34) follows. □

Corollary 1.7.4 *If one of the primal and dual problems is unbounded, no feasible solution exists to the other.*

Definition 1.7.1 The difference between the primal and dual objective values, i.e., $c^{\mathrm{T}} x - b^{\mathrm{T}} y$, is defined as duality gap of the two solutions.

Theorem 1.7.5 (Strong Duality) *If an optimal solution exists to one of the primal and dual problems, an optimal solution exists to the other; the duality gap of the two optimal solutions equals zero.*

Theorem 1.7.6 *If the duality gap of primal and dual feasible solutions equals zero, they are optimal solutions.*

Since it can be converted to the standard form, any LP problem has its dual problem.

There is a 1-1 correspondence between variables/constraints of the primal and dual LP problems, as summarized in the following table:

Primal problem		Dual problem	
Objective function	Min	Objective function	Max
Variables	Nonnegative	Constraints	$\leq$
	Nonpositive		$\geq$
	Free		$=$
Constraints	$\geq$	Variables	Nonnegative
	$\leq$		Nonpositive
	$=$		Free

In practice, the following so-called *variable-bounded* LP model is common:

$$\begin{aligned} &\min\ f = c^{\mathrm{T}}x \\ &\text{s.t.}\ \ Ax = b, \quad l \le x \le u. \end{aligned} \tag{1.35}$$

To derive its dual problem, put it into the following standard form:

$$\begin{aligned} &\min\ c^{\mathrm{T}}x \\ &\text{s.t.}\ \ Ax = b, \\ &\qquad x + s = u, \\ &\qquad -x + t = -l, \\ &\qquad s, t \ge 0. \end{aligned} \tag{1.36}$$

According to the preceding table, the dual problem of (1.35) can be written

$$\begin{aligned} &\max\ b^{\mathrm{T}}y - u^{\mathrm{T}}v + l^{\mathrm{T}}w \\ &\text{s.t.}\ \ A^{\mathrm{T}}y - v + w = c, \quad v, w \ge 0. \end{aligned} \tag{1.37}$$

Clearly, the duality results related to the standard problem, such as the symmetry, weak, and strong duality theorems, are also valid for the variable-bounded problem.

Not surprisingly, an LP method usually solves the pair of primal and dual problems at the same time; e.g., once the simplex algorithm achieves the primal optimal solution

$$\bar{x}_B = B^{-1}b, \qquad \bar{x}_N = 0,$$

it also gives the dual optimal solution below:

$$\bar{y} = B^{-T}c_B, \qquad \bar{z}_B = 0, \quad \bar{z}_N = c_N - N^{\mathrm{T}}\bar{y}.$$

1.8 Optimality Condition

From the duality results in the previous section, it is not difficult to draw optimality condition, which serves as a theoretical basis for developing LP solvers.

At first, we consider the standard LP problem (1.2). Assume that x and (y, z) are primal and dual feasible solutions.

The quantity $x^{\mathrm{T}}z$ is called *complementarity residual* of the two solutions. If the residual is zero, the primal and dual solutions are said to be *complementary*.

By Lemma 1.7.2, the complementarity residual is equivalent to the duality gap. Therefore, the following claim is valid.

Lemma 1.8.1 *The duality gap is zero if and only if the primal and dual solutions are complementary.*

If the primal and dual solutions are feasible, the size of the complementarity residual exhibits how much the solutions approach optimality.

We summarize the optimal condition to the following theorem.

Theorem 1.8.2 (Optimality Condition for the Standard LP Problem) *The solution x is optimal to the standard problem (1.2) if and only if there exist y, z such that*

$$\begin{array}{lll} (i)\ \ Ax = b & x \geq 0 & \text{(primal feasible)} \\ (ii)\ \ A^{\mathrm{T}}y + z = c & z \geq 0 & \text{(dual feasible)} \\ (iii)\ x^{\mathrm{T}}z = 0 & & \text{(complementary)} \end{array} \tag{1.38}$$

For the first time, we introduce the following new concept, in which the feasibility requirement of the primal and dual solutions is dropped to widen the scope of applications.

Definition 1.8.1 The duality gaps associated with indices are defined as

$$\delta_j = \begin{cases} x_j z_j \text{ if } x_j, z_j \geq 0 \\ +\infty \text{ otherwise} \end{cases} \qquad j = 1, \ldots, n.$$

Proposition 1.8.1 *If x and (y, z) are primal and dual feasible solutions, it holds that*

$$c^{\mathrm{T}}x - b^{\mathrm{T}}y = \sum_{j=1,\ldots,n} \delta_j. \tag{1.39}$$

Proof It follows from Lemma 1.7.2 and $x^T z = \sum_{j=1,\ldots,n} \delta_j$, □

Therefore, the duality gap δ_j with a single index j exhibits the portion in the total duality gap. It reflects the improvement in objective value that the pair (x_j, z_j) can provide, potentially. Because of this feature, the duality gaps with indices play an important role in developing face methods

Now, we turn to the variable-bounded LP problem.

Theorem 1.8.3 (Optimal Condition for the Variable-Bounded Problem) *The solution x is optimal to the problem (1.35) if and only if there exist y, v, w such that*

$$\begin{array}{lll} Ax = b & l \leq x \leq u & \text{(primal feasible)} \\ A^{\mathrm{T}}y - v + w - c & v, w, \geq 0 & \text{(dual feasible)} \\ (x - l)^{\mathrm{T}}w = 0 & (u - x)^{\mathrm{T}}v = 0 & \text{(complementary)} \end{array} \tag{1.40}$$

The previous discussions with the standard problem can be transmitted to the variable-bounded problem. We only bring up the following helpful concept.

Definition 1.8.2 Assume that x and z are primal and dual solutions, respectively. The duality gaps with indices are defined as

$$.\delta_j = \begin{cases} (u_j - \bar{x}_j)\bar{z}_j & \text{if } \bar{z}_j < 0 \\ (l_j - \bar{x}_j)\bar{z}_j & \text{if } \bar{z}_j > 0 \\ 0 & \text{if } \bar{z}_j = 0 \end{cases} \qquad j = 1, \ldots, n, \tag{1.41}$$

1.9 Dual Simplex Algorithm

The dual simplex algorithm is famous because it is an efficient LP solver and has long been the primary tool for solving integer/mixed LP problems. In this section, we derive its tableau form first and then its revised version.

Starting from a feasible simplex tableau with the nonnegative right-hand side, the simplex algorithm updates it, iteration by iteration, reducing the corresponding objective value while maintaining the nonnegativity of the right-hand side until the bottom row (reduced costs) becomes nonnegative, and hence optimality is achieved.

Our goal is to gain an optimal simplex tableau, having a nonnegative right-hand side and a nonnegative bottom row, no matter how the tableau came about.

Contrary to the simplex algorithm, therefore, one may start from a tableau having a nonnegative bottom row, and update it in each iteration to increase the associated objective value (as will be clear later) while maintaining the nonnegativity of the bottom row until the right-hand side becomes nonnegative. This idea leads to the so-called dual simplex algorithm.

This approach solves the dual problem, as explained below.

Consider the dual problem represented by the simplex tableau (1.6), i.e.,

$$\begin{array}{ll} \max & g = \bar{b}^{\mathrm{T}} y \\ \text{s.t.} & \begin{bmatrix} I \\ \bar{N}^T \end{bmatrix} y + \begin{bmatrix} z_B \\ z_N \end{bmatrix} = \begin{bmatrix} 0 \\ \bar{z}_N \end{bmatrix}, \qquad z_B, z_N \geq 0. \end{array} \tag{1.42}$$

Substituting $y = 0$ into the equality constraint gives

$$z_B = 0, \qquad z_N = \bar{z}_N.$$

Thereby, the bottom row of the simplex tableau exhibits the dual solution. If it is nonnegative, it is the dual feasible solution, and the tableau is *dual feasible*.

Assume the tableau is dual feasible with $\bar{z}_N \geq 0$.

Firstly, execute the following rule.

Rule 1.9.1 (Dual Row Rule) Select row index by

$$p \in \arg\min\{\bar{b}_i \mid i = 1, \ldots, m\}.$$

If $\bar{b}_p \geq 0$, the tableau is also primal feasible; hence, optimality is achieved. Assume that this is not the case, i.e.,

$$\bar{b}_p < 0. \tag{1.43}$$

Denote the pth row of the tableau by

$$\Delta y_p^T = e_p^T [I \mid \bar{N}]. \tag{1.44}$$

Then, the Δy_p is just the p-indexed column of the dual equality constraint of (1.42).

Taking $-\Delta y_p$ as the direction in y_p-space and Δz as the according direction in the z-space, consider the following update formula:

$$\hat{y}_p = \bar{y}_p - \beta \Delta y_p, \quad \hat{y}_i = 0,\ 1, \ldots, m,\ i \neq p; \quad \hat{z} = \bar{z} + \beta \Delta z, \tag{1.45}$$

where β is termed *dual stepsize*. In order to generate a new dual solution satisfying the equality constraint of the dual problem, it should hold that

$$-\Delta y_p + \Delta z = 0,$$

which with (1.52) gives

$$\Delta z = [e_p^T I, e_p^T \bar{N}]^T. \tag{1.46}$$

Introduce the index set below:

$$J = \{j \in N \mid \bar{a}_{p\,j} < 0\}. \tag{1.47}$$

Lemma 1.9.1 *Assume that $\bar{b}_p < 0$. If the index set J is empty, the primal problem is infeasible.*

Proof Note that emptiness of J implies $e_p^T \bar{N} \geq 0$.

It is verified that formulas (1.45)–(1.54) determine new dual solutions. If the set J is empty, then it holds that $\hat{z} \geq 0$, and hence the new solutions are also dual feasible for all $\beta \geq 0$.

On the other hand, from the dual problem (1.42) we see that the new dual solutions correspond to the dual objective value below:

$$\hat{g} = 0 - \beta \bar{b}_p,$$

which along with (1.49) leads to

$$\hat{g} \geq \bar{g} = 0.$$

Thus, the dual objective value increases infinitely as α goes to infinity.

Therefore, the dual problem is upper unbounded in the dual feasible region; hence, there is no feasible solution to the primal problem. □

If the set J is nonempty, the dual stepsize is limited to maintain the nonnegativity of $\hat{z}$. The following rule gives the maximum possible β and the associated index q.

Rule 1.9.2 (Dual Column Rule) Determine β and column index q such that

$$\beta = -\bar{z}_q/\bar{a}_{pq} = \min_{j \in J} -\bar{z}_j/\bar{a}_{pj} \geq 0. \tag{1.48}$$

Once the pair (p, q) is determined, we perform elementary transformations to turn $\bar{a}_{pq}$ to 1 and eliminate other nonzeros in column q. It is evident that the resulting simplex tableau corresponds to the new dual feasible solution, $\hat{z}_B = 0, \hat{z}_N)$, expressed by (1.45). Updating the associated $\bar{y}$ is unnecessary, as it can be calculabled at the end of the solution process if needed.

Since the primal and dual solutions are complementary, the duality gap is zero, which means that the primal objective value $\bar{f}$, given by the bottom row, equals the dual objective value Lemma (1.8.1). Thus, the $\bar{f}$ increases strictly if $\beta > 0$.

If it holds for some $j \in J$ that $\bar{z}_j = 0$, the $\bar{z}_N$ is said *dual degenerate*. In this case, the β vanishes; hence, the objective value remains unchanged. Beale [1] provided an instance that cannot be solved due to cycling. Fortunately, this case hardly happens in practice.

The following algorithm summarizes the overall steps (Lemke [17], Beale [2]).

Algorithm 1.9.1 (Dual Simplex Algorithm: Tableau Form)
Initial: a dual feasible simplex tableau in form (1.6). This algorithm solves the standard LP problem (1.2).

1. *Select pivot row index* $p \in \arg\min\{\bar{b}_i \mid i = 1, \ldots, m\}$.
2. *Stop if* $\bar{b}_p \geq 0$.
3. *Stop if* $J = \{j \in N \mid \bar{a}_{pj} < 0\} = \emptyset$.
4. *Determined pivot column* $q \in \arg\min_{j \in J} -\bar{z}_j/\bar{a}_{pj}$.
5. *Convert* $\bar{a}_{pq}$ *to* 1, *and eliminate the other nonzeros in the column by elementary transformations.*
6. *Go to Step 1.*

Theorem 1.9.2 *Under the dual nondegeneracy, Algorithm 1.9.1 terminates either at:*

(i) Step 2, achieving a pair of primal and dual optimal solutions.
(ii) Step 3 detects the infeasibility of the problem.

Algorithm 1.9.1 can be revised based on the equivalence between the simplex and revised simplex tableaus.

Let (B, N) be the partition of A. Assume that the factorization $B = LU$ is available.

(a) The basic solution $\bar{x}_B$ $(\bar{x}_N = 0)$:

Initially, it can be determined by solving the following two triangular systems successively:

$$Lu = b, \qquad U\bar{x}_B = u. \tag{1.49}$$

It will be updated subsequently.

(b) The reduced costs $\bar{z}_N$.
Initially, it is determined by solving the following two triangular systems first:

$$U^T u = c_B, \qquad L^T \bar{y} = u. \tag{1.50}$$

Then, compute by

$$\bar{z}_N = c_N - N\bar{y}, \tag{1.51}$$

which is updated subsequently.

(c) The pth row, denoted by v_M^T:
Solve the following two triangular systems for vector w first:

$$U^T u = e_p, \qquad L^T w = u. \tag{1.52}$$

Then, compute by

$$v_N^T = w^T N. \tag{1.53}$$

(d) The pivot column $\bar{a}_q$:
Solve the following two triangular systems:

$$Lu = a_q, \qquad U\bar{a}_q = u. \tag{1.54}$$

Using the preceding expressions, Algorithm 1.9.1 can be revised to the following version.

Algorithm 1.9.2 (Dual Simplex Algorithm)
Initial: (B, N), $B = LU$, $\bar{z}_N \geq 0$, $\bar{x}_B$. This algorithm solves the standard LP problem (1.2).

1. *Select row index $p \in \arg\min\{\bar{x}_{j_i} \mid i = 1, \ldots, m\}$.*
2. *Stop if $\bar{x}_{j_p} \geq 0$ (optimality achieved).*
3. *Solve the two triangular systems (1.52), and compute v_N by (1.53).*
4. *Stop if $J = \{j \in N \mid v_j < 0\} = \emptyset$ (infeasible problem).*
5. *Determine dual stepsize β and column index q such that*
   $\beta = -\bar{z}_q/v_q = \min_{j\in J} -\bar{z}_j/v_j$.
6. *If $\beta \neq 0$, update $\bar{z}_{j_p} = \beta$, $\bar{z}_N = \bar{z}_N + \beta v_N$.*
7. *Solve the two triangular systems (1.54).*

8. *Compute* $\alpha = \bar{x}_{j_p}/v_q$, *an update* $\bar{x}_B = \bar{x}_B - \alpha\bar{a}_q$, $\bar{x}_q = \alpha$.
9. *Adjust* (B, N) *by exchanging* j_p *and* q.
10. *Update* $B = LU$.
11. *Go to Step 1.*

1.10 Double-Phase Dual Simplex Algorithm

The dual simplex algorithms presented previously must start from a dual feasible solution. Therefore, dual-phase is needed, and dual simplex algorithms are used only as dual-phase II.

There are several recommendable dual-phase I procedures (Chapter 15, Pan [36]). In this section, we pick up the simplest and efficient one using the so-called most-obtuse-angle row rule, which is the counterpart of Rule 1.6.2 of the same name.

Let (1.6) be the current simplex tableau. First, execute the following column rule.

Rule 1.10.1 Select the column index q such that

$$q \in \arg\min_{j\in N} \bar{z}_j. \tag{1.55}$$

The solution is dually feasible if $\bar{z}_q \geq 0$. In the other case, execute the following rule.

Rule 1.10.2 (Most-Obtuse-Angle Row Rule) Select the row index

$$p \in \arg\max \{\bar{a}_{i\,q} \mid i = 1, \ldots, m\} \tag{1.56}$$

In case of a tie, select the one corresponding to the minimum component of the right-hand side.

If $\bar{a}_{pq} \leq 0$, then the primal problem is infeasible or lower unbounded, which implies that there is no dual feasible solution.

In the case of $\bar{a}_{pq} > 0$, a geometric interpretation of Rule 1.10.2 is as follows. Introduce vector Δx as follows:

$$\Delta x_B = \bar{a}_q, \quad \Delta x_q = -1; \quad \Delta x_j = 0, \quad j \in N,\ j \neq q.$$

Obviously, $-\Delta x$ is a descent search direction in the x-space, the j_pth component of which can be expressed as

$$\begin{aligned} -\bar{a}_{pq} &= \min_{i=1,\ldots,m} (-\Delta x)^T e_{j_i} = \min_{i=1,\ldots,m} \|\Delta x\| \|e_{j_i}\| \cos\langle -\Delta x, e_{j_i}\rangle \\ &= \|\Delta x\| \min_{i=1,\ldots,m} \cos\langle -\Delta x, e_{j_i}\rangle < 0. \end{aligned}$$

The preceding indicates that the rule selects the row index p so that the gradient of the inequality $x_{j_p} \geq 0$ forms the most obtuse angle with the direction $-\Delta x$, among blocking inequalities. Thus, it seems plausible to kick the index j_p out of the basis.

Once the (p, q) are determined, perform the according elementary transformation and complete the iteration. This process is repeated until the bottom row becomes nonnegative, or it is asserted that there is no dual feasible solution.

Associated steps are summarized into the following procedure.

Subalgorithm 1.10.1 (Tableau Dual-Phase I: Most-Obtuse-Angle Row Rule)
Initial: simplex tableau. This algorithm seeks dual feasibility.

1. *Select column index* $q \in \arg\min_{j \in N} \bar{z}_j$.
2. *Return if* $\bar{z}_q \geq 0$.*(dual feasibility achieved)*
3. *Determine row index* $p \in \arg\max \{\bar{a}_{i\,q} \mid i = 1, \ldots, m\}$.
4. *Return if* $\bar{a}_{p\,q} \leq 0$. *(no dual feasible solution)*
5. *Perform elementary transformations to convert* $\bar{a}_{p\,q}$ *to* 1, *and eliminate all other nonzeros of the column.*
6. *Go to Step 1.*

Now, let us revise Algorithm 1.10.1 for practicable use.

Let $A = (B, N)$ be a partition and let factorization $B = LU$ be available. The solution process will update the LU factors, subsequently.

The following items are computed with the original data.

(a) The reduced costs.

Solve the following two triangular systems successively:

$$U^T w = c_B, \qquad L^T \bar{y} = w. \tag{1.57}$$

Then, compute $\bar{z}_N$ by

$$\bar{z}_N = c_N - N^{\mathrm{T}} \bar{y}. \tag{1.58}$$

(b) The qth column $\bar{a}_q$.

Once the pivot column index q is available, solve the following two triangular systems to obtain the qth column:

$$Lu = a_q, \qquad U\bar{a}_q = u. \tag{1.59}$$

Initially, solve the following two triangular systems for $\bar{x}_B$:

$$Lu = b, \qquad U\bar{x}_B = u.$$

Then, it is straightforward to obtain the following revision.

Subalgorithm 1.10.2 (Dual-Phase I: Most-Obtuse-Angle Row Rule)

Initial: (B, N), $B = LU$, $\bar{x}_B$. *This algorithm finds a dual feasible basic solution to the standard LP problem.*

1. *Solve the two triangular systems (1.57) for* $\bar{y}$, *and compute* $\bar{z}_N$ *by (1.58).*
2. *Select column index* $q \in \arg\min_{j \in N} \bar{z}_j$.
3. *Stop if* $\bar{z}_q \geq 0$. *(dual feasibility is achieved)*
4. *Solve the two triangular systems (1.59) for* $\bar{a}_q$.
5. *Determine row index* $p \in \arg\max_{i=1,\ldots,m} \bar{a}_{i\,q}$.
6. *Stop if* $\bar{a}_{p\,q} \leq 0$. *(there is no dual feasible solution)*
7. *Update LU factors.*
8. *Adjust* (B, N) *by exchanging* j_p *and* q.
9. *Go to Step 1.*

Note: The use of the column rule in Step 2 is only for convenience. For efficiency, the N-M column rule is strongly commendable (Sect. 1.12.3).

The dual algorithm supported by the dual-phase I is called *double-phase dual simplex algorithm.*

We proposed the most-obtuse-angle rule, exciting by its remarkable simplicity, stability, and promising performance with a number of small problems (Pan [22]).

Then, we conducted further computational experiments (Pan [23]). Using MINOS 5.3 as a platform, a code based on the double-phase dual simplex algorithm outperformed MINOS 5.3 on 48 Netlib test problems with a time ratio of **1.37**.

Ten years later, Koberstein and Suhl [16] extended the dual-phase I procedure for solving general real-world LP problems. In their extensive computational experiments, the dual-phase I defeated well-known and commonly used dual-phase-1 procedures.

Finally, it is encouraging that the dual-phase I is still improvable. There are promising variants of the most-obtuse-angle row rule, and the column rule has a much better substitute, as was indicated.

1.11 Getting Optimal Set

Although the simplex method seeks an optimal vertex, it is favorable in applications to find the set of all optimal solutions, providing more possible choices. In this section, we show how to do so from the primal and dual sides, lay a foundation for the face method, and introduce a related fresh bilevel LP model.

Definition 1.11.1 The set of all optimal solutions is termed *optimal set.*

Favorably, it is relatively simple to get the optimal set after the standard LP problem is solved.

Let $A = (B, N)$ be the partition with the optimal basis B.

Assume that $\bar{x}$ and $(\bar{y}, \bar{z})$ are the pair of primal and dual optimal solutions. Using the preceding notation, we cite the following result (Section 5.8, Pan [36]).

Lemma 1.11.1 *If the column partition $N = (B', N')$ is such that*

$$B' = \{j \in N \mid \bar{z}_j = 0\}, \qquad N' = \{j \in N \mid \bar{z}_j > 0\}, \tag{1.60}$$

then, the following expression gives the optimal set:

$$P(B \cup B') = \{x \in \mathcal{R}^n \mid Bx_B + B'x_{B'} = b, \ x_B, x_{B'} \geq 0, \ x_{N'} = 0\}. \tag{1.61}$$

The preceding Lemma gives a constructive proof of the following result.

Theorem 1.11.2 *There exists a unique optimal set if so does an optimal solution. The optimal set is a face.*

The preceding theorem lay a foundation for the face method. The optimal set and maximum optimal face are equivalent.

Getting the optimal set has potential value for applications. If B' is nonempty, and hence the optimal set is infinite, we can choose a solution within the optimal set to fulfill some unique criterion.

Then, the following bilevel LP model arises:

$$\begin{array}{ll} \min \ f_1 = c_1^{\mathrm{T}} x, & \\ \min \ f_2 = c_2^{\mathrm{T}} x, & \\ \text{s.t.} \ \ Ax = b, & \quad x \geq 0. \end{array} \tag{1.62}$$

The preceding finds the optimal set of the first-level problem with the objective f_1 subject to the constraints and then solves the second-level problem with objective function f_2 over the optimal set.

On the other hand, we can get the dual optimal set similarly.

Let (R, R') be the row partition such that

$$R = \{i \mid \bar{x}_{j_i} = 0, \ i = 1, \ldots, m\}, \qquad R' = \{i \mid \bar{x}_{j_i} > 0, \ i = 1, \ldots, m\}. \tag{1.63}$$

Then, there are the following partitions:

$$y = \begin{bmatrix} y_R \\ y_{R'} \end{bmatrix} \qquad z_B = \begin{bmatrix} z_R \\ z_{R'} \end{bmatrix} \qquad \bar{N} = B^{-1}N = \begin{bmatrix} \bar{N}_R^T \\ \bar{N}_{R'}^T \end{bmatrix}.$$

Then, we can state the following result.

Lemma 1.11.3 *The following expression gives the dual optimal set:*

$$D(N) = \{(y, z) \in \mathcal{R}^{m \times n} \mid \bar{N}_R y_R + z_N = \bar{z}_N, \ y_R \leq 0, \ z_N \geq 0, \ y_{R'}, z_{R'} = 0\}. \tag{1.64}$$

1.12 Making Simplex Method Fastest: Pivot Rule

The column rule is a key to the performance of the simplex method since it determines the number of iterations required for solving LP problems.

There are infinitely many choices for the entering index since any index corresponding to negative reduced cost is eligible in theory. For this reason, scholars have suggested various rules in the past to find the best one. From a practical point of view, we ignore the so-called "finite" rules, performing poorly, but only mention practicable ones, such as the steepest-edge rule (Forrest and Goldfarb [8]) and Harris' approximated steepest-edge rule, i.e., Devex rule [12]).

According to our knowledge, the so-called *largest-distance* (renamed *maximum-violation*) and *nested* rules are the two of the most noteworthy new entrants. Therefore, this section only focuses on them and some extensions. For simplicity, we will discuss aiming at the standard problem, as it is a light task to convert to the variable-bounded problem.

1.12.1 Nested Rule

The trick of the nested rule is to destroy the strongest among a group of the most hardened "enemies" in each iteration.

In each iteration, the nested rule calculates reduced costs associated with the working set $N_1 \in N$ and seeks from it the index q with the minimum reduced cost. If the cost is nonnegative, it does the same with the renewed subset $N \backslash N_1$. Optimality is achieved if the resulting $\bar{z}_q$ is still nonnegative. Whenever the minimum $\bar{z}_q$ is negative, it enters the q into the basis and generates a new subset N_1 from current N_1 by kicking out all indices corresponding to nonnegative reduced costs for the next iteration.

Initially, the simplex method takes $N_1 = N$.

We formally state it below, using an optimality tolerance and a percentage.

Rule 1.12.1 (Nested-Dantzig Column Rule)
Input: working set $N_1 \in N$, optimality tolerance ϵ, percentage λ.

1. Calculate $\bar{z}_j,\ j \in N_1$ and select the index q such that $\bar{z}_q = \min_{j \in N_1} \bar{z}_j$.
2. Go to Step 5 if $\bar{z}_q < -\epsilon$.
3. Set $N_1 = N \backslash N_1$, and select the index q such that $\bar{z}_q = \min_{j \in N_1} \bar{z}_j$.
4. Stop if $\bar{z}_q \geq -\epsilon$.
5. Return with the index q and renewed $N_1 = \{j \in N_1 \mid \bar{z}_j < \lambda \bar{z}_q\}$.

It is favorable that the nested rule belongs to the partial pricing category, and it is not difficult to implement.

The nested rule represents a class of rules since one can transform any ordinary rule into its nested form.

We conducted extensive computational experiments, but for the sake of space, we only picked one with with the following five codes (Pan [29]; Section 12.5, [36]):

1. Dantzig: MINOS 5.51 (full pricing option).
2. Devex: Devex rule.
3. P-Dantzig: MINOS 5.51 (default sectional pricing option).
4. N-Devex: Nested Devex rule.
5. N-Dantzig: Rule 1.12.1.

Modified from MINOS 5.51, the last four codes differ only in the column rule used.

Test set: It comprises 80 large-scale sparse problems (Appendix A, Pan [34]), including the 47 biggest Netlib problems (SCRS8–STOCFOR3),[1] 16 Kennington problems,[2] and 17 BPMPD problems with more than 500KB in compressed form.[3] The test set is a nice representative combination. We list the average size of problems below:

Size of test sets

Problem	Netlib (47)	Kennington (16)	BPMPD(17)	Total(80)
Ave. m+n	5867	76, 873	73, 190	34, 374

Code N-Dantzig stands out, and Code Dantzig performed the least ideal, with a time ratio as high as **9.75**! Therefore, we only make a comparison between N-Dantzig and the well-known Devex by listing their time ratios below:

Time ratio for nested rule

Problem	Netlib(47)	Kennington(16)	BPMPD(17)	Total(80)
Devex/N-Dantzig	1.21	5.55	6.54	5.73

N-Dantzig's performance was so excellent that the time ratios for Kennington and BPMPD problems were up to **5.55** and **5.75** although the time ratio is only **1.21** with Netlib problems. So, the nested rule appears more relevant for solving larger problems, favorably.

---

[1] Netlib: http://www.netlib.org/lp/data/.

[2] Kennington: http://www.fp.mcs.anl.gov/otc/Guide/TestProblems/LPtest/.

[3] BPMPD: http://www.sztaki.hu/meszaros/bpmpd/.

1.12.2 Maximum-Violation Rule

The classic Dantzig rule selects the entering index q with the minimum cost. A negative $\bar{z}_q$ indicates its violation of the dual boundary. But, a closer look at the dual problem reveals that it is more logical to take the following rule, involving some *norm*, instead.

Rule 1.12.2 (Maximum-Violation Column Rule) Select the column index q such that

$$q \in \arg\min_{j \in N} \bar{z}_j / \|a_j\|.$$

Then a negative $\bar{z}_q$ corresponds to $\bar{z}_q / \|a_q\|$, which is just the maximum *vertical* violation of the dual boundary.

Surprisingly, the realization of the preceding rule is so simple that only a pre-normalization of columns of A is needed.

The computational experiments involve the following three test codes (Pan [30], Section 12.4, [36]):

1. Devex: Devex rule.
2. MV1: Maximum-violation rule with ∞ norm.
3. MV2: Maximum-violation rule with 2-norm.

All codes were implemented using MINOS 5.51; their differences are only in the column rule used. The test set is the same 80 problems as those for the nested rule. The following table gives the related SPU times of Devex to MV2.

Time ratio for maximum-violation rule

Problem	Netlib(47)	Kennington(16)	BPMPD(17)	Total(80)
Devex/MV2	0.34	4.33	3.64	3.24

We were astonished by the success of the new code, which outperformed Devex with a time ratio of **3.24** overall!, compared with the code yielding from MINOS 5.51 by only adding a line or two for pre-normalization of columns of A. The rule appears more suitable for solving larger problems as it is even inferior on Netlib problems with a time ratio of **0.34**, less than one.

1.12.3 Nested Maximum-Violation (N-M) Rule

Given their outstanding performance, a combination of the nested and maximum-violation rules would be a good idea. This leads to the following rule.

Rule 1.12.3 (N-M Column Rule) The same as Rule 1.12.1 except for pre-normalizing columns of the coefficient matrix.

Computational experiments involve the following three codes (Pan [32], Section 12.6, [36]):

1. Devex: Devex rule.
2. N-M1: Nested maximum-violation rule with ∞ norm.
3. N-M2: Nested maximum-violation rule with 2-norm.

As before, we used MINOS 5.51 as the platform and the same set of 80 test problems. Code N-M2 using 2-norm performed better than code N-M1 using ∞ norm; we therefore make a comparison between the former and Devex. The related time ratios are listed in the following table.

Time ratio for nested maximum-violation (N-M) rule

Problem	Netlib(47)	Kennington(16)	BPMPD(17)	Total(80)
Devex/N-M2	1.16	5.77	10.17	7.27

From the preceding table, we see that the time ratios are all greater than one although it is lower with Netlib problems. As for a comparison between the nest, maximum-violation, and N-M2 rules, the total time ratios are **5.73, 324, and 7.27**, respectively. The outcome responds to our expectations that such a combination produces an additive effect. We would even consider N-M2 making the simplex method the fastest at present.

Finally, we point out that it is straightforward to derive the dual version of the N-M rule similarly from the dual nested and maximum-violation rules. Interested readers may refer to Sections 13.3–13.4, Pan [36].

1.13 Simplex Interior-Point Algorithm

One of the advantages of the interior-point method is that it goes across the feasible region's interior, hence bypassing degenerate vertices. Some algorithms of this type solve large-scale sparse LP problems quickly. But the problem is that it cannot be *warmly started* and applied to handling integer LP problems. In addition, it can only approach an optimal solution on the boundary.

This section presents a fresh LP algorithm featuring both simplex and interior-point methods. It starts from an interior point and generates a series of interior points. But it also pivots in each iteration and finally attains an optimal vertex.

The solver represents a *class* of algorithms. Similarly, one can transform any simplex variant into this type although we only convert the typical simplex algorithm to solve variable-bounded LP problems.

1.13.1 Column Rule and Search Direction

Assume that (B, N) is the current partition, i.e.,

$$B = \{j_1, \ldots, j_m\}, \qquad N = A \backslash B,$$

and that $(\bar{z}_B = 0, \bar{z}_N)$ is the reduced costs or dual solution, i.e.,

$$\bar{z}_N = c_N - N^T B^{-T} c_B.$$

Let $\bar{x}$ be the interior point.

Considering that the duality gaps δ_j, defined by (1.41), exhibit potential objective improvements concerning indices, we take the following rule to select the entering index.

Rule 1.13.1 (Column Rule) Select the index q such that

$$q \in \arg\min_{j \in N} \delta_j. \tag{1.65}$$

Introduce the search direction Δx such that

$$\Delta x_B = \text{sign}(\bar{z}_q) B^{-1} a_q \tag{1.66}$$

$$\Delta x_j = \begin{cases} 0 & j \in N, \; j \neq q \\ -\text{sign}(\bar{z}_q) & j = q \end{cases} \tag{1.67}$$

The following result reveals the eligibility of Δx as the search direction.

Lemma 1.13.1 *Assume that q is determined by Rule 1.13.1. If $\delta_q \neq 0$, the vector Δx is a descent direction in the null of A.*

Proof The proof is omitted. □

1.13.2 Update of Feasible Solution

Consider the following update formula:

$$\hat{x} = \bar{x} + \alpha \Delta x, \tag{1.68}$$

or equivalently,

$$\hat{x}_B = \bar{x}_B + \alpha \Delta x_B \tag{1.69}$$

$$\hat{x}_q = \bar{x}_q - \text{sign}(\bar{z}_q)\alpha \tag{1.70}$$

$$\hat{x}_j = \bar{x}_j \quad j \in N, \ j \neq q, \tag{1.71}$$

where α is the stepsize to be determined.

To maximize the stepsize while keeping the new solution feasible, introduce the following notation:

$$\sigma_i = \begin{cases} (u_{j_i} - \bar{x}_{j_i})/\Delta x_{j_i} & \text{if } \Delta x_{j_i} > 0 \\ (l_{j_i} - \bar{x}_{j_i})/\Delta x_{j_i} & \text{if } \Delta x_{j_i} < 0 \end{cases} \qquad i = 1, \ldots, m, \tag{1.72}$$

which is the maximum range the $\bar{x}_{j_i}$ is allowed to change. Thus, we are led to the following rule for determining the row index.

Rule 1.13.2 (Row Rule) Determine row index p such that

$$\sigma_p = \min_{i=1,\ldots,m} \sigma_i. \tag{1.73}$$

Denote the range that the entering variable x_q is allowed to change, i.e.,

$$\rho_q = |\delta_q/\bar{z}_q|,$$

the maximum stepsize subject to $l \leq \hat{x} \leq u$ is

$$\alpha' = \min\{\rho_q, \sigma_p\} > 0. \tag{1.74}$$

Then, there are the following two cases arising:

(a) $\rho_q \leq \sigma_p$. There is no need for a basis change.
(b) $\rho_q > \sigma_p$. The basis change should be performed with the leaving index j_p and entering index q.

For the new solution $\hat{x}$ to be interior, however, we do not use α' itself but a shortened one instead, i.e.,

$$\alpha = \lambda\alpha', \qquad 0 < \lambda < 1, \tag{1.75}$$

where λ is a predetermined parameter close to 1.

We claim the following result without proof.

Lemma 1.13.2 *Let $\bar{x}$ be an interior solution. If $\delta_q \neq 0$, the $\hat{x}$ defined by (1.69)–(1.71) and (1.75) with (1.72)–(1.74) is an interior solution with the objective value reduced by $\alpha|\bar{z}_q| > 0$.*

1.13.3 Optimality Condition and the Algorithm

However, if δ_q vanishes, the search direction Δx is meaningless. In this case, the following result holds due to the optimal condition for the bounded-variable problem (Theorem 7.3.4, Pan [36]).

Theorem 1.13.3 *The current $\bar{x}$ is an optimal solution if and only if it holds that*

$$\delta_q = 0. \tag{1.76}$$

From a practical point of view, it would be beneficial to use the following *approximate optimality condition* instead.

Definition 1.13.1 The current solution $\bar{x}$ is an ϵ-optimal solution if

$$|\delta_j| \leq \epsilon, \qquad j \in N, \tag{1.77}$$

where $\epsilon > 0$ is a predetermined optimality tolerance.

We summarize the associated steps into the following model.

Algorithm 1.13.1 (Simplex Interior-Point Algorithm)
Initial: $0 < \lambda < 1$, $\epsilon > 0$ *and* $M \gg 1$. *basis* B, *interior solution* $\bar{x}$. *This algorithm solves the bounded-variable problem.*

1. *Solve* $B^{\mathrm{T}}\bar{y} = c_B$ *for* $\bar{y}$.
2. *Compute* $\bar{z}_N = c_N - N^{\mathrm{T}}\bar{y}$.
3. *Compute* δ_j, $j \in N$ *by (1.41).*
4. *Stop if (1.77) is satisfied.*
5. *Select column index* q *by* $q \in \arg\min_{j \in N} \delta_j$.
6. *Solve the system* $B\bar{a}_q = a_q$.
7. *Compute* Δx *by (1.66) and (1.67).*
8. *Determine row index* p *by (1.73) with (1.72).*
9. *Stop if* $\sigma_p > M$.
10. *Set* $\alpha = \lambda\alpha'$, *where* $\alpha' = \min\{\rho_q, \sigma_p\}$
11. *Update* $\bar{x}$ *by (1.69)–(1.71).*
12. *Go to Step 1 if* $\rho_q \leq \sigma_p$.
13. *Update basis* B *by replacing its* p *column with* q*-indexed column.*
14. *Go to Step 1.*

The following is claimed without proof.

Theorem 1.13.4 *Assume the finiteness of Algorithm 1.13.1. It terminates either at:*

(i) Step 4, producing an ϵ-optimal basic solution
(ii) Step 9, declaring the lower unboundedness of the problem

Algorithm 1.13.1 represents a *class* of algorithms since one can convert any simplex variant into such type, e.g., it is not difficult to transit the dual simplex algorithm similarly.

It is the first cross-category algorithm, pushing the boundaries between the simplex and interior-point methods. In contrast to the former, which starts from a vertex, and the latter, which initiates from an interior point, it can start from any *feasible* point. If the first step is nonzero, it generates a series of interior points. As a pivot algorithm, it can be "warmly started"; there is no need for extra purification.

We only report computational experiments with the following two codes (Appendix F, Pan [36]):

1. MINOS: MINOS 5.51 with full pricing.
2. SIP: Algorithm 1.13.1.

The test set includes the 16 Kennington and 17 BPMPD problems, the same as those in Sect. 1.12.1. The results with two BPMPD problems are exceptional and excluded from the report.

The following table gives the CPU time ratio with the two codes.

Time ratio for nested maximum-violation (N-M) rule

Problem	Kennington(16)	BPMPD(17)	Total(80)
MINOS/SIP	3.5	9.0	6.6

The success of Algorithm 1.13.1 is remarkable, winning an unparalleled victory over MINOS 5.51, with a time ratio as high as **6.6** overall. Moreover, it is improvable. The outcome would have been unimaginable if we had incorporated the N-M column rule into Code SIP instead.

Chapter 2
Face Method

The simplex method solves the LP problem by generating a series of vertices to seek an optimal vertex. The so-called face method presented in this chapter follows a fresh new philosophy of moving from face to face until attaining an optimal face together with an optimal solution. In the next section, we introduce the descent search direction. In Sect. 2.2, we discuss the update of the face solution. Section 2.3 addresses the row pivoting operations. In Sect. 2.4, we draw the optimality condition. Then, in Sects. 2.5 and 2.6, we formulate the face algorithm in tableau form and offer a double-phase face algorithm for general-purpose use. We form the face algorithm in revised form in the final section.

Consider the standard LP problem, i.e.,

$$\begin{array}{ll} \min\ f = c^{\mathrm{T}}x & \\ \text{s.t.}\ \ Ax = b, & x \geq 0, \end{array} \tag{2.1}$$

where $A \in \mathcal{R}^{m\times n},\ c \in \mathcal{R}^n,\ b \in \mathcal{R}^m,\ m < n$. It is assumed that Rank $(A) = m$.

2.1 Descent Search Direction

The search direction used is crucial to the success of any LP solver, and seeking an "ideal" search direction is challenging. This section focuses on this topic.

Let $A = (B, N)$ be a partition, where B is the basis. Assume that $\bar{x}$ is a feasible solution, so that

$$B\bar{x}_B + N\bar{x}_N = b, \qquad \bar{x}_B,\ \bar{x}_N \geq 0.$$

and $(\bar{y}, \bar{z})$ is a dual solution such that

$$B^T\bar{y} + \bar{z}_B = c_B, \qquad N^T\bar{y} + \bar{z}_N = c_N.$$

P.-Q. Pan, *Face Method*, https://doi.org/10.1007/978-3-031-93594-7_2

Form the following index set:

$$B' = \{j \in N \mid \bar{z}_j < 0\} \cup \{j \in N \mid \bar{z}_j \geq 0,\ \bar{x}_j > 0\}. \tag{2.2}$$

Reset

$$N = N \backslash B'.$$

Then, for the renewed set N, it holds that

$$\bar{z}_j \geq 0, \qquad \bar{x}_j = 0, \forall\, j \in N.$$

For the so-called face partition $A = (B, B', N)$, introduce several concepts as follows.

Definition 2.1.1 The face is the following subset of the feasible region:

$$P(B \cup B') = \{x \in \mathcal{R}^n \mid Bx_B + B'x_{B'} = b,\ x_{B \cup B'} \geq 0,\ x_N = 0\}. \tag{2.3}$$

The index set B' is said to be*active* and N said to be*nonface/inactive*. The related items, such as matrices, variables, etc., are said similarly.

A solution on the face is called *face solution*. So, the $\bar{x}$ is a face solution.

Any feasible solution can be regarded as a face solution; e.g., even an interior solution is a face solution because it is on the maximum face—the feasible region. It gives us some flexibility. Thereafter, we will not distinguish between a face solution and a feasible solution.

Definition 2.1.2 A face is level face if solutions in it correspond to the same objective value. A level face is optimal face if the corresponding objective value is equal to the optimal value.

Consider minimizing the objective function over the face $P(B \cup B')$, i.e.,

$$\begin{aligned} \min\ & c_B^T x_B + c_{B'}^T x_{B'}, \\ \text{s.t.}\ & Bx_B + B'x_{B'} + N\bar{x}_N = b, \\ & x_{B \cup B'} \geq 0, \end{aligned} \tag{2.4}$$

which is termed *face subprogram.*

Put the face subproblem into the following tableau:

x_B	$x_{B'}$	$\bar{x}_N$	f	RHS
B	B'	N		b
c_B^{T}	$c_{B'}^{\mathrm{T}}$	c_N^{T}	-1	

(2.5)

Assume that Gauss-Jordan elimination transforms the preceding into the so-called face tableau below:

x_B	$x_{B'}$	$\bar{x}_N$
I	$\bar{B}'$	$\bar{N}$
	$\bar{z}_{B'}^{\mathrm{T}}$	$\bar{z}_N^{\mathrm{T}}$

(2.6)

where the right-hand side and f columns were omitted, as will not be used.

We put forward the following direction:

$$\Delta x_{B\cup B'} = \begin{bmatrix} \Delta x_B \\ \Delta x_{B'} \end{bmatrix} = \begin{bmatrix} -\bar{B}'\bar{z}_{B'} \\ \bar{z}_{B'} \end{bmatrix} \tag{2.7}$$

For the sake of space, we omit the cumbersome derivation of the preceding direction but focus on its eligibility to be a descent direction in the face $P(B \cup B')$. It is notable that $\Delta x_{B\cup B'} = 0$ if and only if $\bar{z}_{B'} = 0$.

Theorem 2.1.1 *Assume that B' is nonempty and $\bar{z}_{B'}$ is nonzero. Then, the direction (2.7) satisfies the following condition:*

$$(I, \bar{B}')\Delta x_{B\cup B'} = 0, \tag{2.8}$$

$$(0, \bar{z}_{B'}^{\mathrm{T}})\Delta x_{B\cup B'} > 0. \tag{2.9}$$

Proof The expression (2.8) is verified by using (2.7).

Besides, from (2.7) and the assumption $\bar{z}_{B'} \neq 0$ it follows that

$$(0, \bar{z}_{B'}^{\mathrm{T}})\Delta x_{B\cup B'} = \bar{z}_{B'}^{\mathrm{T}}\bar{z}_{B'} > 0,$$

which confirms the validity of the expression (2.9). □

The preceding theorem guarantees that the vector given by (2.7) is downhill in the null of $(I, \bar{B}')$ and hence is eligible to be as the search direction.

2.2 Improving Face Solution

This section addresses the line search scheme for computing a new face solution.

In each iteration, the face solution $\bar{x}$ is updated by the line search scheme. Since the inactive part $\bar{x}_N = 0$ keeps unchanged, actually only the $\bar{x}_{B\cup B'}$ will be updated. Thereafter, we will not distinguish between the $\bar{x}$ and $\bar{x}_{B\cup B'}$.

Assume that the face solution $\bar{x}_{B\cup B'}$ and the nonzero search direction $\Delta x_{B\cup B'}$ are available.

Introduce the following index set:

$$J = \{j \in B \cup B' \mid \Delta x_j > 0\}. \quad (2.10)$$

If J is nonempty, a new face solution can be determined by the following line search:

$$\hat{x}_{B\cup B'} = \bar{x}_{B\cup B'} - \alpha \Delta x_{B\cup B'}, \quad (2.11)$$

where α is the stepsize such that

$$\alpha = \min_{j\in J} \bar{x}_j/\Delta x_j \geq 0. \quad (2.12)$$

Theorem 2.2.1 *Assume that $\Delta x_{B\cup B'}$ is determined by (2.7), and J defined by (2.10) is nonempty. If $\bar{x}_{B\cup B'}$ is a face solution to subprogram (2.4), then so is the $\hat{x}_{B\cup B'}$ given by (2.11) with (2.12).*

Proof Under the assumption, it holds that

$$(B \cup B')\bar{x}_{B\cup B'} = b, \qquad \bar{x}_{B\cup B'} \geq 0. \quad (2.13)$$

Combining (2.11), (2.8), and the first expression in (2.13) gives

$$(B \cup B')\hat{x}_{B\cup B'} = (B \cup B')\bar{x}_{B\cup B'} - \alpha(B \cup B')\Delta x_{B\cup B'} = b. \quad (2.14)$$

Thus, $\hat{x}_{B\cup B'}$ satisfies the equality constraints.

Note that $\bar{x}_j \geq 0\ \forall j \in B \cup B'$. Without loss of generality, assume that $(B \cup B')\backslash J \neq \emptyset$.

It is obvious by (2.13) that

$$\Delta x_j \leq 0, \quad j \in (B \cup B')\backslash J,$$

from which and (2.11), for any $\alpha \geq 0$, it follows that

$$\hat{x}_j = \bar{x}_j - \alpha \Delta x_j \geq 0, \qquad j \in (B \cup B')\backslash J. \quad (2.15)$$

On the other hand, it holds by (2.12) that

$$\bar{x}_j/\Delta x_j \geq \alpha \geq 0, \qquad j \in J.$$

Multiplying the preceding by $\Delta x_j > 0$ leads to

$$\bar{x}_j \geq \alpha \Delta x_j \geq 0, \qquad j \in J,$$

which together with (2.15) gives

$$\hat{x}_j = \bar{x}_j - \alpha \Delta x_j \geq 0, \qquad j \in J. \quad (2.16)$$

Consequently, (2.14), (2.15), and (2.16) together imply that $\hat{x}_{B\cup B'}$ is a face solution to (2.16). □

From (2.15) and (2.16), we see that only the requirement $\hat{x}_j \geq 0, \forall j \in J$ limits the increase of the stepsize so that α determined by (2.16) is the maximum possible stepsize that can be taken along the search direction. Indices in J and the corresponding variables are often said to be *blocking*.

Let us inspect the progress made in a single iteration. Introduce the following simple result first.

Lemma 2.2.2 *Assume that B' has k columns. For the $(m+k)\times k$ matrix*

$$Z = \begin{bmatrix} -\bar{B}' \\ I_k \end{bmatrix}, \tag{2.17}$$

it holds that

$$(I, \bar{B}')Z = 0, \qquad \text{and} \qquad \operatorname{rank}(Z) = k.$$

Proof The validity can be verified. □

Z is known as a *null space matrix* of $(I, \bar{B}')$, which is the coefficient matrix of the equality constraints of face tableau (2.6). Therefore, the set $\{-Zu \mid u \in R^k\}$ includes all possible search direction candidates.

So, $-Zu$ is an eligible search direction for any given $u \in R^k$. According to the update formula (2.11), the corresponding new solution is

$$\hat{x}_{B\cup B'}(u) = \bar{x}_{B\cup B'} - \alpha Zu,$$

where α is a stepsize. Premultiplying the preceding by $(0, \bar{z}_{B'}^{\mathrm{T}})$ and moving terms, we obtain the corresponding objective decrement, i.e.,

$$\gamma(u) = (0, \bar{z}_{B'}^{\mathrm{T}})\bar{x}_{B\cup B'} - (0, \bar{z}_{B'}^{\mathrm{T}})\hat{x}_{B\cup B'}(u) = \alpha(0, \bar{z}_{B'}^{\mathrm{T}})Zu, \tag{2.18}$$

which not only depends on the direction of u but also on its norm. Therefore, it is reasonable to restrict u by

$$\|u\| = \|\bar{z}_{B'}\|. \tag{2.19}$$

Thereby, we are led to the following problem of maximizing the objective decrement:

$$\begin{array}{ll} \max\ \gamma(u) = \alpha(0, \bar{z}_{B'}^{\mathrm{T}})Zu & \\ \text{s.t. } \|u\| = \|\bar{z}_{B'}\|, & u \in R^k. \end{array} \tag{2.20}$$

Theorem 2.2.3 *Problem (2.20) has the unique optimal solution $u^* = \bar{z}_{B'}$, corresponding to maximum objective decrement $\gamma(u^*) = \alpha\|\bar{z}_{B'}\|^2$.*

Proof From (2.17), (2.18), and the constraint of (2.20), it follows that

$$\begin{aligned}\gamma(u) &= \alpha(0, \bar{z}_{B'}^{\mathrm{T}}) \begin{bmatrix} -\bar{B}' \\ I_k \end{bmatrix} u \\ &= \alpha \bar{z}_{B'}^{\mathrm{T}} u \\ &= \alpha\|\bar{c}_{B'}\| \|u\| \cos\langle \bar{z}_{B'}, u\rangle \\ &= \alpha\|\bar{c}_{B'}\|^2 \cos\langle \bar{z}_{B'}, u\rangle.\end{aligned}$$

It is obvious that $u^* = \bar{z}_{B'}$ is the unique optimal solution to problem (2.20), corresponding to $\gamma(u^*) = \alpha\|\bar{c}_{B'}\|^2$. □

It is then obvious that $-Zu^*$ with $u^* = \bar{z}_{B'}$ is just the search direction we used. Theorem 2.2.3 ensures that it leads to the maximum possible objective decrement, i.e.,

$$\gamma(u^*) = \alpha\|\bar{z}_{B'}\|^2.$$

This is a fundamental advantage of the face method. We all know that the simplex method attains the objective decrement $\alpha|\bar{z}_q|$ only.

Nevertheless, in the case when J is empty, the update formula is meaningless. Then we are done, as the following says.

Theorem 2.2.4 *Assume that $\bar{x}_{B\cup B'}$ is a face solution and $\Delta x_{B\cup B'}$ defined by (2.7) is nonzero. If J defined by (2.10) is empty, problem (2.1) is lower unbounded.*

Proof The emptiness of J implies $\Delta x_{B\cup B'} \le 0$, which along with $\bar{x}_{B\cup B'} \ge 0$ leads to

$$\hat{x}_{B\cup B'} = \bar{x}_{B\cup B'} - \alpha\Delta x_{B\cup B'} \ge 0, \qquad \forall\, \alpha \ge 0. \tag{2.21}$$

From the preceding, (2.8) and $B \cup B'\bar{x}_B = b$, it follows that

$$B\hat{x}_B = B\bar{x}_B - \alpha B\Delta x_B = b, \qquad \forall\, \alpha \ge 0.$$

Therefore, $\hat{x}_B$ is a face solution to (2.4) for all $\alpha \ge 0$.

Moreover, by (2.21) and (2.9), it holds that

$$c_B^{\mathrm{T}}\bar{x}_B - c_B^{\mathrm{T}}\hat{x}_B = \alpha c_B^{\mathrm{T}}\Delta x_B > 0, \qquad \forall\, \alpha > 0.$$

Consequently, the decrement in objective value goes to infinity, as so does α. □

2.3 Row Pivoting Operation

Let $A = (B, B', N)$ be the current face partition, and let $\bar{x}$ be the face solution. Assume that the new face solution $\hat{x}_B$ was computed by (2.11). This section discusses pivot operations for updating the face tableau.

Introduce the following subset:

$$J_\alpha = \{j \in J \mid \bar{x}_j/\Delta x_j = \alpha\}. \tag{2.22}$$

It is obvious that

$$\hat{x}_j = 0, \qquad \forall\, j \in J_\alpha. \tag{2.23}$$

There are the following two cases arising:

(a) Simple iteration: $B' \cap J_\alpha$ is nonempty.
In this case, there exist blocking indices in B', and no basis change is needed. What to do is only adjust (B', N) by moving them to N:

$$B' = B' \backslash (B' \cap J_\alpha), \qquad N = N \cup (B' \cap J_\alpha) \tag{2.24}$$

and go to the next iteration.

(b) $B' \cap J_\alpha$ is empty.
In this case, all the blocking indices belong to B, and hence, a pivoting operation is required.

Designate by j_i the index of the i-th column, e_i, of $\bar{B} \triangleq I$, so that the variable x_{j_i} corresponds to the i-th row of the tableau (2.6).

Select the pivot row using the following rule first.

Rule 2.3.1 (Row Rule) Determine the row index p such that

$$j_p \in \arg \max_{j_i \in J_\alpha} \Delta x_{j_i}, \tag{2.25}$$

which determines the index j_p to leave the basis B.

The geometric interpretation of the preceding rule is as follows.

It is evident that the cosine of the angle between the search direction $-\Delta x_B$ and the gradient of a constraint $x_j \geq 0$ equals

$$\alpha_j = -\Delta x_j / \|\Delta x_B\|, \qquad \forall j \in B. \tag{2.26}$$

Based on the heuristic characteristic of optimal solution (Proposition 1.3.1), for each $j_i \in B$, the constraint $x_{j_i} \geq 0$, corresponding to the minimum α_{j_i} or the maximum Δx_{j_i}, tends to be active at the optimal solution. It indicates that one should choose the according index to enter N and take the associated row as the pivot row.

In the following, we bring up matching column rules.
Note that $\Delta x_{j_p} > 0$ since $j_p \in J_\alpha \in J$.
Extract the p-th row of $\bar{B}'$ from tableau (2.6), i.e.,

$$v_{B'}^{\mathrm{T}} = e_p^{\mathrm{T}} \bar{B}'. \tag{2.27}$$

Then, from (2.7) it follows that

$$\Delta x_{j_p} = -e_p^{\mathrm{T}} \bar{B}' \bar{z}_{B'} = -\sum_{j \in B'} v_j \bar{z}_j > 0,$$

which implies that there exists $j \in B'$ such that $v_j \bar{z}_j < 0$. Thus, we can propose the following plausible column rule to matching Rule 2.3.1.

Rule 2.3.2 (Column Rule) Select a column index q to enter B such that

$$q \in \arg\min\{v_j \bar{z}_j \mid j \in B'\}. \tag{2.28}$$

If there is a tie, take one corresponding to the maximum $\hat{x}_q$

Once the row index p and column index q are determined, make the basis change accordingly. To do so, update the face tableau by premultiplying the p-th row by $1/\bar{a}_{p,q}$, and adding $-\bar{a}_{i,q}$ times of the row to the i-th row, for $i = 1, \ldots, m,\ i \neq p$. Accordingly, adjust (B, B') by

$$B = B \backslash \{j_p\} \cup \{q\}, \qquad B' = B' \backslash \{q\}, \qquad N = N \cup \{j_p\}. \tag{2.29}$$

If the B' is nonempty and $\bar{z}_{B'}$ is nonzero, carry out the next iteration.

2.4 Optimality Condition

Obviously, the size of B' reduces iteration by iteration in a nested manner. Once the B' becomes empty or nonempty but $\bar{z}_{B'}$ vanishes, the search direction is meaningless. Then, it is time to test for optimality.

We claim the following result.

Theorem 2.4.1 *Assume that the B' is empty or nonempty, but $\bar{z}_{B'}$ equals 0. If the following set is empty, then the $\bar{x}$ and $(\bar{y}, \bar{z})$ are a pair of primal and dual optimal solutions:*

$$T = \{j \in N \mid \bar{z}_j < 0\} \cup \{j \in N \mid \bar{z}_j \geq 0,\ \ \bar{x}_j > 0\}. \tag{2.30}$$

Proof Since $\bar{x}$ and $\bar{z}$ are primal and dual solutions, respectively, it holds that

$$A\bar{x} = b, \qquad A^{\mathrm{T}}\bar{y} + \bar{z} = c, \quad \bar{y} = B^{-T}c_B. \tag{2.31}$$

The emptiness of the T implies that the set N is empty or it is nonempty, but it holds that

$$\bar{z}_j \geq 0, \qquad \bar{x}_j = 0, \qquad \forall\, j \in N. \tag{2.32}$$

Therefore, indices in the N exhibit complementarity. In addition, it is evident that the same is true for the sets B and B' if the latter is nonempty but $\bar{z}B'$ is zero. Therefore, the claim is valid. □

According to Lemma 1.11.1, the following claim is valid.

Corollary 2.4.2 *Under the same assumption of Theorem 2.4.1, the face $P(B \cup B')$ is optimal.*

If the set T is nonempty, the face is level face but not optimal. Adjust the set by

$$B' = B' \cup T, \qquad N = N \backslash T. \tag{2.33}$$

Consequently, (2.32) holds for the renewed set N if it is nonempty, and the renewed B' is such that

$$\bar{z}_j < 0 \quad \text{or} \qquad \bar{z}_j \geq 0,\;\; \bar{x}_j \geq 0, \qquad \forall\, j \in B' \tag{2.34}$$

Thereby, we can start the next iteration.

2.5 Face Algorithm: Tableau Form

In this section, we form the dual algorithm in tableau form.

The face tableau is manipulated, iteration by iteration, with the $\bar{x}_{B \cup B'}$ updated and (B, B', N) adjusted, unless lower unboundedness is detected. The size of the face $P(B \cup B')$ decreases strictly until the active set B' becomes empty or $\bar{z}B'$ vanishes, reaching a level face. Such a series of continuous iterations is called *face contraction (process)*. If set T is empty, the optimal face is achieved; otherwise, bring the T from N to B' and proceed with another round of face contraction.

Initially, the dual solution $(\bar{y}, \bar{z})$ can be determined by

$$\bar{z}_B = 0, \qquad \bar{z}_N = c_N - N^T\bar{y}, \qquad \bar{y} = B^{-T}c_B,$$

which is taken as the bottom line in the face tableau.

We summarize the overall steps into the following model.

Algorithm 2.5.1 (Face Algorithm: Tableau Form)
Initial : face tableau, feasible solution $\bar{x}$. This algorithm solves the standard LP problem (2.1).

1. *Compute $\Delta x_{B\cup B'}$ by (2.7).*
2. *Stop if J defined by (2.10) is empty.*
3. *Determine α by (2.12) and J_α by (2.22).*
4. *If $\alpha > 0$, update $\bar{x}_{B\cup B'}$ by (2.11).*
5. *Go to Step 8 if $B' \cap J_\alpha$ is empty.*
6. *Adjust (B', N) by (2.24).*
7. *Go to Step 1 if $\bar{z}_{B'}$ is nonzero; else, to Step 13.*
8. *Determine row index p by (2.25).*
9. *Determine column index q by (2.28).*
10. *Multiply the p-th row by $1/\bar{a}_{p,q}$, and add $-\bar{a}_{i,q}$ times of the p-th row to the i-th row for all $i = 1, \ldots, m+1,\ i \neq p$.*
11. *Adjust (B, B', N) by (2.29).*
12. *Go to Step 1 if B' is nonempty and $\bar{z}_{B'}$ is nonzero.*
13. *Stop if T defined by (2.30) is empty.*
14. *Adjust $B' = B' \cup T,\ N = N \setminus T$.*
15. *Go to Step 1.*

Theorem 2.5.1 *Assume nondegeneracy in a single iteration for each contraction process. Algorithm 2.5.1 terminates either at:*

(i) Step 13, giving a pair of primal and dual optimal solutions
(ii) Step 2, declaring lower unboundedness

Proof Obviously, the dimension of the face in any contraction process strictly decreases, and the algorithm reaches a level face at the end of each contraction process within a finite number of iterations unless detecting lower unboundedness.

Assume that Algorithm 2.5.1 does not terminate. Then, it must carry out infinitely many contraction processes; each includes an iteration with nonzero stepsize. According to Theorem 2.2.3, the traversed level faces correspond to strictly reduced objective values. It contradicts the finiteness of the number of level faces. Therefore, the algorithm terminates.

By Theorem 2.4.1, exit Step 13 gives a pair of primal and dual optimal solutions. By Theorem 4.3.4, exit Step 2 indicates lower unboundedness. □

Although degeneracy is still a problem, the nondegeneracy assumption in Algorithm 2.5.1 seems to be eased. It is unimaginable that a contraction process does not include any iteration involving nonzero stepsize.

It is noted that if stepsize is nonzero, components of $\hat{x}_{B'}$ are all strictly positive, and hence any chosen entering $\hat{x}_q$ will never vanish. It is also evident that the leaving index j_p will never re-enter B until another face contraction is conducted.

2.6 Double-Phase Face Algorithm

The face method can only start from a feasible solution. In practice, a phase I process is indispensable if a feasible solution is unknown. In this section, we recommend such a phase I procedure.

The resulting double-phase face algorithm is suitable for general-purpose use. Phase I solves an auxiliary problem using the face method to get an initial feasible solution. Then, phase II solves the original problem using the face method.

The auxiliary problem is formed as follows.

Let h be any nonnegative n-dimensional vector. Set

$$a_{n+1} = b - Ah \tag{2.35}$$

and construct the following auxiliary problem:

$$\begin{aligned} &\min\ g = x_{n+1} \quad \text{to} \quad 0, \\ &\text{s.t.}\ \ [A, a_{n+1}] \begin{bmatrix} x \\ x_{n+1} \end{bmatrix} = b, \qquad x \geq 0. \end{aligned} \tag{2.36}$$

It is verified that the following is a feasible solution to the preceding problem:

$$\begin{bmatrix} h \\ 1 \end{bmatrix} \geq 0.$$

Further, it seems attractive to have the added column normalized. Thus, we set the auxiliary column instead of (2.35):

$$a_{n+1} = (b - Ah)/\|b - Ah\|, \tag{2.37}$$

where $\|\square\|$ denotes the 2-norm of vector $\square$. Then, the problem (2.36) has the following feasible solution:

$$\begin{bmatrix} h \\ \|b - Ah\| \end{bmatrix} \geq 0.$$

Thereby, we can obtain a face solution to the original problem once the phase I problem is solved with the objective value 0.

Example 2.6.1 Solve the following standard problem by double-phase face algorithm:

$$A = \begin{bmatrix} 1 & & & 4 & & 3 & 5 \\ & 1 & & -2 & 8 & -5 & 4 \\ & & 1 & -8 & -4 & & -3 \end{bmatrix} \qquad b = \begin{bmatrix} 15 \\ -18 \\ 9 \end{bmatrix}$$

$$c = [-6, -9, -3, 4, -41, 9, 2]^T$$

Set $B = \{1, 2, 3\}$, $N = \{4, 5, 6, 7\}, h = [1, 1, 1, 0, 0, 0, 0]^T$,
$a_8 = (b - Bh)/\|b - Bh\| = [2159/3843, -337/442, 287/894]^T$,
$\bar{x}_8 = \|b - Bh\| = 7775/312$.
Phase I: Call Algorithm 2.5.1 to solve the following auxiliary problem:

$$A = \begin{bmatrix} 1 & & & 4 & & 3 & 5 & 2159/3843 \\ & 1 & & -2 & 8 & -5 & 4 & -337/442 \\ & & 1 & -8 & -4 & & -3 & 287/894 \end{bmatrix} \qquad b = \begin{bmatrix} 15 \\ -18 \\ 9 \end{bmatrix}$$

$c^T = [0, 0, 0, 0, 0, 0, 0, 1]$[1]
Initial: $B = \{1, 2, 3\}$, $B' = \{8\}, N = \{4, 5, 6, 7\}$,
The available feasible solution to the auxiliary problem: $\bar{x} = [1, 1, 1, 0, 0, 0, 0, 7775/312]^T$.
Put the auxiliary problem into the following tableau:

x_1	x_2	x_3	x_4	x_5	x_6	x_7	x_8
1	0	0	4	0	3	5	2159/3843
0	1	0	−2	+8	−5	4	−337/442
0	0	1	−8	−4	0	−3	287/894
0	0	0	0	0	0	0	1
1	1	1	0	0	0	0	7775/312

Iteration 1:

1. $\Delta x_{B \cup B'} = [-2159/3843, 337/442, -287/894, 1]^{\mathrm{T}}$.
2. $J = \{2, 8\} \neq \emptyset$.
3. $\alpha = \min\{1/(337/442), (7775/312)/1\} = 442/337$.
4. $\bar{x}_{B \cup B'} = [1, 1, 1, 7775/312]^{\mathrm{T}} - (442/337)$
   $[-2159/3843, 337/442, -287/894, 1]^{\mathrm{T}} = [33/19, 0, 27/19, 5123/217]^T$.
5. $B' \cap J_\alpha = \emptyset$. (turn to Step 8)
8. $p = 2, j_p = 2$.
9. $q = 8$.
10.

[1] The auxiliary objective variable is also lower bounded from 0.

x_1	x_2	x_3	x_4	x_5	x_6	x_7	x_8
1	14/19	0	$\frac{48}{19}$	$\frac{112}{19}$	$\frac{-22,150}{32,373}$	$\frac{151}{19}$	0
0	−442/337	0	$\frac{884}{337}$	$\frac{-3536}{337}$	$\frac{2210}{337}$	$\frac{-1768}{337}$	1
0	8/19	1	$\frac{-168}{19}$	$\frac{-11,897}{18,837}$	$\frac{-33,295}{30,124}$	$\frac{-49,546}{37,655}$	0
0	442/337	0	$\frac{-884}{337}$	$\frac{3536}{337}$	$\frac{1768}{337}$	$\frac{-2210}{337}$	0
33/19	0	27/19	0	0	0	0	5123/217

11. $B = \{1, 8, 3\},\ B' = \emptyset,\ N = \{2, 4, 5, 6, 7\}$.
12. $B' = \emptyset$.
13. $T = \{4, 6\}$.
14. $B' = \{4, 6\}$.

Iteration 2:

1. $\Delta x_B = -\bar{B}'\bar{z}_{B'} = -\begin{bmatrix} 48/19 & -22,150/32,373 \\ 884/337 & 2210/337 \\ -168/19 & -33,295/30,124 \end{bmatrix}\begin{bmatrix} -884/337 \\ -2210/337 \end{bmatrix} =$
$\begin{bmatrix} 1667/779 \\ 11,424/229 \\ -1583/52 \end{bmatrix}$, $\Delta x_{B'} = [-884/337, -2210/337]^{\mathrm{T}}$.
2. $J = \{1, 8\} \neq \emptyset$.
3. $\alpha = \min\{(33/19)/(1667/779), (5123/217)/(11,424/229)\} = 1291/2728$.
4. $\bar{x}_{B \cup B'} = [33/19, 5123/217, 27/19, 0, 0]^{\mathrm{T}} - (1291/2728)$ $[1667/779, 11,424/229, -1583/52, -884/337, -2210/337]^{\mathrm{T}} = [4145/5724,$ $0, 459/29, 36/29, 90/29]^T$.
5. $B' \cap J_\alpha = \emptyset$. (turn to Step 8)
8. $p = 2,\ j_p = 6$.
9. $\max[|-884/337|, |-2210/337|] = 2210/337,\ q = 6$.
10.

x_1	x_2	x_3	x_4	x_5	x_6	x_7	x_8
1	3/5	0	14/5	24/5	0	37/5	337/3230
0	−1/5	0	2/5	−8/5	1	−4/5	337/2210
0	1/5	1	−42/5	−12/5	0	−11/5	2174/12,899
0	0	0	0	0	0	0	1
4145/5724	0	459/29	36/29	0	90/29	0	0

11. $B = \{1, 6, 3\}$, $B' = \{4\}$, N = {2,5,7}.
   The auxiliary variable x_8 left the basis. The feasible solution achieved is $\bar{x} = [4145/5724, 0, 459/29, 36/29, 0, 90/29, 0]$

Phase II:

To get the initial tableau for the original problems, drop the x_8 column and replace the objective row by the following original costs:

$c^T = [-6, -9, -3, 4, -41, 9, 2]$.

Then eliminate basic costs by elementary transformations, leading to

x_1	x_2	x_3	x_4	x_5	x_6	x_7
1	3/5	0	14/5	24/5	0	37/5
0	−1/5	0	2/5	−8/5	1	−4/5
0	1/5	1	−42/5	−12/5	0	−11/5
0	−3	0	−8	−5	0	47
4145/5724	0	459/29	36/29	0	90/29	0

$B = \{1, 6, 3\}$, $B' = \{2, 4, 5\}$, $N = \{7\}$.

Iteration 3:

1. $\Delta x_B = -\bar{B}'\bar{z}_{B'} = -\begin{bmatrix} 3/5 & 14/5 & 24/5 \\ -1/5 & 2/5 & -8/5 \\ 1/5 & -42/5 & -12/5 \end{bmatrix} \begin{bmatrix} -3 \\ -8 \\ -5 \end{bmatrix} = \begin{bmatrix} 241/5 \\ -27/5 \\ -393/5 \end{bmatrix}$,

   $\Delta x_{B'} = [-3, -8, -5]^{\mathrm{T}}$.
2. $J = \{1\} \neq \emptyset$.
3. $\alpha = \min\{(4145/5724)/(241/5)\} = 57/3794$.
4. $\bar{x}_{B \cup B'} = [4145/5724, 90/29, 459/29, 0, 36/29, 0]^{\mathrm{T}} - (57/3794)$ $[241/5, -27/5, -393/5, -3, -8, -5]^{\mathrm{T}}$

   $= [0, 2519/791, 6038/355, 171/3794, 659/484, 285/3794]^T$.
5. $B' \cap J_\alpha = \emptyset$. (turn to Step 8)
8. $p = 1$, $j_p = 1$.
9. $\min[|-3|, |-8|, |-5|] = 8$, $q = 4$.
10.

x_1	x_2	x_3	x_4	x_5	x_6	x_7
5/14	3/14	0	1	12/7	0	37/14
−1/7	−2/7	0	0	−16/7	1	−13/7
3	2	1	0	12	0	20
20/7	−9/7	0	0	61/7	0	477/7
0	171/3794	6038/355	659/484	285/3794	2519/791	0

11. $B = \{4, 6, 3\}$, $B' = \{2, 5\}$, $N = \{1, 2, 5, 7\}$.
12. $B' \neq \emptyset$.

Iteration 4:

1. $\Delta x_B = -\bar{B}'\bar{z}_{B'} = -\begin{bmatrix} 3/14 & 12/7 \\ -2/7 & -16/7 \\ 2 & 12 \end{bmatrix}\begin{bmatrix} -9/7 \\ 61/7 \end{bmatrix} = \begin{bmatrix} -1437/98 \\ 958/49 \\ -102 \end{bmatrix}$, $\Delta x_{B'} = [-9/7, 61/7]^{\mathrm{T}}$.
2. $J = \{6, 5\} \neq \emptyset$.
3. $\alpha = \min\{(2519/791)/(958/49), (285/3794)/(61/7)\} = 143/16{,}589$.
4. $\bar{x}_{B\cup B'} = [659/484, 2519/791, 6038/355, 171/3794, 285/3794]^{\mathrm{T}} - (143/16{,}589)$ $[-1437/98, 958/49, -102, -9/7, 61/7]^{\mathrm{T}} = [1979/1330, 564/187, 3345/187, 99/1763, 0]^T$.
   $\bar{x} = [0, 99/1763, 3345/187, 1979/1330, 0, 564/187, 0]$
5. $B' \cap J_\alpha = \{5\}$.
6. $B' = \{2\} \neq \emptyset$.
7. (go to Step 1)

Iteration 5:

1. $\Delta x_B = -\bar{B}'\bar{z}_{B'} = -[3/14, -2/7, 2]^T[-9/7] = [27/98, -18/49, 18/7]^T$, $\Delta x_{B'} = [-9/7]$.
2. $J = \{4, 3\} \neq \emptyset$.
3. $\alpha = \min\{(1979/1330)/(27/98), (3345/187)/(18/7)\} = 1388/257$.
4. $\bar{x}_{B\cup B'} = [1979/1330, 564/187, 3345/187, 99/1763]^{\mathrm{T}} - (1388/257)$ $[27/98, -18/49, 18/7, -9/7]^{\mathrm{T}} = [0, 5, 269{,}131/67{,}283, 584{,}249/83{,}464]^T$.
5. $B' \cap J_\alpha = \emptyset$. (turn to Step 8)
8. $p = 1$, $j_p = 4$.
9. $q = 2$.
10.

x_1	x_2	x_3	x_4	x_5	x_6	x_7
5/3	1	0	14/3	8	0	37/3
1/3	0	0	4/3	0	1	5/3
−1/3	0	1	−28/3	−4	0	−14/3
5	0	0	6	19	0	84
0	584,249/83,464	269,131/67,283	0	0	5	0

11. $B - \{2, 6, 3\}$, $B' = \emptyset$, $N = \{1, 4, 5, 7\}$.
13. $T = \emptyset$. Optimality achieved.

Primal and dual optimal solution:
$\bar{x} = [0, 584{,}249/83{,}464, 269{,}131/67{,}283, 0, 0, 5, 0]^T$,
$\bar{z} = [5, 0, 0, 6, 19, 0, 84]^T$, $f = -30$.
As the end $\bar{z}_N > 0$, the problem has the unique optimal vertex solution $\bar{x}$.

2.7 Face Algorithm

In this section, we convert Algorithm 2.5.1 to a revised version, using original data instead of the face tableau.

Assume that $B = LU$ is available.

Involved items can be calculated as follows:

(i) The reduced costs $\bar{z}_(B')$ and $\bar{z}_N$.

Firstly, solve the following two $m \times m$ triangular systems successively for $\bar{y} = B^{-T} c_B$:

$$U^{\mathrm{T}} w = c_B, \qquad L^{\mathrm{T}} y = w. \tag{2.38}$$

Then, compute

$$\bar{z}_{B'} = c_{B'} - (B')^{\mathrm{T}} \bar{y}, \tag{2.39}$$

or

$$\bar{z}_N = c_N - N^{\mathrm{T}} \bar{y}. \tag{2.40}$$

(ii) Search direction $\Delta x_{B \cup B'}$. Assuming $\Delta x_{B'} = \bar{z}_{B'}$ being available, only $\Delta x_B = -B^{-1} B' \bar{z}_{B'}$ is computed.

Solve the following two $m \times m$ triangular systems:

$$Lw = -B' \bar{z}_{B'}, \qquad U \Delta x_B = w. \tag{2.41}$$

The following column rule would be helpful for illustration.

Rule 2.7.1 (Column Rule) Select an entering index q such that

$$q \in \arg \max_{j \in B'} |\bar{z}_j|. \tag{2.42}$$

Practically, the following column rule is recommendable because of using duality gaps δ_j, reflecting the potential objective value improvement.

To compute duality gaps for $j \in B'$ by (1.41), set the lower bounds $l_j = 0$ and sufficiently large upper bounds, like

$$u_j = 10^{20}.$$

Therefore, the following rule is relevant.

Rule 2.7.2 (Column Rule) Select a column index q to enter B such that

$$q \in \arg\max_{j \in B'} |\delta_j|. \tag{2.43}$$

The preceding column rule is preferable since it removes from B' index q corresponding to the maximum duality gap. The following two column rules are pivot row free. The pivot size can be monitored while updating LU factors.

We revised Algorithm 2.5.1 into the following model: Rule 2.7.1 is used for column selection.

Algorithm 2.7.1 (Face algorithm)
Initial: partition $A = (B, B', N)$ with basis B and empty B', feasible solution $\bar{x}$, factorization $B = LU$. This algorithm solves the standard LP problem (2.1).

1. *Solve the two triangular systems (2.38) for $\bar{y}$.*
2. *Go to Step 5 if B' is empty.*
3. *Compute $\bar{z}_{B'}$ by(2.39).*
4. *Go to Step 8 if $\bar{z}_{B'}$ is nonzero.*
5. *Compute $\bar{z}_N$ by (2.40).*
6. *Stop if the set T defined by (2.30) is empty; (optimality achieved)*
7. *Adjust $B' = B' \cup T$, $N = N \backslash T$.*
8. *Set $\Delta x_{B'} = \bar{z}_{B'}$, and solve the two triangular systems (2.41) for Δx_B.*
9. *Stop if J defined by (2.10) is empty. (lower unbounded)*
10. *Determine α by (2.12) and J_α by (2.22).*
11. *If $\alpha > 0$, update $\bar{x}_{B \cup B'}$ by (2.11).*
12. *If $B' \cap J_\alpha$ is nonempty, then:*

    (1) adjust (B', N) by (2.24);
    (2) go to Step 5 if B' is empty; else go to Step 8.

13. *Determine row index p by (2.25).*
14. *Determine column index q by (2.42).*
15. *Adjust (B, B', N) by (2.29).*
16. *Update L and U.*
17. *Go to Step 1.*

Note: This algorithm solves four triangular systems in each iteration (in Steps 1 and 12(i)).

Example 2.7.1 Solve the following standard problem by Algorithm 2.7.1:

$$A = \begin{bmatrix} -1 & -2 & 8 & -5 & 4 & 0 & 8 & 1 & 6 \\ -2 & 5 & 0 & -3 & 0 & -4 & 0 & -8 & 5 \\ 8 & 7 & 4 & 0 & 3 & 5 & 1 & 0 & -2 \\ 3 & 0 & -6 & -2 & -2 & -8 & -6 & 7 & -6 \end{bmatrix} \qquad b = \begin{bmatrix} 16 \\ -10 \\ 13 \\ -2 \end{bmatrix}$$

$c = [0, 0, 0, 0, -8, 2, -7, -6, 7]^T$.

Initial: $B = \{1, 2, 3, 4\}$, $B' = \emptyset$, $N = \{5, 6, 7, 8, 9\}$,

$\bar{x} = [1, 0, 1, 0, 0, 0, 1, 1, 0]^T$.

Permuted LU factors:

$$L = \begin{bmatrix} -1/8 & -1/6 & 1 & 0 \\ -1/4 & 1 & 0 & 0 \\ 1 & 0 & 0 & 0 \\ 3/8 & -7/18 & -32/39 & 1 \end{bmatrix}, \quad U = \begin{bmatrix} 8 & 7 & 4 & 0 \\ 0 & 27/4 & 1 & -3 \\ 0 & 0 & 26/3 & -11/2 \\ 0 & 0 & 0 & -599/78 \end{bmatrix}.$$

Iteration 1:

1. $\bar{y} = 0$.
2. $B' = \emptyset$.
5. $\bar{z}_N = [-8, 2, -7, -6, 7]^T$.
6. $T = \{5, 7, 8\}$.
7. $B' = \{5, 7, 8\}, \ N = \{6, 9\}$.
8. $\Delta x_{B'} = [-8, -7, -6]^T, \ u = [94, -48, 31, -16]^T$, $w = [31, -161/4, 547/6, 2459/78]^T, \ \Delta x_B = [1955/252, -5367/599, 1749/221, -2459/599]^T$.
9. $J = \{1, 3\}$.
10. $\alpha = \min[1/(1955/252), 1/(1749/221)] = 221/1749, \ J_\alpha = \{3\}$.
11. $\bar{x}_{B \cup B'} = [187/9481, 257/227, 0, 568/1095, 1954/1933, 669/355, 1025/583]^T$.
    $\bar{x}[187/9481, 257/227, 0, 568/1095, 1954/1933, 0, 669/355, 1025/583, 0]^T$.
12. $B' \cap J_\alpha = \emptyset$.
13. $p = 3$.
14. $\max[|-8|, |-7|, |-6|] = 8, \ q = 5$.
15. $B = \{1, 2, 5, 4\}, \ B' = \{7, 8\}$.
16. $$L = \begin{bmatrix} -1/8 & -1/6 & 1 & 0 \\ -1/4 & 1 & 0 & 0 \\ 1 & 0 & 0 & 0 \\ 3/8 & -7/18 & -17/27 & 1 \end{bmatrix}, \quad U = \begin{bmatrix} 8 & 7 & 3 & 0 \\ 0 & 27/4 & 3/4 & -3 \\ 0 & 0 & 9/2 & -11/2 \\ 0 & 0 & 0 & -179/27 \end{bmatrix}.$$

Iteration 2:

1. $w = [0, 0, -16/9, 264/179]^T, \ \bar{y} = [-152/179, 232/537, -296/537, 264/179]^T$.
2. $\bar{z}_{B'} = [2611/284, -1935/161]^T$.
8. $\Delta x_{B'} = [2611/284, -1935/161]^T, \ u = [-11{,}014/179, -4519/47, -2611/284, 9054/65]^T$,
   $w = [-2611/284, -3741/38, -11{,}705/148, 4646/85]^T$,
   $\Delta x_B = [57{,}397/2551, -4811/317, -3733/135, -1921/233]^T$.
9. $J = \{1, 7\}$.
10. $\alpha = \min[(187/9481)/(57{,}397/2551), (669/355)/(2611/284)] = 121/138{,}031, \ J_\alpha = \{7\}$.
11. $\bar{x}_{B \cup B'} = [1023/947, 97/12{,}693, 3953/916, 901/1936, 0]^T$,
    $\bar{x} = [0, 97/12{,}693, 0, 901/1936, 3953/916, 0, 0, 1023/947, 0]^T$.
12. $B' \cap J_\alpha \neq \emptyset$.
13. $p = 1$

14. $\max[|2611/284|, |-1935/161|] = 1935/161,\ q = 8.$
15. $B = \{8, 2, 5, 4\},\ B' = \{7\}.$
16. $L = \begin{bmatrix} -1/8 & -11/56 & 1 & 0 \\ 1 & 0 & 0 & 0 \\ 0 & 1 & 0 & 0 \\ -7/8 & 5/8 & -217/257 & 1 \end{bmatrix}, \quad U = \begin{bmatrix} -8 & 5 & 0 & -3 \\ 0 & 7 & 3 & 0 \\ 0 & 0 & 257/56 & -43/8 \\ 0 & 0 & 0 & -2355/257 \end{bmatrix}.$

Iteration 3:

1. $w = [3/4, -15/28, -358/257, 675/1181]^T,$
   $\bar{y} = [-2144/2355, 892/785, -687/641, 675/1181]^T.$
2. $\bar{z}_{B'} = [3837/802]^T.$
8. $\Delta x_{B'} = [3837/802]^T,\ u = [-4325/113, 0, -3837/802, 11{,}511/401]^T,$
   $w = [0, -3837/802, -13{,}921/355, -1777/1256]^T,$
   $\Delta x_B = [1169/666, 3049/1051, -3195/382, 237/1535]^T.$
9. $J = \{8, 2, 4, 7\}.$
10. $\alpha = \min[497/281)/(1169/666), (2575/2248)/(3049/1051),$
    $(679/1291)/(237/1535).(486/259)/(3837/802)] = 433/1104,\ J_\alpha = \{7\}.$
11. $\bar{x}_{B\cup B'} = [0, 2575/2248, 1091/1054, 679/1291, 486/259, 497/281]^T.$
    $\bar{x} = [0, 97/12{,}693, 0, 901/1936, 3953/916, 0, 0, 1023/947, 0]^T.$
12. $B' \cap J_\alpha = \emptyset.$

    (1) $B' = \emptyset,\ N = \{1, 3, 6, 7, 9\};$
    (2) (to Step 5)

    Iteration 4:
5. $\bar{z}_N = c_N - N^T\bar{y} = [1595/194, 35{,}324/2355, 3147/191, 3837/802, 121/15]^T.$
6. $T = \emptyset$. Optimality achieved.

Primal and dual optimal solutions are as follows:
$\bar{x} = [0, 97/12{,}693, 0, 901/1936, 3953/916, 0, 0, 1023/947, 0]^T,\ f = c^T\bar{x} = -7422/181,$
$\bar{z}[1595/194, 0, 35{,}324/2355, 0, i, 3147/191, 3837/802, 0, 121/15]^T.$

Chapter 3
Generalized Face Method

In this chapter, we generalize the face method for solving the variable-bounded problem (1.35).

We first put forward the optimality condition in Sect. 3.1. In Sect. 3.2, we address the update of the feasible solution. Section 3.3 focuses on the pivoting operations. Section 3.4 formulates the generalized face algorithm in tableau and revised forms. Section 3.5 offers a practicable version of the generalized face algorithm. In the final section, we discuss the face versus the simplex method.

We are concerned with the variable-bounded problem, i.e.,

$$\begin{aligned} &\min\ c^T x \\ &\text{s.t.}\quad Ax = b, \quad l \leq x \leq u, \end{aligned} \tag{3.1}$$

where $A \in R^{m\times n}$ $(m < n)$, $rank(A) = m, b \in R^m, c, l, u \in R^n$.

Let $A = (B, B', N)$ be a partition, where B is the basis. Assume that $\bar{x}$ is a feasible solution such that

$$l_{B\cup B'} \leq \bar{x}_{B\cup B'} \leq u_{B\cup B'}; \qquad \bar{x}_j = l_j/u_j, \quad j \in N. \tag{3.2}$$

Introduce the following concepts:

Definition 3.0.1 The face is defined as the following subset of the feasible region:

$$\begin{aligned} P(B\cup B') = \{x \in \mathcal{R}^n \mid Bx_B + B'x_{B'} = b,\ l_{B\cup B'} \leq x_{B\cup B'} \leq u_{B\cup B'}; \\ x_j = l_j/u_j,\ j \in N\}, \end{aligned} \tag{3.3}$$

The index set B' is *active* and N *nonface/inactive*. A face is a *level face* if solutions in it correspond to the same objective value. A level face is *optimal face* if the corresponding objective value equals the optimal value.

P.-Q. Pan, *Face Method*, https://doi.org/10.1007/978-3-031-93594-7_3

3.1 Optimality Condition

In this section, we discuss the optimality in the current context.

Let x be a primal solution and let $(\bar{y}, \bar{z})$ be a dual solution such that

$$\bar{z}_B = 0, \qquad N^{\mathrm{T}}\bar{y} + \bar{z}_N = c_N, \tag{3.4}$$

where $B^T\bar{y} = c_B$.

Introduce the following index set:

$$T = \{j \in N \mid \bar{z}_j < 0,\ u_j > \bar{x}_j\} \cup \{j \in N \mid \bar{z}_j > 0,\ \bar{x}_j > l_j\}, \tag{3.5}$$

which is composed of indices with nonzero duality gaps (Definition 1.8.2).

We have the following result.

Theorem 3.1.1 *Let $\bar{x}$ be a feasible solution and let $\bar{z}$ satisfy (3.4). Assume that B' is empty or $\bar{z}_{B'}$ vanishes. If the T is empty, $(\bar{x}, \bar{z})$ are a pair of primal and dual optimal solutions.*

Proof Emptiness of T implies that

$$\bar{x}_j = u_j, \qquad \text{s.t.} \quad \bar{z}_j < 0, \quad j \in N. \tag{3.6}$$

and that

$$\bar{x}_j = l_j, \qquad \text{s.t.} \quad \bar{z}_j > 0, \quad j \in N. \tag{3.7}$$

Set

$$v_B, w_B = \bar{z}_B = 0, \tag{3.8}$$

and introduce the following two vectors:

$$v_j = \begin{cases} -\bar{z}_j & \text{if } \bar{z}_j < 0 \\ 0 & \text{if } \bar{z}_j \geq 0 \end{cases} \qquad w_j = \begin{cases} 0 & \text{if } \bar{z}_j \leq 0 \\ \bar{z}_j & \text{if } \bar{z}_j > 0 \end{cases} \qquad j \in N \tag{3.9}$$

it is evident that

$$-v + w = \bar{z}, \qquad v, w \geq 0,$$

which along with (3.4) and (3.8) implies dual feasibility satisfied. Then, (3.6)–(3.7) and (3.9) reveal that $\bar{x}$ and $(\bar{y}, v, w)$ satisfy complementarity slackness. Therefore, the optimality condition (1.40) is satisfied, and the claim is valid. □

Corollary 3.1.2 *Under the same assumption of Theorem 3.1.1, the face $P(B \cup B')$ is optimal.*

If T is nonempty, reset

$$B' = B' \cup T, \qquad N = N \backslash B'. \tag{3.10}$$

Further, proceed to the next iteration.

3.2 Updating Face Solution

The $\Delta x_{B \cup B'}$ given by (2.7) is still an applicable search direction in the face.

Let $\bar{x}$ be the current face solution. With $\bar{x}_N$ fixed, we only update $\bar{x}_{B \cup B'}$ by the following line search scheme:

$$\hat{x}_{B \cup B'} = \bar{x}_{B \cup B'} - \alpha \Delta x_{B \cup B'}, \tag{3.11}$$

where α is a stepsize.

To maintain feasibility, the new solution $\hat{x}_{B \cup B'}$ must satisfy

$$l_{B \cup B'} \leq \hat{x}_{B \cup B'} \leq u_{B \cup B'}.$$

This restriction leads to the maximum stepsize below:

$$\alpha = \min \alpha_j, \tag{3.12}$$

where

$$\alpha_j = \begin{cases} (\bar{x}_j - u_j)/\Delta x_j & \text{if } \Delta x_j < 0 \\ (\bar{x}_j - l_j)/\Delta x_j & \text{if } \Delta x_j > 0 \end{cases} \qquad j \in B \cup B' \tag{3.13}$$

It can be shown that the new solution is feasible, and the corresponding objective value strictly decreases if the stepsize α is positive.

As for the merit of the search direction, Theorem 2.2.3 is still correct. Thus, the search direction leads to the maximum possible objective decrement, proportional to $\|\bar{z}_{B'}\|^2$.

3.3 Pivoting Operation

Let us focus on pivoting operations and related index set adjustments.

Assume that the new feasible solution $\hat{x}_{B \cup B'}$ has been computed. Then, the following index set is well-defined:

$$J_\alpha = \{j \in B \cup B' \mid \alpha_j = \alpha\}, \tag{3.14}$$

where α and α_j are defined by (3.12) and (3.13), respectively. For each $j \in J_\alpha$, the $\hat{x}_j$ reaches one of its bounds. Therefore, indices/variables associated with J_α are said to be *blocking*.

There are the following two types of iterations.

(a) Simple iteration: $B' \cap J_\alpha$ is nonempty.
Since blocking indices exist in B' in this case, there is no need for a basis change, so the involved computing cost is low. Before carrying out the next iteration, what to do is to adjust the following sets:

$$B' = B' \backslash (B' \cap J_\alpha), \qquad N = N \cup (B' \cap J_\alpha). \tag{3.15}$$

(b) Full iteration: $B' \cap J_\alpha = \emptyset$.
This case implies $J_\alpha \subset B$. Then, there is a need for pivot selection and basis change. Note that the ith basic variable x_{j_i} corresponds to the i-th row of tableau (2.6).

First, determine the pivot row using the following rule.

Rule 3.3.1 (Row Rule) Assume that $\Delta x_{B \cup B'}$ and J_α are defined by (2.7) and (3.14), respectively. Select row index p such that

$$j_p \in \arg\max_{j_i \in J_\alpha} |\Delta x_{j_i}|. \tag{3.16}$$

Once the blocking index $j_p \in B$ is determined to leave B, what to do next is to select an entering index from B'. Among multiple possible choices, the following column rule is recommendable as it removes the index corresponding to the maximum potential objective improvement.

Rule 3.3.2 (Column Rule) Denote the p-th row of $\bar{B}'$ by $v_{B'}^T$. Select the entering index q such that

$$q \in \arg\max \{|\delta_j| \mid |v_j| > \epsilon, \ j \in B'\}, \tag{3.17}$$

where δ_j, $j \in B'$ denote duality gaps defined by (1.41), and ϵ is a predetermined tolerance for pivot size.

Once the pivot row and column indices are determined, update the canonical face tableau (2.6) by elementary transformations, and adjust the following index sets (B, B', N) accordingly:

$$B = (B \backslash \{j_p\}) \cup \{q\}, \ B' = B' \backslash \{q\}, \ N = N \cup \{j_p\}. \tag{3.18}$$

Note that the size of B' decreases strictly until B' vanishes. Then, it is time to test for the optimality.

3.4 Generalized Face Algorithm

Initially, assume that the following set B' is nonempty:

$$B' = \{j \in N \mid \bar{z}_j < 0,\ u_j > \bar{x}_j\} \cup \{j \in N \mid \bar{z}_j \geq 0,\ \bar{x}_j > l_j\}. \tag{3.19}$$

Assume that $\bar{z}_{B'}$ is nonzero.

We summarize the overall steps into the following model.

Algorithm 3.4.1 (Generalized Face Algorithm: Tableau Form)
Initial : Partition $A = (B, B', N)$, face tableau in form (2.6), feasible solution $\bar{x}$, big number σ. This algorithm solves the variable-bounded problem (3.1).

1. *Compute $\Delta x_{B \cup B'}$ by (2.7).*
2. *Determine α by (3.12)–(3.13) and J_α.*
3. *Stop if $\alpha > \sigma$.*
4. *If $\alpha > 0$, update $\bar{x}_{B \cup B'}$ by (3.11).*
5. *If $J_\alpha \cap B'$ is nonempty, then:*

   (1) adjust index sets by (3.15);
   (2) go to Step 1 if B' is nonempty, else go to Step 11.

6. *Determine row index p such that $j_p \in \arg\max_{j_i \in J_\alpha} |\Delta x_{j_i}|$.*
7. *Select entering index q by (3.17).*
8. *Multiply the p-th row by $1/\bar{a}_{p,q}$, and add $-\bar{a}_{i,q}$ times of the p-th row to the i-th row for $i = 1, \ldots, m+1;\ i \neq p$.*
9. *Adjust (B, B', N) by (3.18).*
10. *Go to Step 1 if B' is nonempty and $\bar{z}_{B'}$ is nonzero.*
11. *Stop if T defined by (3.5) is empty.*
12. *Adjust $B' = B' \cup T$, $N = N \backslash T$.*
13. *Go to Step 1.*

We state the following result with the proof omitted.

Theorem 3.4.1 *If each face contraction involves a nondegenerate feasible solution, Algorithm 3.4.1 terminates either at:*

(i) Step 11, giving an optimal face together with an optimal solution
(2) Step 3, declaring lower unboundedness

Let us turn to the revision of the preceding algorithm.

Instead, we recommend using duality gaps for column pivoting in Rule 2.7.2.

Initially, partition $A = (B, B', N)$ is available, where B' is assumed to be an empty set.

Using expressions before Algorithm 2.7.1, it is not difficult to revise Algorithm 3.4.1 to the following model.

Algorithm 3.4.2 (Generalized Face Algorithm)
Initial : partition $A = (B, B', N)$, factorization $B = LU$, feasible solution $\bar{x}$, big number σ. This algorithm solves the variable-bounded problem (3.1).

1. *Solve the two triangular systems (2.38) for $\bar{y}$.*
2. *Go to Step 4 if B' is empty; else, compute $\bar{z}_{B'}$ by (2.39).*
3. *Go to Step 7 if $\bar{z}_{B'}$ is nonzero*
4. *Compute $\bar{z}_N$ by (2.40).*
5. *Stop if the set T defined by (2.30) is empty; (optimality achieved)*
6. *Adjust $B' = B' \cup T, N = N \backslash T$.*
7. *Set $\Delta x_{B'} = \bar{z}_{B'}$, and solve triangular systems (2.41) for Δx_B.*
8. *Determine α by (3.12)–(3.13) and J_α.*
9. *Stop if $\alpha > \sigma$ (lower unboundedness declared).*
10. *If $\alpha > 0$, update $\bar{x}_{B \cup B'}$ by (3.11).*
11. *If $J_\alpha \cap B' \neq \emptyset$, then:*

    (1) adjust index sets by (3.15);
    (2) go to Step 1 if B' is empty, else go to Step 2.

12. *Determine row index p such that $j_p \in \arg\max_{j_i \in J_\alpha} |\Delta x_{j_i}|$.*
13. *Select column index $q \in \arg\max_{j \in B'} \delta_j$, where δ_j is defined by (1.41).*
14. *Adjust (B, B', N) by (3.18).*
15. *Update L and U.*
16. *Go to Step 1.*

Example 3.4.1 Solve the following variable-bounded problem by Algorithm 3.4.1:

$$A = \begin{bmatrix} 1 & & & 4 & -2 & 0 & 3 & -5 \\ & 1 & & -1 & 3 & 6 & 2 & 8 \\ & & 1 & 2 & -1 & -6 & 5 & -4 \end{bmatrix} \qquad b = \begin{bmatrix} 1 \\ 19 \\ -3 \end{bmatrix}$$

$c = [0, 0, 0, 5, 7, 3, -8, -2]^T$
$u = [5, 2, 4, 8, 2, 6, 3, 4]^T$
$\bar{x} = [1, 1, 1, 1, 1, 1, 1, 1]^T$
$l = [0, 0, 0, 0, 0, 0, 0, 0]^T$

Put the problem into the following face tableau:

x^T	x_1	x_2	x_3	x_4	x_5	x_6	x_7	x_8
	1			4	−2	0	3	−5
		1		−1	3	6	2	8
			1	2	−1	−6	5	−4
				5	7	3	−8	−2
u^T	5	2	4	8	2	6	3	4
$\bar{x}^T$	1	1	1	1	1	1	1	1
l^T	0	0	0	0	0	0	0	0

where the bottom 3 rows are for reference.

Initial: $\sigma = 10^{14}$, $B' = \{4, 5, 6, 7, 8\}$, $N = \emptyset$.
Iteration 1:

1. $\Delta x_{B'} = \bar{z}_{B'} = [5, 7, 3, -8, -2]^T$,

$$\Delta x_B = -\bar{B}'\bar{z}_{B'} = -\begin{bmatrix} 4 & -2 & 0 & 3 & -5 \\ -1 & 3 & 6 & 2 & 8 \\ 2 & -1 & -6 & 5 & -4 \end{bmatrix} [5, 7, 3, -8, -2]^T =$$

$[8, -2, 47]^T$.

2. $\alpha = \min[(\bar{x}(1) - 0)/(8), (\bar{x}(2) - u(2))/(-2), (\bar{x}(3) - 0)/(47), (\bar{x}(4) - 0)/(5), (\bar{x}(5) - 0)/(7), (\bar{x}(6) - 0)/(3), (\bar{x}(7) - u(7))/(-8), (\bar{x}(8) - u(8))/(-2)]$
$= \min[1/8, 1/2, 1/47, 1/5, 1/7, 1/3, 1/4, 3/2] = 1/47$, $J_\alpha = \{3\}$.
4. $\bar{x}_{B \cup B'} = \bar{x}_{B \cup B'} - \alpha \Delta x_{B \cup B'} = [39/47, 49/47, 0, 42/47, 40/47, 44/47, 55/47, 49/47]^T$
5. $J_\alpha \cap B' = \emptyset$.
6. $p = 3$, $j_p = 3$.
7. $\max[(\bar{x}(4) - 0)(5), (\bar{x}(5) - 0)(7), (\bar{x}(6) - 0)(3), (\bar{x}(7) - u(7))(-8), (\bar{x}(8) - u(8))(-2)]$
$= \max[210/47, 280/47, 132/47, 688/47, 278/47] = 688/47$, $q = 7$.
8.

x^T	x_1	x_2	x_3	x_4	x_5	x_6	x_7	x_8
	1		−3/5	14/5	−7/5	18/5		−13/5
		1	−2/5	−9/5	17/5	42/5		48/5
			1/5	2/5	−1/5	−6/5	1	−4/5
			8/5	41/5	27/5	−33/5		−42/5
u^T	5	2	4	8	2	6	3	4
$\bar{x}^T$	39/47	49/47	0	42/47	40/47	44/47	55/47	49/47
l^T	0	0	0	0	0	0	0	0

9. $B = \{1, 2, 7\}$, $B' = \{4, 5, 6, 8\}$, $N = \{3\}$.
10. $\bar{z}_{B'} \neq 0$.

Iteration 2:

1. $\Delta x_{B'} = \bar{z}_{B'} = [41/5, 27/5, -33/5, -42/5]^T$,

$$\Delta x_B = -\bar{B}'\bar{z}_{B'} = -\begin{bmatrix} 4 & -2 & 0 & 3 & -5 \\ -1 & 3 & 6 & 2 & 8 \\ 2 & -1 & -6 & 5 & -4 \end{bmatrix} \bar{z}_{B'} = [-337/25, 3312/25, -421/$$

$25]^T$.

2. $\alpha = \min[(\bar{x}(1) - u(1))/(-337/25), (\bar{x}(2) - 0)/(3312/25), (\bar{x}(7) - u(7))/(-421/25), (\bar{x}(4) - 0)/(41/5), (\bar{x}(5) - 0)/(27/5), (\bar{x}(6) - u(6))/(-33/5), (\bar{x}(8) - u(8))/(-42/5)] = \min[641/2072, 55/6989, 123/1132, 210/1927, 200/1269, 1190/1551, 695/1974] = 55/6989$, $p = 2$, $J_\alpha = \{2\}$.

4. $\bar{x}_{B\cup B'} = \bar{x}_{B\cup B'} - \alpha\Delta x_{B\cup B'} = [1649/1762, 0, 667/512, 781/942, 887/1097, 831/841, 1255/1132]^T$
   $\bar{x} = [1649/1762, 0, 0, 781/942, 887/1097, 831/841, 667/512, 1255/1132]^T$.
5. $J_\alpha \cap B' = \emptyset$.
6. $p = 2,\ j_p = 2$.
7. $\max[(\bar{x}(4) - 0)(41/5), (\bar{x}(5) - 0)(27/5), (\bar{x}(6) - u(6))(-33/5), (\bar{x}(8) - u(8))(-42/5)]$
   $= \max[911/134, 3314/759, 14{,}753/446, 6679/275] = 14{,}753/446,\ q = 6$
8.

x^T	x_1	x_2	x_3	x_4	x_5	x_6	x_7	x_8
	1	−3/7	−3/7	25/7	−20/7			−47/7
		5/42	−1/21	−3/14	17/42	1		8/7
		1/7	1/7	1/7	2/7		1	4/7
		11/14	9/7	95/14	113/14			−6/7
u^T	5	2	4	8	2	6	3	4
$\bar{x}^T$	39/47	49/47	0	42/47	40/47	44/47	55/47	49/47
l^T	0	0	0	0	0	0	0	0

9. $B = \{1, 6, 7\},\ B' = \{4, 5, 8\},\ N = \{2, 3\}$.
10. $\bar{z}_{B'} \neq 0$.

Iteration 3:

1. $\Delta x_{B'} = \bar{z}_{B'} = [95/14, 113/14, -6/7]^T$,
   $$\Delta x_B = -\bar{B}'\bar{z}_{B'} = -\begin{bmatrix} 25/7 & -20/7 & -47/7 \\ -3/14 & 17/42 & 8/7 \\ 1/7 & 2/7 & 4/7 \end{bmatrix} \bar{z}_{B'} = [-97/14, -5/6,$$
   $-39/14]^T$.
2. $\alpha = \min[(\bar{x}(1) - u(1))/(-97/14), (\bar{x}(6) - u(6))/(-5/6), (\bar{x}(7) - u(7))/(-39/14),$
   $(\bar{x}(4) - 0)/(95/14), (\bar{x}(5) - 0)/(113/14), (\bar{x}(8) - u(8))/(-6/7)] = \min[603/1028,$
   $5058/841, 1235/2027, 298/2439, 567/5660, 4058/1203] = 567/5660,\ J_\alpha = \{5\}$.
4. $\bar{x}_{B\cup B'} = \bar{x}_{B\cup B'} - \alpha\Delta x_{B\cup B'}$
   $= [947/581, 1422/1327, 1460/923, 601/4025, 0, 1265/1059]^T$,
   $\bar{x} = [947/581, 0, 0, 601/4025, 0, 1422/1327, 1460/923, 1265/1059]^T$.
5. $J_\alpha \cap B'$ is nonempty.

   (1) $B' = \{4, 8\},\ N = \{2, 3, 5\}$.
   (2) (switch to the Step 1)

Iteration 4:

1. $\Delta x_{B'} = \bar{z}_{B'} = [95/14, -6/7]^T$,

$$\Delta x_B = -\bar{B}'\bar{z}_{B'} = -\begin{bmatrix} 25/7 & -47/7 \\ -3/14 & 8/7 \\ 1/7 & 4/7 \end{bmatrix} \bar{z}_{B'} = [-2939/98, 477/196, -47/98]^T.$$

2. $\alpha = \min[(\bar{x}(1) - u(1))/(-2939/98), (\bar{x}(6) - 0)/(477/196),$
$(\bar{x}(7) - u(7))/(-47/98), (\bar{x}(4) - 0)/(95/14), (\bar{x}(8) - u(8))/(-6/7)]$
$= \min[277/2465, 166/377, 1999/676, 173/7862, 1810/553] = 173/7862,$
$J_\alpha = \{4\}.$
4. $\bar{x}_{B\cup B'} = \bar{x}_{B\cup B'} - \alpha \Delta x_{B\cup B'}$
$= [5443/2377, 903/887, 1957/1229, 0, 961/792]^T,$
$\bar{x} = [5443/2377, 0, 0, 0, 0, 903/887, 1957/1229, 961/792]^T.$
5. $J_\alpha \cap B'$ is nonempty.

    (1) $B' = \{8\}, \ N = \{2, 3, 4, 5\}.$
    (2) (switch to the next Step 1)

Iteration 5:

1. $\Delta x_{B'} = \bar{z}_{B'} = [-6/7]^T,$
$\Delta x_B = -\bar{B}'\bar{z}_{B'} = [-282/49, 48/49, 24/49]^T.$
2. $\alpha = \min[(\bar{x}(1) - u(1))/(-282/49), (\bar{x}(6) - 0)/(48/49),$
$(\bar{x}(7) - 0)/(24/49), (\bar{x}(8) - u(8))/(-6/7)]$
$= \min[429/911, 662/637, 3095/952, 2318/713] = 429/911, \ J_\alpha = \{1\}.$
4. $\bar{x}_{B\cup B'} = \bar{x}_{B\cup B'} - \alpha \Delta x_{B\cup B'}$
$= [5, 157/282, 64/47, 76/47]^T,$
$\bar{x} = [5, 0, 0, 0, 0, 157/282, 64/47, 76/47]^T.$
5. $J_\alpha \cap B'$ is empty.
6. $p = 1, \ j_p = 1.$
7. $q = 8$
8.

x^T	x_1	x_2	x_3	x_4	x_5	x_6	x_7	x_8
	−7/47	3/47	3/47	−25/47	20/47			1
	8/47	13/282	−17/141	37/94	−23/282	1		
	4/47	5/47	5/47	21/47	2/47		1	
	−6/47	79/94	63/47	595/94	793/94			
u^T	5	2	4	8	2	6	3	4
$\bar{x}^T$	5	0	0	0	0	157/282	64/47	76/47
l^T	0	0	0	0	0	0	0	0

9. $B = \{8, 6, 7\}, \ B' = \emptyset, \ N = \{1, 2, 3, 4, 5\}.$
10. Nothing to do.
11. $T = \emptyset$. Optimality is achieved.

The pair of primal and dual optimal face solution:
$\bar{x} = [5, 0, 0, 0, 0, 157/282, 64/47, 76/47]^T,$

$\bar{z} = [-6/47, 79/94, 63/47, 595/94, 793/94, 0, 0, 0]^T$, $\bar{f} = c^T\bar{x} = -1171/94$.

3.5 Practicable Face Algorithm

Note that the computing cost for the search direction of the face algorithm may become enormous when the number of columns of the active matrix B' is large in solving large-scale LP problems, let alone zero components of the $\bar{z}_{B'}$. As a countermeasure, this section introduces a practicable face algorithm, taking into account duality gaps (1.41) in conjunction with the N-M pivot rule (Sect. 1.12.3).

Using the lower bound $l_j = 0$ and a sufficiently large upper bound on x_j, like

$$u_j = 10^{20},$$

the duality gaps δ_j defined by (1.41) can be determined.

The trick is to take a certain percent of the top components of the B' to calculate the search direction (based on duality gaps).

Let $\epsilon > 0$ be a given optimality tolerance and μ be a given proportion, like $0.6 - 0.9$.

There are the following two cases:

(a) The B' is empty or $\bar{z}_{B'}$ vanishes.
Optimality is achieved if

$$\max_{j\in N} |\delta_j| < \epsilon. \tag{3.20}$$

In the other case, determine the following threshold:

$$\sigma_1 = \mu \max_{j\in N} \delta_j, \tag{3.21}$$

and set

$$B' = \{j \in N \mid \delta_j > \sigma_1\}, \qquad N\backslash B'. \tag{3.22}$$

(b) B' is nonempty and $\bar{z}_{B'}$ is nonzero.
Determine the following threshold:

$$\sigma = \mu \max_{j\in B'} \delta_j, \tag{3.23}$$

and reset

$$B' = \{j \in B' \mid \delta_j > \sigma\}, \qquad N \backslash B'. \tag{3.24}$$

Actualizing the preceding considerations leads to the following algorithm.

Algorithm 3.5.1 (Practicable Face Algorithm)
Initial: partition $A = (B, B', N)$ with basis B and empty B', feasible solution $\bar{x}$, factorization $B = LU$, optimality tolerance ϵ and proportion μ. This algorithm is the same as Algorithm 3.4.2 except for Steps 2-6 replace by

2. *Go to Step 4 if B' is empty; else, compute $\bar{z}_{B'}$ by (2.39).*
3. *If $\bar{z}_{B'}$ is nonzero, reset B' by (3.24) and go to Step 7.*
4. *Compute $\bar{z}_N$ by (2.40).*
5. *Stop if the condition (3.20) is satisfied; (optimality achieved)*
6. *Adjust B' and N by (3.22).*

3.6 Remarks: Face Method vs. Simplex Method

In summary, let us examine the features of the face versus simplex method.

(i) First, the face method no longer moves from vertex to vertex but from face to face instead, with a sequence of associated face solutions, until reaching an optimal face together with an optimal solution. One benefit of this fresh philosophy is that the face method would have more chance to gain success against degeneracy.

(ii) Unlike the simplex method, which uses the search direction along a descent edge of the underlying polyhedron, the face method takes a combination of edges corresponding to nonzero active reduced costs. It would be the source of the unique feature that it gains the remarkable objective decrement per iteration. The number of iterations required for solving LP problems should decrease significantly.

(iii) The face method determines the pivot row first and selects the pivot column after, which is the opposite of the simplex method. While the pivot row is deterministic, the pivot column is selectable. The column rule considering duality gaps is especially promising, which should significantly contribute to the method's efficiency. Moreover, we strongly recommend incorporating the N-M Rule 1.12.3 into the face method.

(iv) The face method is stable, relative to the simplex method, because its pivot is selectable, while the latter's pivot can be arbitrarily close to zero. As we know, the latter sometimes cannot maintain a usable basis and fails to solve large-scale and challenging problems.

(v) As for computing cost in a single iteration, the face method is similar to the simplex method. It also involves four triangular systems. LU factorization can be implemented similarly to the simplex context.

Chapter 4
Dual Face Method

In this chapter, we present the dual face method for solving the standard LP problem (2.1) but achieve the goal from the dual side.

Section 4.1 addresses the basic facts essential for understanding the dual face method. Section 4.2 introduces the ascent search direction. Section 4.3 discusses the update of the dual face solution. Section 4.4 handles the pivoting operations. Section 4.5 focuses on the optimality condition. Then, we formulate the dual face algorithm in tableau form in Sect. 4.6, which must start from a dual feasible solution. We present the so-called hybrid of primal/dual face methods in Sect. 4.7, which can start from any solution. The last section formulates the dual face algorithm in the revised form.

To solve the standard LP problem, i.e.,

$$\begin{array}{ll} \min\ f = c^{\mathrm{T}}x & \\ \text{s.t.}\ \ Ax = b, & x \geq 0, \end{array} \tag{4.1}$$

we will handle its dual problem, i.e.,

$$\begin{array}{ll} \max\ g = b^{\mathrm{T}}y & \\ \text{s.t.}\ \ A^{\mathrm{T}}y + z = c, & z \geq 0, \end{array} \tag{4.2}$$

where $A \in R^{m\times n},\ b \in R^m,\ c \in R^n,\ \mathrm{Rank}(A) = m.$

The classical basis is a submatrix consisting of m independent columns of A. To allow fewer columns, we generalize it as follows (Pan [24, 31]).

Definition 4.0.1 Deficient basis is a submatrix consisting of independent columns of A whose column space includes the right-hand side b.

The deficient basis is generalized further (Pan [27, 28]):

Definition 4.0.2 Pseudo-basis is a submatrix consisting of independent columns of A.

P.-Q. Pan, *Face Method*, https://doi.org/10.1007/978-3-031-93594-7_4

Introduce the following concept.

Definition 4.0.3 For the partition $A = (B, N)$ with pseudo-basis B, the dual face is defined as

$$D(N) = \{(y, z_N) \in \mathcal{R}^m \times \mathcal{R}^{n-k} \mid B^{\mathrm{T}} y = c_B,\ N^{\mathrm{T}} y + z_N = c_N,\ z_B = 0,\ z_N \geq 0\}. \tag{4.3}$$

Definition 4.0.4 A dual face is a level dual face if the dual objective value is constant over it. A level dual face is *optimal* if the constant equals the dual optimal value.

The deficient basis/pseudo-basis is especially suitable for attacking the problem from its dual side.

4.1 Source of Dual Face Method

This section derives basic facts, which serves as the source of the dual face method.

Let B be an $m \times k$ pseudo-basis. Consider the following least-squares problem:

$$\min_{x_B \in \mathcal{R}^k} \|b - Bx_B\|. \tag{4.4}$$

The unique solution to the preceding problem is

$$\bar{x}_B = (B^{\mathrm{T}} B)^{-1} B^{\mathrm{T}} b. \tag{4.5}$$

The residual of (4.4) at the solution $\bar{x}_B$ is the key item, that is,

$$\Delta y = b - B\bar{x}_B = b - B(B^{\mathrm{T}} B)^{-1} B^{\mathrm{T}} b. \tag{4.6}$$

It is evident that the Δy vanishes if and only if B is a deficient basis. In the other case, it holds that

$$B^{\mathrm{T}} \Delta y = 0, \qquad b^{\mathrm{T}} \Delta y > 0. \tag{4.7}$$

If it is nonzero, the Δy is in the null of B^{T} and is an ascent direction concerning the dual objective function. Note that Δy is just the orthogonal projection of b onto the null of B^{T}. It should be a desirable search direction as the closest in the null to the dual objective gradient b.

Nevertheless, the Δy given by (4.6) is unsuitable for application. Imagine that the $(b,\ B)$ was in the following form:

$$b = \begin{bmatrix} b_1 \\ b_2 \end{bmatrix} \qquad B = \begin{bmatrix} I \\ 0 \end{bmatrix} \qquad \begin{matrix} k \\ m-k \end{matrix}.$$

Then, the according least-squares problem becomes

$$\min_{x_B \in \mathcal{R}^k} \left\| \begin{bmatrix} b_1 \\ b_2 \end{bmatrix} - \begin{bmatrix} I \\ 0 \end{bmatrix} x_B \right\|. \tag{4.8}$$

The solution to the preceding problem is obtained directly, i.e.,

$$\bar{x}_B = b_1, \qquad \Delta y = \begin{bmatrix} 0 \\ b_2 \end{bmatrix}.$$

We will realize this idea using the canonical form of $Ax = b$.

Without loss of generality, assume that B consists of the first k columns and R the first k rows such that the $k \times k$ submatrix B has full rank.

Using notation

$$N = A \backslash B, \qquad R' = \{1, \ldots, m\} \backslash R,$$

the primal problem, (4.1), can be put into the tableau below:

x_B	x_N	f	rhs
B_R	N_R		b_R
$B_{R'}$	$N_{R'}$		$b_{R'}$
c_B^{T}	c_N^{T}	-1	

(4.9)

Performing Gauss-Jordan elimination to the preceding, we convert B_R into the unit matrix, yielding the following tableau:

x_B	x_N	f	rhs
I	$\bar{N}_R$		$\bar{b}_R$
	$\bar{N}_{R'}$		$\bar{b}_{R'}$
	$\bar{z}_N^{\mathrm{T}}$	-1	$-\bar{g}$

(4.10)

The dual subproblem corresponding to the (4.10) is then

$$\begin{aligned} &\max\ \bar{b}_R^{\mathrm{T}} y_R + \bar{b}_{R'}^{\mathrm{T}} y_{R'} \\ &\text{s.t.}\ \begin{bmatrix} I & \\ \bar{N}_R^{\mathrm{T}} & \bar{N}_{R'}^{\mathrm{T}} \end{bmatrix} \begin{bmatrix} y_R \\ y_{R'} \end{bmatrix} + \begin{bmatrix} z_B \\ z_N \end{bmatrix} = \begin{bmatrix} c_B \\ c_N \end{bmatrix}, \quad z_B = \bar{z}_B,\ z_N \geq 0. \end{aligned} \tag{4.11}$$

For the preceding problem, the counterpart of the least-squares problem (4.4) is simpler, that is,

$$\min_{x_B \in \mathcal{R}^k} \left\| \begin{bmatrix} \bar{b}_R \\ \bar{b}_{R'} \end{bmatrix} - \begin{bmatrix} I \\ 0 \end{bmatrix} x_B \right\| \tag{4.12}$$

which gives its unique solution $\bar{x}_B = \bar{b}_R$ with residual

$$\Delta y = \begin{bmatrix} 0 \\ \bar{b}_{R'} \end{bmatrix}. \tag{4.13}$$

If $\bar{b}_{R'}$ is nonzero, therefore, Δy is eligible to be the search direction in y-space. The matching search direction in z-space is then

$$\Delta z \equiv \begin{bmatrix} \Delta z_B \\ \Delta z_N \end{bmatrix} = \begin{bmatrix} 0 \\ -\bar{N}_{R'}^{\mathrm{T}} \bar{b}_{R'} \end{bmatrix}. \tag{4.14}$$

The zero-valued vector Δz_B ensures the $\bar{z}_B = 0$ being fixed, so is the following item:

$$y_R = c_B - \bar{z}_B. \tag{4.15}$$

Dropping such inactive parts, the subproblem (4.11) is simplified as

$$\begin{aligned} &\max \ \bar{b}_R^{\mathrm{T}}(c_B - \bar{z}_B) + \bar{b}_{R'}^{\mathrm{T}} y_{R'} \\ &\text{s.t.} \ \ \bar{N}_{R'}^{\mathrm{T}} y_{R'} + z_N = c_N - \bar{N}_R^{\mathrm{T}}(c_B - \bar{z}_B), \qquad z_N \geq 0. \end{aligned} \tag{4.16}$$

Note that the first term of the preceding objective is constant.

The feasible region of the preceding program can be expressed as

$$D' = \{(y_{R'}, z_N) \in \mathcal{R}^{m-k} \times \mathcal{R}^{n-k} \mid \bar{N}_{R'}^{\mathrm{T}} y_{R'} + z_N = c_N - \bar{N}_R^{\mathrm{T}}(c_B - \bar{z}_B), \quad z_N \geq 0\}. \tag{4.17}$$

Note that for any given $(y_{R'}, z_N) \in D'$, there is the $(\tilde{y}_R, \tilde{y}_{R'}, \tilde{z}_B, \tilde{z}_N)$ such that

$$\tilde{y}_R = B_R^{-T}(c_B - \bar{z}_B - B_{R'}^{\mathrm{T}} y_{R'}), \quad \tilde{y}_{R'} = y_{R'}, \quad \tilde{z}_B = \bar{z}_B, \quad \tilde{z}_N = z_N.$$

Then, we claim the following result revealing the relationship between D' and the dual face $D(N)$ defined by (4.3).

Lemma 4.1.1 *If $(y_{R'}, z_N)$ belongs to D', then*

$$(B_R^{-T}(c_B - \bar{z}_B - B_{R'}^{\mathrm{T}} y_{R'}), \ y_{R'}, \bar{z}_B, \ z_N) \tag{4.18}$$

belongs to $D(N)$, and vice versa.

Proof Premultiplying the first row by B_R^{-1}, and then adding $-B_{R'}$ times of the first row to the second row, we transform (4.9) into the following tableau:

x_B	x_N	f	rhs
I	$B_R^{-1}N_R$		$B_R^{-1}b_R$
	$N_{R'} - B_{R'}B_R^{-1}N$		$b_{R'} - B_{R'}B_R^{-1}b_R$
	$\bar{z}_N^{\mathrm{T}}$	-1	

(4.19)

From equivalence between (4.10) and (4.19), it follows that

$$\bar{N}_R = B_R^{-1}N_R, \qquad \bar{N}_{R'} = N_{R'} - B_{R'}B_R^{-1}N_R. \tag{4.20}$$

Assume that the former solution is in D'. From the equality constraint of (4.16), (4.20), and (4.15), it follows that

$$\begin{aligned} z_N &= c_N - \bar{N}_{R'}^{\mathrm{T}}y_{R'} - \bar{N}_R^{\mathrm{T}}(c_B - \bar{z}_B) \qquad (4.21)\\ &= c_N - (N_{R'}^{\mathrm{T}} - N_R^{\mathrm{T}}B_R^{-T}B_{R'}^{\mathrm{T}})y_{R'} - N_R^{\mathrm{T}}B_R^{-T}(c_B - \bar{z}_B)\\ &= c_N - N_{R'}^{\mathrm{T}}y_{R'} + N_R^{\mathrm{T}}B_R^{-T}B_{R'}^{\mathrm{T}}y_{R'} - N_R^{\mathrm{T}}B_R^{-T}(c_B - \bar{z}_B)\\ &= c_N - N_{R'}^{\mathrm{T}}y_{R'} - N_R^{\mathrm{T}}B_R^{-T}(c_B - \bar{z}_B - B_{R'}^{\mathrm{T}}y_{R'}). \qquad (4.22)\end{aligned}$$

Substituting

$$y_R = B_R^{-T}(c_B - \bar{z}_B - B_{R'}^{\mathrm{T}}y_{R'}) \tag{4.23}$$

into the preceding leads to

$$N_R^{\mathrm{T}}y + N_{R'}^{\mathrm{T}}y_{R'} + z_N = c_N,$$

which implies that $(y = (y_R^{\mathrm{T}}, y_{R'}^{\mathrm{T}})^{\mathrm{T}}, z_N)$ satisfies

$$N_R^{\mathrm{T}}y_R + B_{R'}y_{R'} = c_B - \bar{z}_B,$$

which implies that

$$B_R^{\mathrm{T}}y_R - \bar{z}_B = c_B.$$

Therefore, the solution (4.18) is in $D(N)$.

Conversely, assume that the solution is in $D(N)$. From the definition of $D(N)$ and (4.23), it follows that

$$\begin{aligned} z_N &= c_N - N_R^{\mathrm{T}}y_R\\ &= c_N - N_R^{T}y_R - N_{R'}^{T}y_{R'}\\ &= c_N - N_R^{\mathrm{T}}B_R^{-T}(c_B - \bar{z}_B - B_{R'}^{\mathrm{T}}y_{R'}) - N_{R'}^{\mathrm{T}}y_{R'}\end{aligned}$$

$$= c_N - N_R^{\mathrm{T}} B_R^{-T}(c_B - \bar{z}_B) - (N_{R'}^T - N_R^{\mathrm{T}} B_R^{-T} B_{R'}^{\mathrm{T}}) y_{R'},$$

which along with (4.20) leads to the equality constraint of (4.16). Therefore, the former point is in D', and the proof is complete. □

Theorem 4.1.2 *If B is a deficient basis, then the $D(N)$ and D' are level dual faces.*

Proof Without loss of generality, assume that $k < m$.

According to the definition of deficient basis, the residual Δy given by (4.13) is zero, and so is $\bar{b}_{R'}$. Therefore, D' is a level dual face, over which the objective of (4.16) equals $\bar{b}^{\mathrm{T}}(c_B - \bar{z}_B)$.

By Lemma 4.1.1, there is the one-to-one correspondence between D' and $D(N)$, i.e.,

$$(y_{R'}, z_N \geq 0) \quad \longleftrightarrow \quad (z_B = \bar{z}_B,\ y_R = B_R^{-T}(c_B - \bar{z}_B - B_{R'}^{\mathrm{T}} y_{R'}),\ y_{R'},\ z_N \geq 0).$$

The objective value of the dual problem (4.2) at the latter solution equals

$$\begin{aligned} b_R^{\mathrm{T}} y_R + b_{R'}^{\mathrm{T}} y_{R'} &= b_R^{\mathrm{T}} B_R^{-T}(c_B - \bar{z}_B - B_{R'}^{\mathrm{T}} y_{R'}) + b_{R'}^{\mathrm{T}} y_{R'} \\ &= b_R^{\mathrm{T}} B_R^{-T}(c_B - \bar{z}_B) - y_{R'}^{\mathrm{T}} B_{R'} B_R^{-1} b + y_{R'}^{\mathrm{T}} b_{R'} \\ &= b_R^{\mathrm{T}} B_R^{-T}(c_B - \bar{z}_B) + y_{R'}^{\mathrm{T}}(b_{R'} - B_{R'} B_R^{-1} b_R). \end{aligned} \tag{4.24}$$

On the other hand, a comparison between the right hands of (4.10) and (4.19) gives that

$$b_{R'} - B_{R'} B_R^{-1} b_R = \bar{b}_{R'} = 0,$$

substituting which into (4.24) leads to

$$b_R^{\mathrm{T}} y_R + b_{R'}^{\mathrm{T}} y_{R'} = b_R^{\mathrm{T}} B_R^{-T}(c_B - \bar{z}_B).$$

Therefore, the objective value of problem (4.2) is constant over $D(N)$. □

Corollary 4.1.3 *The dual face $D(N)$ is a level dual face if and only if the $\bar{b}_{R'}$ in the tableau (4.10) vanishes.*

Based on discussions of this section, solving the subproblem (4.16) corresponds to maximizing the dual objective function over the dual face $D(N)$. Similarly, we can derive the following counterpart of the search direction (4.13)–(4.14):

$$\Delta y_{R'} = \bar{b}_{R'}, \qquad \Delta z_N = -\bar{N}_{R'}^{\mathrm{T}} \bar{b}_{R'}. \tag{4.25}$$

4.2 Ascent Search Direction

Assume that $(\bar{y}, \bar{z})$ is a dual feasible solution such that

$$A^T \bar{y} + \bar{z} = c, \qquad \bar{z} \geq 0.$$

Let $(\tilde{B}, N)$ be a partition with normal basis $\tilde{B}$. Put the canonical system of $Ax = b$ into the following tableau with $\bar{z}$ as the bottom line:

$x_{\tilde{B}}$	x_N	f	rhs
I	$\bar{N}$		$\bar{b}$
$\bar{z}_{\tilde{B}}^{\mathrm{T}}$	$\bar{z}_N^{\mathrm{T}}$	-1	$-\bar{g}$

(4.26)

As a normal basis, the $\tilde{B}$'s ith index j_i corresponds to its ith column, i.e.,

$$\tilde{B} = [a_{j_1}, \ldots, a_{j_m}].$$

So, x_{j_i} is the ith basic variable, corresponding to the ith row in tableau (4.26). Then, the preceding tableau determines the following basic solution:

$$\bar{x}_{\tilde{B}} = \bar{b}, \qquad \bar{x}_N = 0. \tag{4.27}$$

Now, generate the subset N' such that

$$N' = \{j_i \in \tilde{B} \mid \bar{x}_{j_i} < 0\} \cup \{j_i \in \tilde{B} \mid \bar{x}_{j_i} \geq 0, \ \ \bar{z}_{j_i} > 0\}, \tag{4.28}$$

and set

$$B = \tilde{B} \backslash N'.$$

Then, it is obvious that

$$\bar{x}_B \geq 0, \qquad \bar{z}_B = 0,$$

and the tableau (4.26) can be rewritten (with f column omitted) as

x_B	$x_{N'}$	x_N	rhs
I		$\bar{N}_R$	$\bar{b}_R$
	I'	$\bar{N}_{R'}$	$\bar{b}_{R'}$
	$\bar{z}_{N'}^{\mathrm{T}}$	$\bar{z}_N^{\mathrm{T}}$	$-\bar{g}$

(4.29)

which is termed *(canonical) dual face tableau*. The sets B and N' are said to be *inactive* and *active*, respectively (as will be clear later).

Assume that B is $m \times k$ pseudo-basis with $0 \leq k \leq m$[1].

According to Definition 4.0.3, for the partition $A = (B, N', N)$, the dual face is

$$\begin{aligned} D(N' \cup N) = \{(y, z_{N' \cup N}) \in \mathcal{R}^m \times \mathcal{R}^{n-k} \mid B^{\mathrm{T}} y = c_B, \\ (N' \cup N)^{\mathrm{T}} y + z_{N' \cup N} = c_{N' \cup N}, z_B = 0,\ z_{N' \cup N} \geq 0\}, \end{aligned} \tag{4.30}$$

where the N' is said to be *active* and B is *inactive*, as $z_B = 0$ is fixed.

Corollary 4.2.1 *If N' is empty or nonempty but $\bar{x}_{N'}$ vanishes, the $D(N' \cup N)$ is a level dual face.*

Proof Under the assumption, the B is a deficient basis. Then, the validity comes from Corollary 4.1.3. □

The tableau (4.29) exhibits the primal solution $\bar{x}$, i.e.,

$$\bar{x}_B = \bar{b}_R, \qquad \bar{x}_{N'} = \bar{b}_{R'}, \qquad \bar{x}_N = 0. \tag{4.31}$$

Note that the solution $\bar{x}$ is not feasible. Even if feasible, the pair $(\bar{x}, \bar{z})$ may not be optimal because the complementarity condition may not be satisfied.

Considering (4.31), the tableau (4.29) is associated with the following dual problem:

$$\begin{aligned} &\max\ \bar{b}_R^{\mathrm{T}} y_R + \bar{b}_{R'}^{\mathrm{T}} y_{R'} \\ &\text{s.t.} \quad \begin{bmatrix} I & \\ & I \\ \bar{N}_R^{\mathrm{T}} & \bar{N}_{R'}^{\mathrm{T}} \end{bmatrix} \begin{bmatrix} y_R \\ y_{R'} \end{bmatrix} + \begin{bmatrix} z_B \\ z_{N'} \\ z_N \end{bmatrix} = \begin{bmatrix} 0 \\ \bar{z}_{N'} \\ \bar{z}_N \end{bmatrix}, \qquad z_B = 0;\ z_{N'}, z_N \geq 0. \end{aligned}$$

As $z_B = 0$ is fixed, it follows from the first equation of the preceding that $y = 0$. Thereby, the preceding problem is transformed to

$$\begin{aligned} &\max\ \bar{b}_{R'}^{\mathrm{T}} y_{R'} \\ &\text{s.t.} \quad \begin{bmatrix} I \\ \bar{N}_{R'}^{\mathrm{T}} \end{bmatrix} y_{R'} + \begin{bmatrix} z_{N'} \\ z_N \end{bmatrix} = \begin{bmatrix} \bar{z}_{N'} \\ \bar{z}_N \end{bmatrix}, \qquad z_{N'}, z_N \geq 0, \end{aligned} \tag{4.32}$$

which is referred to as *dual face subprogram*. Note that it only involves the active part of the tableau (4.29).

The gradient $\bar{x}_{N'}$ of the dual objective is a favorable search direction in the $y_{R'}$-space. Consequently, we have the following dual search direction:

[1] It is helpful to have the pseudo-basis including empty case ($k = 0$).

$$\Delta y_{R'} = \bar{b}_{R'} \tag{4.33}$$

$$\Delta z_{N' \cup N} \triangleq \begin{bmatrix} \Delta z_{N'} \\ \Delta z_N \end{bmatrix} = \begin{bmatrix} -\bar{b}_{R'} \\ -\bar{N}_{R'}^{\mathrm{T}} \bar{b}_{R'} \end{bmatrix}. \tag{4.34}$$

The preceding formula coincides with (4.25), corresponding to the residual of the associated least-squares problem. Note that the search direction vanishes if and only if so does the residual $\bar{b}_{R'}$.

The following claim confirms the eligibility of the search directions.

Lemma 4.2.2 *Assume that the direction* $(\Delta y_{R'}, \Delta z_{N' \cup N})$ *is defined by (4.33)–(4.34) and that* $\Delta y_{R'}$ *is nonzero. Then,* $\Delta y_{R'}$ *is the steepest ascent direction in the* $y_{R'}$*-space, and the search direction* $(\Delta y_{R'}, \Delta z_{N' \cup N})$ *is in the null of the coefficient matrix of the subprogram (4.32).*

Proof Obviously, the $\Delta y_{R'}$ is the steepest ascent direction in the $y_{R'}$-space, as it is the gradient of the objective function.

In addition, the following equation is easily verified:

$$\begin{bmatrix} I \\ \bar{N}_{R'}^{\mathrm{T}} \end{bmatrix} \Delta y_{R'} + \begin{bmatrix} \Delta z_{N'} \\ \Delta z_N \end{bmatrix} = \begin{bmatrix} 0 \\ 0 \end{bmatrix}. \tag{4.35}$$

Therefore, the claim is valid. □

4.3 Improving Dual Face Solution

Assume that $(\bar{z}_B, \bar{z}_{N' \cup N})$ is the current dual face solution, as exhibited by the bottom of the dual face tableau (4.29) and that the search direction $(\Delta y, \Delta z_{N' \cup N})$ is available. Our task is to update the dual face solutions to make progress. Note that the inactive part $\bar{z}_B = 0$ is fixed, and hence only $\bar{z}_{N' \cup N}$ will be updated.

Introduce the following index set:

$$J = \{j_i \in N' \mid \Delta z_{j_i} < 0\} \cup \{j \in N \mid \Delta z_j < 0\}. \tag{4.36}$$

If J is nonempty, the following update formula is well-defined:

$$\hat{y}_{R'} = \beta \Delta y_{R'}, \tag{4.37}$$

$$\hat{z}_{N' \cup N} = \bar{z}_{N' \cup N} + \beta \Delta z_{N' \cup N}, \tag{4.38}$$

where β is the dual stepsize. To maintain the dual feasibility of the new solution, the maximum possible β is determined by

$$\beta = \min_{j \in J} -\bar{z}_j / \Delta z_j \geq 0. \tag{4.39}$$

At the same time, the following set can be determined:

$$J_\beta = \{j \in J \mid -\bar{z}_j/\Delta z_j = \beta\}. \tag{4.40}$$

The related variables are said to be blocking, as it holds that

$$\hat{z}_j = 0, \qquad \forall\, j \in J_\beta.$$

Therefore, the associated components of $\hat{z}_j$ reach the boundary.

Obviously, updating $\bar{z}_{N'\cup N}$ by the formula (4.38) amounts to adding $-\beta\bar{b}_{R'}^T$ times of rows corresponding to R' to the bottom line in tableau (4.29).

Using the preceding notation, we claim the following result.

Theorem 4.3.1 *If J defined by (4.36) is nonempty, then (4.37)–(4.39) give a new dual face solution to subprogram (4.32).*

Proof Since the old solution satisfies the equality constraint of (4.32) and directions (4.33)–(4.34) satisfy (4.35), the new solution fulfills the equality constraint.

From (4.38)–(4.39), it follows that

$$\beta \le -\bar{z}_j/\Delta z_j, \qquad j \in J. \tag{4.41}$$

Besides, (4.36) ensures that

$$\Delta z_j < 0, \qquad j \in J.$$

Thus, multiplying (4.41) by Δz_j gives

$$\beta\Delta z_j \ge -\bar{z}_j, \qquad j \in J,$$

which along with (4.38) leads to

$$\hat{z}_j = \bar{z}_j + \beta\Delta z_j \ge 0, \qquad j \in J. \tag{4.42}$$

In the case of $(N' \cup N)\backslash J \ne \emptyset$, on the other hand, (4.36) implies that

$$\Delta z_j \ge 0, \quad j \in (N' \cup N)\backslash J,$$

combining which, $\bar{z}_{N'\cup N} \ge 0$ and $\beta \ge 0$, results in

$$\hat{z}_j = \bar{z}_j + \beta\Delta z_j \ge 0, \qquad j \in (N' \cup N)\backslash J. \tag{4.43}$$

Therefore, it holds that $\hat{z}_{N'\cup N} \ge 0$, and the proof is complete. □

The following claim is valid based on Lemma 4.1.1.

Corollary 4.3.2 *The new solution* $(\bar{z}_B, \hat{z}_{N' \cup N})$ *is a dual face solution to the original problem (4.1).*

The final dual solution in y-space can be calculated cheaply at the end of the solution process if needed. Therefore, we will update the dual solution $\bar{z}$ only.

The following reveals the increment in the objective value in a single iteration.

Theorem 4.3.3 *If search direction is determined by (4.33)–(4.34), the new solution (4.37)–(4.38) increases the dual objective value by*

$$\Delta f = \beta \, \|\bar{b}_{R'}\|^2 \tag{4.44}$$

with respect to dual subprogram (4.32).

Proof By (4.37), the old zero-valued solution in $y_{R'}$-space is updated to the new solution $\beta \bar{b}_{R'}$, and hence the corresponding objective value increases from zero up to $\beta \bar{b}_{R'}^{\mathrm{T}} \bar{b}_{R'}$, which is equal to (4.44). □

The progress made seems impressive.

However, if the stepsize β vanishes, no improvement is made, so the so-called new solution is not new. This case occurs if and only if some component of $\bar{z}_J$ is 0. The related dual face solution is said to be *dual degenerate*.

As for the case of J empty when the stepsize β is not well-defined, the following claim applies.

Theorem 4.3.4 *Assume that* $\bar{b}_{R'}$ *is nonzero. If* $J = \emptyset$, *there is no feasible solution to the standard problem.*

Proof Emptiness of J means that

$$\Delta z_j \geq 0, \qquad j \in N' \cup N,$$

from which, $\bar{z}_{N' \cup N} \geq 0$ and $\beta \geq 0$, it follows that

$$\hat{z}_j = \bar{z}_j + \beta \Delta z_j \geq 0, \qquad j \in N' \cup N. \tag{4.45}$$

Therefore, the new solution is dual feasible to (4.32) for all $\beta > 0$. Under the condition $\bar{b}_{R'} \neq 0$, the corresponding objective increment, given by Theorem 4.3.3, goes to ∞, as so is β. It means that the dual subprogram (4.32) is upper unbounded, and hence there exists no feasible solution to the corresponding primal subprogram and the primal standard problem. □

4.4 Pivoting Operation

The dual face tableau (4.29) needs to be updated in each iteration. In this section, we address associated pivoting operations and index set adjustments.

Assume that the J_β was determined by (4.40) and the new dual solution $\hat{z}_{N' \cup N}$ was calculated.

The following two cases arise:

(a) $N' \cap J_\beta$ is nonempty.
Assume that the row index set $\tilde{R}$ corresponds to the column index set $N' \cap J_\beta$. In this case, the N' includes the indices associated with binding inequalities. To move them to N, execute the following adjustment:

$$\begin{aligned} &B = B \cup (N' \cap J_\beta),\ R = R \cup (N' \cap J_\beta), \\ &N' = N' \backslash (N' \cap J_\beta),\ R' = R' \backslash \tilde{R}. \end{aligned} \tag{4.46}$$

Then, pivoting is unnecessary and the involved computational cost is relatively cheap.

(b) $N' \cap J_\beta$ is empty.
It is clear that J_β belongs to N in this case. Then, the following column rule applies.

Rule 4.4.1 (Column Rule) Select the column index q such that

$$q \in \arg \min_{j \in J_\beta} \Delta z_j. \tag{4.47}$$

It is noted that the Δz_q reaches its maximum possible absolute value.

Assume that the column index q is determined. There could be multiple choices for row pivoting.

Let column index q be available. Consider the following row rule first.

Rule 4.4.2 (Row Rule) Select the row index p such that

$$p \in \arg \min\{-\bar{b}_i \bar{a}_{i,q} \mid i \in R'\}. \tag{4.48}$$

The following gives the plausibility of the preceding rule.

Proposition 4.4.1 *Δz_q and $-\bar{b}_p \bar{a}_{p,q}$ are both negative.*

Proof It is evident that $\bar{a}_{i'q} = 0$ implies $\bar{N}_{R',q} = 0$, and hence

$$(\bar{N}_{R',q})^{\mathrm{T}} \bar{b}_{R'} = 0. \tag{4.49}$$

In addition, by (4.36) and $q \in J$, it is known that $\Delta z_q < 0$. Hence, combining (4.47) and (4.49) leads to

$$\Delta z_q = -\bar{N}_{R',q}^{\mathrm{T}} \bar{b}_{R'} = \sum_{i \in R'} -\bar{b}_i \bar{a}_{i,q} < 0.$$

The preceding along with (4.48) gives $-\bar{b}_p\bar{a}_{p,q} < 0$, which has a minus sign as Δz_q. □

Obviously, the pivot $\bar{a}_{p,q}$ determined by Rule 4.4.2 is nonzero. It is more than that. The rule also conforms to the heuristic characteristic of the optimal solution in the dual space.

Once the pivot is determined, what to do next is to update the tableau by multiplying the p-th row by $1/\bar{a}_{p,q}$, and adding $-\bar{a}_{i,q}$ times of row p to the i-th row, for $i = 1, \ldots, m,\ i \neq p$.

Then, adjust the index sets as follows:

$$\begin{aligned} &B = B \cup q, \quad R = R \cup p, \\ &N' = N' \backslash j_p, \ R' = R' \backslash p, \ N = N \backslash q \cup j_p. \end{aligned} \tag{4.50}$$

Related column and row permutations are ignored in conjunction with the tableau. After all, it is not necessary to move around the entire column/row of entries in implementation.

4.5 Optimality Condition

The size of N' reduces iteration by iteration in a nested manner. Once the N' becomes empty, or nonempty but $\bar{x}_{N'}$ vanishes, the search direction is not well-defined. According to Corollary 4.1.3, the reached $D(N' \cup N)$ is level dual face. Then, it is time to check if it is optimal.

We claim the following result.

Theorem 4.5.1 *Assume that the N' is empty or it is nonempty but $\bar{x}_{N'} = 0$. If the following set is empty, then the $\bar{z}$ and $\bar{x}B$ are a pair of primal and dual optimal solutions:*

$$T = \{j_i \in B \mid \bar{x}_{j_i} < 0\} \cup \{j_i \in B \mid \bar{x}_{j_i} \geq 0,\ \ \bar{z}_{j_i} > 0\}. \tag{4.51}$$

Theorem 4.5.2 *Let $\bar{z}$ be a dual face solution and let B be a deficient basis. If set T defined by (4.51) is empty, then $(\bar{x}, \bar{z})$ is a pair of primal and dual optimal solutions with objective value $\bar{f}$, corresponding to dual optimal face $D(N' \cup N)$.*

Proof Note that deficiency of B ensures

$$\bar{x}_{N'} = \bar{b}_{R'} = 0. \tag{4.52}$$

Emptiness of T ensures that

$$\bar{x}_B \geq 0,$$

from which, (4.31) and (4.52), it follows that the solution

$$\bar{x} = (\bar{x}_B, \bar{x}_{N'}, \bar{x}_N)$$

is a feasible (face) solution to the primal problem (4.1).

Furthermore, emptiness of T also implies that

$$\bar{x}_{j_i}\bar{z}_{j_i} = 0, \qquad j_i \in B,$$

and hence the dual complimentarity condition is satisfied, i.e., $\bar{x}^T\bar{z} = 0$. Therefore, the pair $(\bar{x}, \bar{z})$ is a primal and dual optimal solution.

By Corollary 4.2.1, $D(N' \cup N)$ is a level dual face if B is a deficient basis. Since $\bar{z}$ is an optimal dual solution and it belongs to $D(N' \cup N)$, the latter is an optimal dual face. The bottom line of tableau (4.29) exhibits the optimal objective value $\bar{f}$. □

Corollary 4.5.3 *Theorem 4.5.2 is valid if B is a full basis.*

If T is nonempty, adjust the sets as follows:

$$\begin{array}{ll} B = B \backslash T & R = R \backslash T' \\ N' = N' \cup T & R' = R' \cup T' \end{array}. \tag{4.53}$$

Then, carry out the next iteration.

4.6 Dual Face Algorithm: Tableau Form

Based on discussions in the previous sections, we formulate the dual face algorithm in tableau form in this section.

Iteration by iteration, the active set N' reduces, while inactive set B expands; in other words, the associated dual face is contracted. Once the dual face vanishes, the process is interrupted because the search direction is not well-defined. Such a series of iterations is termed *dual face contraction*. Optimality is tested at the end of it. If the optimality condition is not fulfilled, a new dual face is formed, and another round of dual face contraction is performed.

Initially, we take a partition $A = (B, N)$ with full basis B.

There is a dual feasible solution $(\bar{y}, \bar{z})$ such that

$$A^T\bar{y} + \bar{z} = c, \qquad \bar{z} \geq 0,$$

and a primal solution $\bar{x}$ such that

$$B\bar{x}_B = b, \qquad \bar{x}_N = 0.$$

Form index sets

$$N' = \{j_i \in B \mid \bar{x}_{j_i} < 0\} \cup \{j_i \in B \mid \bar{x}_{j_i} \geq 0,\ \bar{z}_{j_i} > 0\}, \qquad B = B \backslash N',$$

to provide the partition $A = (B, N', N)$ and related dual face tableau.

Overall steps are summarized into the following algorithm.

Algorithm 4.6.1 (Dual Face Algorithm: Tableau Form)
Initial : partition (B, N', N) and (R, R'), dual face tableau in form (4.29). This algorithm solves the standard LP problem (4.1).

1. *Compute search direction $\Delta z_{N' \cup N}$ by (4.34).*
2. *Stop if $J = \{j \in N' \cup N \mid \Delta z_j < 0\}$ is empty.*
3. *Determine stepsize β and J_β by (4.39)–(4.40).*
4. *If $\beta > 0$, update $\bar{z}_{N' \cup N}$ by (4.38).*
5. *Go to Step 8 if $N' \cap J_\beta$ is empty.*
6. *Adjust (N', B) by (4.46).*
7. *Go to Step 1 if $\bar{x}_{N'}$ is nonzero; else, to Step 13.*
8. *Select the column index q by (4.47).*
9. *Determine row index p by (4.48).*
10. *Premultiply the p-th row by $1/\bar{a}_{p,q}$, and add $-\bar{a}_{i,q}$ times of the row to the i-th row, for $i = 1, \ldots, m,\ i \neq p$.*
11. *Adjust sets by (4.50).*
12. *Go to Step 1 if N' is nonempty and $\bar{x}_{N'}$ is nonzero.*
13. *Stop if T defined by (4.51) is empty.*
14. *Adjust sets by (4.53).*
15. *Go to Step 1.*

We make the following claim with the proof omitted.

Theorem 4.6.1 *If each dual face contraction involves a nondegenerate dual face solution, Algorithm 4.6.1 terminates either at:*

(i) Step 13, giving a pair of primal and dual optimal solutions
(ii) Step 2, declaring primal infeasibility

Example 4.6.1 Solve the following standard LP problem by Algorithm 4.6.1:

$$A = \begin{bmatrix} 1 & & & & 4 & -6 & 3 & 8 \\ & 1 & & & 2 & -11 & -4 & -5 \\ & & 1 & & -8 & 4 & -12 & 12 \\ & & & 1 & 6 & -12 & 3 & 1 \end{bmatrix} \qquad b = \begin{bmatrix} 6 \\ 9 \\ -17 \\ -12 \end{bmatrix}$$

$c = [0, 0, 5, 4, 2, 1, 2, 0]^T$

Initially, $\bar{z} = [0, 0, 5, 4, 2, 1, 2, 0]^T$, $\bar{f} = 0$.
$B = \{1, 2\},\ R = \{1, 2\},\ N' = \{3, 4\},\ R' = \{3, 4\}$.
Iteration 1:

1. $\bar{b}_{R'} = [-17, -12]^T$,

$\Delta z_{N'}^T = -\bar{b}_{R'}^T = [17, 12]$, $\Delta z_N^T = -\bar{b}_{R'}^T \bar{N}_{R'} = [-64, -76, -168, 216]$.

2. $J = \{5, 6, 7\}$.
3. $\beta = \min[-2/(-64), -1/(-76), -2/(-168)] = \min[1/32, 1/76, 1/84] = 1/187$, $J_\beta = \{7\}$.
4. $\bar{z}_{N' \cup N}^T = [437/84, 29/7, 26/21, 2/21, 0, 18/7]$, $\bar{f} = 433/84$.
5. $N' \cap J_\beta = \emptyset$.
8. $q = 7$.
9. $p \in \arg\min[-(-17)(-12), -(-12)(3)] = 3$.
10.

x_1	x_2	x_3	x_4	x_5	x_6	x_7	x_8	rhs
1		1/4		2	−5		11	7/4
	1	−1/3		14/3	−37/3		−9	44/3
		−1/12		2/3	−1/3	1	−1	17/12
		1/4	1	4	−11		4	−65/4
		437/84	29/7	26/21	2/21		18/7	433/84

11. $B = \{1, 2, 7\}$, $R = \{1, 2, 3\}$, $N' = \{4\}$, $R' = \{4\}$, $N = \{3, 5, 6, 8\}$.
12. (go to Step 1)

Iteration 2:

1. $\Delta z_{N'}^T = -\bar{b}_{R'}^T = [65/4]$, $\Delta z_N^T = -\bar{b}_{R'}^T \bar{N}_{R'} = [65/16, 65, -715/4, 65]$.
2. $J = \{6\}$.
3. $\beta = \min[-(2/21)/(-715/4)] = 8/15{,}015$, $J_\beta = \{6\}$.
4. $\bar{z}_{N' \cup N}^T = [137/33, 229/44, 14/11, 0, 86/33]$, $\bar{f} = 233/44$.
5. $N' \cap J_\beta = \emptyset$.
8. $q = 6$.
9. $p = 4$.
10.

x_1	2	x_3	x_4	x_5	x_6	x_7	x_8	rhs
1		3/22	−5/11	2/11			101/11	201/22
	1	−27/44	−37/33	2/11			−445/33	1447/44
		−1/11	−1/33	6/11		1	−37/33	21/11
		−1/44	−1/11	−4/11	1		−4/11	65/44
		229/44	137/33	14/11			86/33	233/44

11. $B = \{1, 2, 7, 6\}$, $R = \{1, 2, 3, 4\}$, $N' = \emptyset$, $R' = \emptyset$, $N = \{3, 4, 5, 8\}$.
13. $T = \emptyset$. Optimality achieved.

The pair of primal and dual optimal face solutions is as follows:
$\bar{x} = [201/22, 1447/44, 0, 0, 0, 65/44, 21/11, 0]^T$,
$\bar{z} = [0, 0, 229/44, 137/33, 14/11, 0, 0, 86/33]^T$, $\bar{f} = 233/44$.

4.7 Hybrid of Primal/Dual Face Methods

The face (dual face) method must start from a feasible (dual feasible) solution. Therefore, it is futile if no face (dual face) solution is available.

There are multiple options to resolve the problem, but our trick is to use hybrids of the primal/dual face methods, in conjunction with the perturbation techniques (Pan [25, 26, 36]). We prefer this approach as it is probably the best.

4.7.1 *Primal-Dual Face Algorithm*

The hybrid comprises two phases. The phase I uses the face (dual face) algorithm to achieve dual (primal) feasibility, while phase II uses the dual (primal) face algorithm to achieve optimality.

Assume that $\bar{x} \in \mathcal{R}^n$ is a nonnegative vector, satisfying the following constraints:

$$Ax = \tilde{b}, \qquad x \geq 0,$$

where $\tilde{b} = A\bar{x}$ is a perturbation of the right-hand side b. Therefore, the face Algorithm 2.5.1 can get started to solve the perturbed problem. Once it is solved, a dual feasible solution to the original problem is obtained. Then, the dual face Algorithm 4.6.1 is applicable for achieving optimality.

Nevertheless, it is unnecessary to solve the perturbed problem completely since our goal is only to gain a nonnegative dual solution. For the same sake, the formula (2.34) is better replaced by the simpler expression below:

$$\bar{z}_j < 0, \qquad \forall\ j \in B'. \tag{4.54}$$

The goal is achieved once B' is empty.

The approach is demonstrated through the following instance.

Example 4.7.1 Solve the following standard LP problem by primal-dual face algorithm:

$$A = \begin{bmatrix} 1 & & & & -2 & -5 & 4 & -8 \\ & 1 & & & 4 & 2 & -6 & -3 \\ & & 1 & & 1 & 0 & 0 & 2 \\ & & & 1 & -5 & 3 & 4 & -5 \end{bmatrix} \qquad b = \begin{bmatrix} -11 \\ -2 \\ 3 \\ -2 \end{bmatrix}$$

$c = [0, 0, 0, 0, 2, -7, -5, -3]^T$.

To form primal-phase problem, perturb the right-hand side to the closed nonnegative one below:

$$B = [1, 1, 3, 1]^T.$$

It can be verified that the following solution:

$$\bar{x} = [1, 1, 3, 1, 0, 0, 0, 0]^T$$

satisfies $Ax = B$.

Primal phase I: call Algorithm 2.7.1 to handle the perturbed problem to achieve dual feasibility.

Put the problem into the following face tableau (the perturbed right-hand side is omitted):

x_1	x_2	x_3	x_4	x_5	x_6	x_7	x_8	rhs
1				−2	−5	4	−8	−11
	1			4	2	−6	−3	−2
		1		1			2	3
			1	−5	3	4	−5	−2
				2	−7	−5	−3	
1	1	3	1					−

where the bottom line exhibits $\bar{x}$ as reference.

Iteration 1: $B = \{1, 2, 3, 4\},\ B' = \{6, 7, 8\},\ N = \{5\}$.

1. $\bar{B}' = \begin{bmatrix} -5 & 4 & -8 \\ 2 & -6 & -3 \\ 0 & 0 & 2 \\ 3 & 4 & -5 \end{bmatrix}$, $\Delta x_{B'} = [-7, -5, -3]^T$,
   $\Delta x_B = -\bar{B}' \Delta x B' = [-39, -25, 6, 26]^{\mathrm{T}}$.
2. $J = \{3, 4\} \neq \emptyset$.
3. $\alpha = \min[3/6, 1/26] = 1/26,\ p = 4,\ J_\alpha = \{4\}$.
4. $\bar{x}_{B\cup B'} = \bar{x}_{B\cup B'} - \alpha \Delta x_{B\cup B'} = [5/2, 51/26, 36/13, 0, 7/26, 3/26]^T$,
   $\bar{x} = [5/2, 51/26, 36/13, 0, 0, 7/26, 5/26, 3/26]^T$.
5. $B' \cap J_\alpha = \emptyset$.
8. $p = 4,\ j_p = 4$.
9. $q = \arg\min[3(-7), 4(-5), (-5](-3)] = 6$.
10.

x_1	x_2	x_3	x_4	x_5	x_6	x_7	x_8	rhs
1			5/3	−31/3		32/3	−49/3	−43/3
	1		−2/3	22/3		−26/3	1/3	−2/3
		1		1			2	3
			1/3	−5/3	1	4/3	−5/3	−2/3
			7/3	−29/3		13/3	−44/3	−14/3
5/2	51/26	36/13			7/26	5/26	3/26	−

11. $B = \{1, 2, 3, 6\}$, $B' = \{8\}$, $N = \{4, 5\}$.
12. (switch to Step 1)

Iteration 2:

1. $\Delta x_{B'} = [-44/3]$, $\Delta x_B = -\bar{B}'\Delta x B' = [-2156/9, 44/9, 88/3, -220/9]^{\mathrm{T}}$.
2. $J = \{2, 3\} \neq \emptyset$.
3. $\alpha = \min[x(2)/(44/9), x(3)/(88/3)] = \min[459/1144, 27/286] = 27/286$, $p = 3,\ J_\alpha = \{3\}$.
4. $\bar{x}_{B\cup B'} = \bar{x}_{B\cup B'} - \alpha\Delta x_{B\cup B'} = [653/26, 3/2, 0, 67/26, 3/2]^T$, $\bar{x} = [653/26, 3/2, 0, 0, 0, 67/26, 5/26, 3/2]^T$.
5. $B' \cap J_\alpha$ is empty.
8. $p = 3,\ j_p = 3$.
9. $q = 8$.
10.

x_1	x_2	x_3	x_4	x_5	x_6	x_7	x_8	rhs
1		49/6	5/3	−13/6		32/3		61/6
	1	−1/6	−2/3	43/6		−26/3		−7/6
		1/2		1/2			1	3/2
		5/6	1/3	−5/6	1	4/3		11/6
		22/3	7/3	−7/3		13/3		52/3
653/26	3/2				67/26	5/26	3/2	–

11. $B = \{1, 2, 8, 6\}$, $B' = \emptyset$, $N = \{3, 4, 5, 7\}$.
13. $T = \{5\}$.
14. $B' = \{5\},\ N = \{3, 4, 7\}$.

Iteration 3:

1. $\bar{B}' = [-13/6, 43/6, 1/2, -5/6]6T,\ \Delta x_{B'} = [-7/3]^T$, $\Delta x_B = -\bar{B}'\Delta x B' = [-91/18, 301/18, 7/6, -35/18]^{\mathrm{T}}$.
2. $J = \{2, 8\} \neq \emptyset$.
3. $\alpha = \min[x(2)/(301/18), x(8)/(7/6)] = \min[27/301, 9/7] = 27/301$, $p = 2,\ J_\alpha = \{2\}$.
4. $\bar{x}_{B\cup B'} = \bar{x}_{B\cup B'} - \alpha\Delta x_{B\cup B'} = [4270/167, 0, 60/43, 1538/559, 9/43]^T$, $\bar{x} = [4270/167, 0, 0, 0, 9/43, 1538/559, 5/26, 60/43]^T$.
5. $B' \cap J_\alpha = \emptyset$.
8. $p = 2,\ j_p = 2$.
9. $q = 5$.
10.

x_1	x_2	x_3	x_4	x_5	x_6	x_7	x_8	rhs
1	13/43	349/43	63/43			346/43		422/43
	6/43	−1/43	−4/43	1		−52/43		−7/43
	−3/43	22/43	2/43			26/43	1	68/43
	5/43	35/43	11/43		1	14/43		73/43
	14/43	313/43	91/43			65/43		729/43

The tableau exhibits the dual feasible solution.
Dual phase II: call Algorithm 4.6.1 to achieve optimality.
The initial sets:

$$B = \{1, 8, 6\}, \quad R = \{1, 3, 4\}, \quad N' = \{5\}, \quad R' = \{2\}, \quad N = \{2, 3, 4, 7\}.$$

The primal solution is

$$\bar{x}_B = [422/43, -7/43, 68/43, 73/43]^T, \qquad \bar{x}_N = 0, \qquad \bar{f} = -729/43.$$

Iteration 4:

1. $\Delta z_{N'} = [7/43]$, $\Delta z_N = [42/1849, -7/1849, -28/1849, -364/1849]^T$.
2. $J = \{3, 4, 7\}$.
3. $\beta = \min[-z(3)/(-7/1849), -z(4)/(-28/1849), -z(7)/(-364/1849)]$
   $= \min[13{,}459/7, 559/4, 215/28] = 215/28$, $J_\beta = \{7\}$.
4. $\hat{z}_{N' \cup N} = \bar{z}_{N' \cup N} + \beta \Delta z_{N' \cup N} = [5/4, 1/2, 29/4, 2, 0]$,
   $\bar{f} = \bar{f} + \beta(-7/43)^2 = -67/4$. $\bar{z} = [0, 1/2, 29/4, 2, 5/4, 0, 0, 0]^T$.
5. $N' \cap J_\beta = \emptyset$.
8. $q = 7$.
9. $p = 2$, $j_p = 5$.
10.

x_1	x_2	x_3	x_4	x_5	x_6	x_7	x_8	rhs
1	16/13	207/26	11/13	173/26				227/26
	−3/26	1/52	1/13	−43/52		1		7/52
		1/2		1/2			1	3/2
	2/13	21/26	3/13	7/26	1			43/26
	1/2	29/4	2	5/4				67/4

11. $B = \{1, 7, 8, 6\}$, $R = \{1, 2, 3, 4\}$, $N' = R' = \emptyset$, $N = \{2, 3, 4, 5\}$.
13. $T = \emptyset$. Optimality is achieved.

The primal and dual optimal face solutions are
$\bar{x} = [227/26, 0, 0, 0, 0, 43/26, 7/52, 3/2]^T$
$\bar{z} = [0, 1/2, 29/4, 2, 5/4, 0, 0, 0]^T, \qquad \bar{f} = -67/4.$

4.7.2 Dual-Primal Face Algorithm

Contrary to the preceding approach, it may achieve the primal feasibility in phase I and achieve optimality in phase II. We still take the preceding example to demonstrate the process.

Put the problem into the following tableau:

x_1	x_2	x_3	x_4	x_5	x_6	x_7	x_8	rhs
1				-2	-5	4	-8	-11
	1			4	2	-6	-3	-2
		1		1		0	2	3
			1	-5	3	4	-5	-2
				2	-7	-5	-3	0
				2	1	1	1	$-$

where the bottom line lists the perturbed nonnegative dual solution, i.e.,

$$\bar{z} = [0, 0, 0, 0, 2, 1, 1, 1]^T.$$

Dual phase I: call Algorithm 4.6.1 to achieve primal feasibility.

We don't have to solve the perturbed problem to the end, as what is wanted is only a nonnegative primal solution. In addition, the formula (4.51) is replaced by

$$T = \{j_i \in B \mid \bar{x}_{j_i} < 0\}, \qquad T' = \{i \in R \mid \bar{x}_{j_i} < 0\}. \tag{4.55}$$

Obviously, the goal is achieved if T is empty.

Initial: $B = \{3\}$, $R = \{3\}$, $N' = \{1, 2, 4\}$, $R' = \{1, 2, 4\}$, $N = \{5, 6, 7.8\}$.

Iteration 1:

1. $\Delta z_{N'} = -\bar{b}_{R'} = [11, 2, 2,]^T$,
   $\Delta z_N = -\bar{N}^{\mathrm{T}}_{R'}\bar{b}_{R'}) = [-24, -45, 40, -104]^T$.
2. $J = \{5, 6, 8\}$.
3. $\beta = \min[-z(5)/(-24), -z(6)/(-45), -z(8)/(-104)]$
   $= \min[1/12, 1/45, 1/104] = 1/104, \ J_\beta = \{8\}$.
4. $\hat{z}_{N' \cup N} = \bar{z}_{N' \cup N} + \beta \Delta z_{N' \cup N}$
   $= [11/104, 1/52, 1/52, 23/13, 59/104, 18/13, 0]^T$,
   $\bar{z} - [11/104, 1/52, 0, 1/52, 23/13, 59/104, 18/13, 0]^T$.
   $\bar{f} = \bar{f} + \beta \bar{b}^T_{R'} \bar{b}_{R'} = -5332/315$.
5. $N' \cap J_\beta = \emptyset$.
8. $q = 8$.
9. $p = \arg\min[-A(1, 9) * A(1, 8), -A(2, 9) * A(2, 8), -A(4, 9) * A(4, 8)] = \arg\min[-88, -6, -10] = 1, \ j_p = 1$.

10.

x_1	x_2	x_3	x_4	x_5	x_6	x_7	x_8	rhs
− 1/8				1/4	5/8	−1/2	1	11/8
−3/8	1			19/4	31/8	−15/2		17/8
1/4		1		1/2	−5/4	1		1/4
−5/8			1	−15/4	49/8	3/2		39/8
−3/8				11/4	−41/8	−13/2		33/8

The right-hand side $\bar{b}$ becomes nonnegative; hence primal feasibility is achieved. Primal phase II: call Algorithm 2.7.1.

$B = \{8, 2, 3, 4\}$, $B' = \{1, 6, 7\}$, $N = \{5\}$, $\bar{x} = [0, 17/8, 1/4, 39/8, 0, 0, 0, 11/8]^T$.

1. $\bar{B}' = \begin{bmatrix} -1/8 & 5/8 & -1/2 \\ -3/8 & 31/8 & -15/2 \\ 1/4 & -5/4 & 1 \\ -5/8 & 49/8 & 3/2 \end{bmatrix}$, $\Delta x_{B'} = [-3/8, -41/8, -13/2]^T$,

   $\Delta x_B = -\bar{B}' \Delta x B' = [-3/32, -929/32, 3/16, 1309/32]^{\mathrm{T}}$.
2. $J = \{3, 4\} \neq \emptyset$.
3. $\alpha = \min[(1/4)/(3/16), (39/8)/(1309/32)] = \min[4/3, 156/1309] = 156/1309$,

   $p = 4,\ J_\alpha = \{4\}$.
4. $\bar{x}_{B \cup B'} = [11/8, 17/8, 1/4, 39/8, 000]^T - \alpha[-3/32, -929/32, 3/16, 1309/32, -3/8, -41/8, -13/2]^T = [822/593, 955/171, 298/1309, 0, 117/2618, 499/817, 1014/1309]^T$, $\bar{x} = [117/2618, 955/171, 298/1309, 0, 0, 499/817, 1014/1309, 822/593]^T$.
5. $B' \cap J_\alpha = \emptyset$.
8. $p = 4,\ j_p = 4$.
9. $\min[(-5/8)(-3/8), (49/8)(-41/8), (3/2](-13/2)] = -2009/64,\ q = 6$.
10.

x_1	x_2	x_3	x_4	x_5	x_6	x_7	x_8	rhs
− 3/49			−5/49	31/49		−32/49	1	43/49
1/49	1		−31/49	349/49		−414/49		−47/49
6/49		1	10/49	−13/49		64/49		61/49
−5/49			8/49	−30/49	1	12/49		39/49
−44/49			41/49	−19/49		−257/49		402/49

11. $B = \{8, 2, 3, 6\}$, $B' = \{1, 5, 7\}$, $N = \{4\}$.

Iteration 3:

1. $\bar{B}' = \begin{bmatrix} -3/49 & 31/49 & -32/49 \\ 1/49 & 349/49 & -414/49 \\ 6/49 & -13/49 & 64/49 \\ -5/49 & -30/49 & 12/49 \end{bmatrix}$, $\Delta x_{B'} = [-44/49, -19/49, -257/49]^T$,

   $\Delta x_B = -\bar{B}' \Delta x B' = [-482/149, -4278/103, 2359/344, 879/920]^{\mathrm{T}}$.
2. $J = \{3, 6\} \neq \emptyset$.
3. $\alpha = \min[(298/1309)/(2359/344), (499/817)/(879/920)]$
   $= \min[106/3193, 622/973] = 106/3193$, $p = 3$, $J_\alpha = \{3\}$.
4. $\bar{x}_{B \cup B'} = [822/593, 955/171, 298/1309, 499/817, 117/2618, 0, 1014/1309]^T$
   $-\ \alpha[-482/149, -4278/103, 2359/344, 879/920, -44/49, -19/49, -257/49]^T$
   $= [348/233, 10{,}146/1457, 0, 868/1499, 97/1302, 111/8623, 648/683]^T$,
   $\bar{x} = [97/1302, 10{,}146/1457, 0, 0, 111/8623, 868/1499, 648/683, 348/233]^T$.
5. $B' \cap J_\alpha = \emptyset$.
8. $p = 3$, $j_p = 3$.
9. $\min[(6/49)(-44/49), (-13/49)(-19/49), (64/49)(-257/49)] = -5727/836$, $q = 7$.
10.

x_1	x_2	x_3	x_4	x_5	x_6	x_7	x_8	rhs
		1/2		1/2			1	3/2
13/16	1	207/32	11/16	173/32				227/32
3/32		49/64	5/32	−13/64		1		61/64
−1/8		−3/16	1/8	−9/16	1			9/16
−13/32		257/64	53/32	−93/64				845/64

11. $B = \{8, 2, 7, 6\}$, $B' = \{1, 5\}$, $N = \{3, 4\}$.

Iteration 4:

1. $\bar{B}' = \begin{bmatrix} 0 & 1/2 \\ 13/16 & 173/32 \\ 3/32 & -13/64 \\ -1/8 & -9/16 \end{bmatrix}$, $\Delta x_{B'} = [-13/32, -93/64]^T$,

   $\Delta x_B = -\bar{B}' \Delta x B' = [93/128, 16{,}765/2048, -1053/4096, -889/1024]^{\mathrm{T}}$.
2. $J = \{8, 2\} \neq \emptyset$.
3. $\alpha = \min[x(8)/(93/128), x(2)/(16{,}765/2048)] = \min[1145/557, 507/596] = 507/596$,
   $p = 2$, $J_\alpha = \{2\}$.
4. $\bar{x}_{B \cup B'} = [348/233, 10{,}146/1457, 648/683, 868/1499, 97/1302, 111/8623]^T$
   $- \alpha[93/128, 16{,}765/2048, -1053/4096, -889/1024, -13/32, -93/64]^T$
   $= [661/755, 0, 1248/1069, 1282/973, 1075/2559, 627/502]^T$,
   $\bar{x} = [1075/2559, 0, 0, 0, 627/502, 1282/973, 1248/1069, 661/755]^T$.

5. $B' \cap J_\alpha = \emptyset$.
8. $p = 2,\ j_p = 2$.
9. $\min[(13/16)(-13/32), (173/32)(-93/64)] = \min[-169/512, -2836/361]$ $= -2836/361,\ q = 5$.
10.

x_1	x_2	x_3	x_4	x_5	x_6	x_7	x_8	rhs
−13/173	−16/173	−17/173	−11/173				1	146/173
26/173	32/173	207/173	22/173	1				227/173
43/346	13/346	349/346	63/346			1		211/173
−7/173	18/173	84/173	34/173		1			225/173
−65/346	93/346	1991/346	637/346					2614/173

11. $B = \{8, 5, 7, 6\}, B' = \{1\}, N = \{2, 3, 4\}$.

Iteration 5:

1. $\bar{B}' = [-13/173, 26/173, 43/346, -7/173]^T$, $\Delta x_{B'} = [-13/32]$, $\Delta x_B = -\bar{B}' \Delta x B' = [-169/5536, 169/2768, 559/11{,}072, -91/5536]^{\mathrm{T}}$.
2. $J = \{5, 7\} \neq \emptyset$.
3. $\alpha = \min[x(5)/(169/2768), x(7)/(559/11{,}072)] = \min[6669/326, 3561/154]$ $= 6669/326,\ p = 2,\ J_\alpha = \{5\}$.
4. $\bar{x}_{B \cup B'} = [661/755, 627/502, 1248/1069, 1282/973, 1075/2559]^T - \alpha$ $[-169/5536, 169/2768, 559/11{,}072, -91/5536, -13/32]^T$ $= [3/2, 0, 4247/31{,}549, 43/26, 227/26]^T$, $\bar{x} = [227/26, 0, 0, 0, 0, 43/26, 4247/31{,}549, 3/2]^T$.
5. $B' \cap J_\alpha = \emptyset$.
8. $p = 2,\ j_p = 5$.
9. $q = 1$.
10.

x_1	x_2	x_3	x_4	x_5	x_6	x_7	x_8	rhs
		1/2		1/2			1	3/2
1	16/13	207/26	11/13	173/26				227/26
	−3/26	1/52	1/13	−43/52		1		7/52
	2/13	21/26	3/13	7/26	1			43/26
	1/2	29/4	2	5/4				67/4

11. $B = \{8, 1, 7, 6\}, B' = \emptyset, N = \{2, 3, 4, 5\}$.
13. $T = \emptyset$. Optimality is achieved. Primal and dual optimal solutions:

$$\bar{x} = [227/26, 0, 0, 0, 0, 43/26, 7/52, 3/2]^T, \qquad \bar{f} = -67/4$$
$$\bar{z} = [0, 1/2, 29/4, 2, 5/4, 0, 0, 0]^T.$$

4.8 Dual Face Algorithm

In this section, we convert Algorithm 4.6.1 to the revised form.

Initially, take some partition $A = (B, N', N)$ with full basis B together with the following sets:

$$R = \{1, \ldots, m\}, \qquad N' = R' = \emptyset. \tag{4.56}$$

Let $\bar{z}$ be a dual feasible solution such that

$$\bar{z}_B = 0, \qquad \bar{z}_N = c_N - N^T \bar{y}, \qquad B^T \bar{y} = c_B.$$

Assume that the factorization $\tilde{B} \triangleq B \cup N' = LU$ is available.

The following items are to be computed:

(a) Search direction: $\Delta z_N = -\bar{N}_{R'}^T \bar{b}_{R'}$ (4.34).
It can be expressed as

$$\Delta z_N^T = -[0^T, \bar{b}_{R'}^T]\bar{N} = -([0^T, \bar{b}_{R'}^T](B \cup N')^{-1})N.$$

Thereby, solve the following two triangular systems:

$$U^T u = -[0^T, \bar{b}_{R'}^T]^T, \qquad L^T w = u, \tag{4.57}$$

and calculate

$$\Delta z_N = N^T w. \tag{4.58}$$

(b) Primal solution: $\bar{x}_{B \cup N'}$.
It can be obtained by solving the following two triangular systems:

$$Lw = b, \qquad U\bar{b} = w.$$

The primal solution is then

$$\bar{x}_{B \cup N'} = \bar{b}, \qquad \bar{x}_N = 0. \tag{4.59}$$

(c) Updating dual objective value.

$$\bar{f} = \bar{f} + \beta \bar{x}_{N'}^T \bar{x}_{N'}. \tag{4.60}$$

For the sake of simplicity, the following row rule would be useful for illustration.

Rule 4.8.1 (Row Rule) Select the row index p such that

$$p \in \arg\max_{i\in R'} |\bar{x}_{j_i}|. \tag{4.61}$$

Practically, we recommend the following row rule, which uses the so-called duality gaps, as introduced below:

Note that

$$\bar{x}_{N'} = \bar{b}_{R'}.$$

Set the lower bound on z_j to $l_j = 0$ and upper bound to some sufficiently large number, like

$$u = 10^{20}.$$

Then the duality gaps (in the dual context) can be defined as

$$\delta_{j_i} = \begin{cases} (u_j - \bar{z}_{j_i})\bar{x}_{j_i}, & \text{if } \bar{x}_{j_i} < 0, \\ (l_j - \bar{z}_{j_i})\bar{x}_{j_i}, & \text{if } \bar{x}_{j_i} \geq 0, \end{cases} \qquad i \in R', \tag{4.62}$$

which reflects the potential improvement in the objective value with respect to indices $j_i \in N'$.

Rule 4.8.2 (Row Rule) Select the row index p such that

$$p \in \arg\max_{i\in R'} |\delta_{j_i}|. \tag{4.63}$$

The preceding rule is advantageous since it removes from N' the index j_p, corresponding to the maximum duality gap.

Then, Algorithm 4.6.1 can be converted to the following revised form.

Algorithm 4.8.1 (Dual Face Algorithm)

Initial: partition (B, N', N) with empty N', dual feasible solution $\bar{z}$, $B = LU$. This algorithm solves the standard LP problem (4.1).

1. *Solve the two triangular systems (4.59) for $\bar{x}_{B\cup N'}$.*
2. *Go to Step 5 if N' is nonempty and $\bar{x}_{N'}$ is nonzero.*
3. *Stop if the set T defined by (4.51) is empty (optimality achieved).*
4. *Adjust sets by (4.53).*
5. *Set $\Delta z_{N'} = -\bar{b}_{R'}$.*
6. *Solve the two triangular systems (4.57) and compute Δz_N by (4.58).*
7. *Stop if $J = \{j \in N' \cup N \mid \Delta z_j < 0\}$ is empty (infeasible problem).*
8. *Determine stepsize β and J_β by (4.39)–(4.40).*
9. *If $\beta > 0$, update $\bar{z}_{N'\cup N}$ by (4.38) and $\bar{f}$ by (4.60).*
10. *If $N' \cap J_\alpha$ is nonempty, then:*

(1) Adjust sets by (4.46)
(2) Go to Step 3 if N' is empty; else, go to Step 5.

11. *Determine the column index q by (4.47).*
12. *Determine row index p by (4.61).*
13. *Adjust sets by (4.50).*
14. *Update $\tilde{B} = B \cup N' = LU$.*
15. *Go to Step 1.*

Example 4.8.1 Solve the following standard LP problem by Algorithm 4.6.1:

$$A = \begin{bmatrix} 1 & 0 & 0 & 1 & -2 & -14 & 9 & -4 \\ 0 & 2 & -3 & 0 & -18 & -21 & 32 & -12 \\ -3 & 0 & 1 & -3 & 6 & 49 & -35 & 14 \\ 0 & -2 & 3 & 8 & 34 & -51 & -32 & -20 \end{bmatrix} \qquad b = \begin{bmatrix} 4 \\ 1 \\ -17 \\ -17 \end{bmatrix}$$

$c = [2, 1, 3, 0, 1, 2, 0, 1]^T$.

Initial: $B = \{1, 2, 3, 4\},\ R = \{1, 2, 3, 4\},\ N' = R' = \emptyset,\ N = \{5, 6, 7, 8\}$.
$\bar{z} = [2, 0, 3, 0, 1, 2, 0, 1]^T, \bar{f} = 0$.

$$B \cup N' = LU, \text{where } L = \begin{bmatrix} 1 & 0 & 0 & 0 \\ 0 & -1 & 0 & 0 \\ -3 & 0 & 1 & 0 \\ 0 & 1 & 0 & -2 \end{bmatrix}, \qquad U = \begin{bmatrix} 1 & 0 & 0 & 1 \\ 0 & -2 & 3 & 0 \\ 0 & 0 & 1 & 0 \\ 0 & 0 & 0 & -4 \end{bmatrix}.$$

Iteration 1:

1. $\bar{x}_{B \cup N'} = [6, -7 - 5, -2]^T$.
3. $T = \{1, 2, 3, 4\},\ T' = \{1, 2, 3, 4\}$.
4. $N' = \{1, 2, 3, 4\},\ R' = \{1, 2, 3, 4\}\ B = R = \emptyset$.
5. $\Delta z_{N'} = [-6, 7, 5, 2]^T$.
6. $u = [-6, -7/2, 31/2, -2]^T,\ w = [81/2, 9/2, 31/2, 1]^T,\ \Delta z_N = [-35, 47, -66, -19]^T$.
7. $J = \{1, 5, 7, 8\}$.
8. $\beta = \min[-\bar{z}(1)/(-6), -\bar{z}(5)/(-35), -\bar{z}(7)/(-66), -\bar{z}(8)/(-19)] = 0,\ J_\beta = \{7\}$.
9. Nothing to do.
10. $N' \cap J_\beta = \emptyset$.
11. $q = 7$.
12. $p = \arg\max[|6|, |-7||-5|, |-2|] = 2,\ j_2 = 2,\ \bar{x}_{j_2} = -7$.
13. $B = \{7\},\ R = \{2\},\ N' = \{1, 3, 4\},\ R' = \{1, 3, 4\},\ , N = \{2, 5, 6, 8\}$.
14. $\tilde{B} = [A(:, 1), A(:, 7), A(:, 3), A(:, 4)]$,
permutated LU factors:

$$L = \begin{bmatrix} -1/3 & -1/12 & 1 & 0 \\ 0 & 1 & 0 & 0 \\ 1 & 0 & 0 & 0 \\ 0 & -1 & 0 & 1 \end{bmatrix} \qquad U = \begin{bmatrix} -3 & -35 & 1 & -3 \\ 0 & 32 & -3 & 0 \\ 0 & 0 & 1/12 & 0 \\ 0 & 0 & 0 & 8 \end{bmatrix}.$$

Iteration 2:

1. $\bar{x}_{B\cup N'} = [87/4, -7/4, -19, -2]^T$.
2. (switch to Step 5)
5. $\Delta z_{N'} = [-87/4, 19, 2]^T$.
6. $u = [29/4, 1015/128, 13{,}647/32, 95/32]^T$, $w = [13{,}647/32, 743/16, 4781/32, 95/32]^T$, $\Delta z_N = [1391/16, -11{,}063/16, 895/4, -3693/16]^T$.
7. $J = \{1, 5, 8\}$.
8. $\beta = \min[-\bar{z}(1)/(-87/4), -\bar{z}(5)/(-11{,}063/16), \bar{z}(8)/(-3693/16)]$ $= \min[8/87, 16/11{,}063, 16/3693] = 16/11{,}063,\ J_\beta = \{5\}$.
9. $\bar{z}_{N'\cup N} = \bar{z}_{N'\cup N} + \beta\Delta z_{N'\cup N}$ $= [1189/604, 2534/837, 32/11{,}063, 958/851, 0, 955/411, 461/692]^T$. $\bar{z} = [1189/604, 958/851, 2534/837, 32/11{,}063, 0, 955/411, 0, 461/692]$, $\bar{f} = \bar{f} + \beta\bar{x}_{N'}^T\bar{x}_{N'} = 583/481$.
10. $N' \cap J_\beta = \emptyset$.
11. $q = 5$.
12. $p = \arg\max[|87/4|, |-19|, |-2|] = 3,\ j_3 = 3,\ \bar{x}_{j_3} = -19$.
13. $B = \{5, 7\},\ R = \{3, 2\},\ N' = \{1, 4\},\ R' = \{1, 4\},\ ,\ N = \{2, 3, 6, 8\}$.
14. $B \cup N' = [A(:, 1), A(:, 7), A(:, 5), A(:, 4)]$,

permutated LU factors:

$$L = \begin{bmatrix} -1/3 & -1/12 & -3/32 & 1 \\ 0 & 1 & 0 & 0 \\ 1 & 0 & 0 & 0 \\ 0 & -1 & 1 & 0 \end{bmatrix} \qquad U = \begin{bmatrix} -3 & -35 & 6 & -3 \\ 0 & 32 & -18 & 0 \\ 0 & 0 & 16 & 8 \\ 0 & 0 & 0 & 3/4 \end{bmatrix}.$$

Iteration 3:

1. $\bar{x}_{B\cup N'} = [331/72, 5/8, 19/18, -37/9]^T$.
2. (switch to Step 5)
5. $\Delta z_{N'} = [-331/72, 37/9]^T$.
6. $u = [331/216, 1604/957, 4895/3734, -4099/1728]^T$, $w = [-4099/1728, 2531/986, 1766/2381, 209/192]^T$, $\Delta z_N = [479/162, -1699/460, -6658/167, -1864/57]^T$.
7. $J = \{1, 3, 6, 8\}$.
8. $\beta = \min[-\bar{z}(1)/(-331/72), -\bar{z}(3)/(-1699/460), -\bar{z}(6)/(-6658/167), -\bar{z}(8)/(-1864/57)]$ $= \min[498/1163, 1291/1575, 747/12{,}817, 227/11{,}143] = 227/11{,}143$, $J_\beta = \{8\}$.
9. $\bar{z}_{N'\cup N} = \bar{z}_{N'\cup N} + \beta\Delta z_{N'\cup N}$ $= [2143/1143, 731/8437, 1065/898, 989/335, 860/569, 0]^T$. $\bar{z} = [2143/1143, 1065/898, 989/335, 731/8437, 0, 860/569, 0, 0]^T$. $\bar{f} = \bar{f} + \beta\bar{x}_{N'}^T\bar{x}_{N'} = 2124/1069$.
10. $N' \cap J_\beta = \emptyset$.
11. $q = 8$.
12. $p = \arg\max[|331/72|, |-37/9|] = 4,\ j_p = 4,\ \bar{x}_4 = -37/9$.
13. $B = \{5, 7, 8\},\ R = \{3, 2, 4\},\ N' = \{1\},\ R' = \{1\},\ ,\ N = \{2, 3, 4, 6\}$.

14. $B \cup N' = [A(:,1), A(:,7), A(:,5), A(:,8)]$,

$$L = \begin{bmatrix} -1/3 & -1/12 & -3/32 & 1 \\ 0 & 1 & 0 & 0 \\ 1 & 0 & 0 & 0 \\ 0 & -1 & 1 & 0 \end{bmatrix} \qquad U = \begin{bmatrix} -3 & -35 & 6 & 14 \\ 0 & 32 & -18 & -12 \\ 0 & 0 & 16 & -32 \\ 0 & 0 & 0 & -10/3 \end{bmatrix}.$$

Iteration 4:

1. $\bar{x}_{B \cup N'} = [271/160, 137/160, 17/20, 37/40]^T$.
2. (switch to Step 5)
5. $\Delta z_{N'} = [-271/160]$.
6. $u = [271/480, 1897/3072, 667/1381, -1553/346]^T$,
   $w = [-1553/346, 674/2205, -912/979, 139/2235]$.
   $\Delta z_{N'} = [-271/160]$, $\Delta z_N = [1269/2606, -1185/713, -884/739, 2151/283]^T$.
7. $J = \{1, 3, 4\}$.
8. $\beta = \min[-\bar{z}(1)/(-271/160), -\bar{z}(3)/(-1185/713), -\bar{z}(4)/(-884/739)]$
   $= \min[1004/907, 2311/1301, 284/3921] = 284/3921$, $J_\beta = \{4\}$.
9. $\bar{z}_{N' \cup N} = \bar{z}_{N' \cup N} + \beta \Delta z_{N' \cup N}$
   $= [198/113, 138/113, 320/113, 0, 233/113]^T$.
   $\bar{z} = [198/113, 138/113, 320/113, 0, 0, 233/113, 0, 0]^T$,.
   $\bar{f} = \bar{f} + \beta \bar{x}_{N'}^T \bar{x}_{N'} = 248/113$.
10. $N' \cap J_\beta = \emptyset$.
11. $q = 4$.
12. $p = 1$, $j_4 = 1$, $\bar{x}_{j_1} = 271/160$.
13. $B = \{4, 7, 5, 8\}$, $R = \{1, 2, 3, 4\}$, $N' = R' = \emptyset$, , $N = \{1, 2, 3, 6\}$.
14. $B \cup N' = [A(:,4), A(:,7), A(:,5), A(:,8)]$,

$$L = \begin{bmatrix} 1/8 & -13/47 & 25/123 & 1 \\ 0 & -32/47 & 1 & 0 \\ -3/8 & 1 & 0 & 0 \\ 1 & 0 & 0 & 0 \end{bmatrix} \qquad U = \begin{bmatrix} 8 & -32 & 34 & -20 \\ 0 & -47 & 75/4 & 13/2 \\ 0 & 0 & -246/47 & -356/47 \\ 0 & 0 & 0 & 226/123 \end{bmatrix}$$

Iteration 5:

1. $\bar{x}_B = [271/113, 112/113, 165/226, 331/226]^T$.
2. $N' = \emptyset$.
3. $T = \emptyset$. Optimality achieved.

The primal and dual optimal basic solutions:

$$\bar{x} = [0, 0, 0, 271/113, 165/226, 0, 112/113, 331/226]^T,$$

$$\bar{z} = [198/113, 138/113, 320/113, 0, 0, 233/113, 0, 0]^T, \qquad \bar{f} = 248/113.$$

Note: It is interesting to compare the face method and dual face method. Surprisingly enough, it turns out they are just a pigeon pair. It seems natural to use the latter to achieve dual feasibility, and then use the former to attain primal feasibility, or vice versa.

This algorithm solves four triangular systems in Steps 2 and 7. The algorithm's computing cost in a single iteration is comparable to the simplex method, highlighting its efficiency and practicality.

The dual face method offers a distinct advantage in its stability, consistently selecting a significant pivot in size, thereby enhancing its reliability.

The most promising feature of the dual method would be the dual objective increment $\beta\|\bar{b}_{R'}\|^2$ achieved in a single iteration. The number of iterations required for solving LP problems would be significantly reduced.

Chapter 5
Generalized Dual Face Method

This chapter is devoted to the generalized dual face method for solving the variable-bounded LP problem. In the next section, we address the update of the dual face solution and related column rule. Section 5.2 discusses the row rule and updates the primal solution. Section 5.3 addresses the optimality condition. Then, Sects. 5.4 and 5.5 formulate the generalized dual face algorithm in tableau and revised forms. Section 5.6 presents the practicable dual face algorithm. We propose a tentative algorithm in the final section.

To solve the variable-bounded LP problem (3.1), we handle its dual problem (1.37), which can be rewritten

$$\begin{array}{ll} \max & b^{\mathrm{T}}y + u_{\Gamma}^{\mathrm{T}}z_{\Gamma} + l_{\Pi}^{\mathrm{T}}z_{\Pi} \\ \text{s.t.} & A^{\mathrm{T}}y + z = c, \end{array} \tag{5.1}$$

where $\Gamma = \{j \mid z_j < 0,\ j = 1, \ldots, n\}$, $\quad \Pi = \{j \mid z_j > 0,\ j = 1, \ldots, n\}$.

Introduce the following concept.

Definition 5.0.1 For the partition $A = (B, N)$ with pseudo-basis B, the dual face is a subset of the feasible region defined as

$$D(N) = \{(y, z_N) \in \mathcal{R}^m \times \mathcal{R}^{n-k} \mid B^{\mathrm{T}}y = c_B,\ N^{\mathrm{T}}y + z_N = c_N,\ z_B = 0\}. \tag{5.2}$$

Definition 5.0.2 A dual face is a dual-level face if the dual objective value is constant over it. A level dual face is *optimal* if the constant equals the dual optimal value.

5.1 Column Rule and Updating Dual Solution

Let $A = (B, N', N)$ be a partition, where B is pseudo-basis.

Assume that $\bar{z}$ is a dual solution such that

P.-Q. Pan, *Face Method*, https://doi.org/10.1007/978-3-031-93594-7_5

$$A^T\bar{y} + \bar{z} = c$$

for some $\bar{y}$. Initially, we may set $\bar{z} = c$ ($\bar{y} = 0$). Note that only $\bar{z}$ needs to be updated, as the final dual solution in y-space can be calculated cheaply at the end of the solution process if needed.

Let $(\tilde{B}, N)$ be a partition with the normal basis $\tilde{B}$. Put the canonical system of $Ax = b$ into the following tableau with $\bar{z}$ as the bottom line:

$x_{\tilde{B}}$	x_N	f	rhs
I	$\bar{N}$		$\bar{b}$
$\bar{z}_{\tilde{B}}^{\mathrm{T}}$	$\bar{z}_N^{\mathrm{T}}$	-1	$-\bar{g}$

(5.3)

Let $\bar{x}$ be a primal solution such that

$$\bar{x}_j = \begin{cases} l_j, & \text{if } \bar{z}_j \geq 0, \\ u_j, & \text{if } \bar{z}_j < 0. \end{cases} \qquad j \in N. \tag{5.4}$$

Then, the basics can be calculated by

$$\bar{x}_{\tilde{B}} = \bar{b} - \bar{N}\bar{x}_N. \tag{5.5}$$

5 Form the subset N' such that

$$N' = \{j_i \in \tilde{B} \mid \bar{z}_{j_i} \neq 0\} \cup \{j_i \in \tilde{B} \mid \bar{x}_{j_i} < l_{j_i}\} \cup \{j_i \in \tilde{B} \mid \bar{x}_{j_i} > u_{j_i}\}, \tag{5.6}$$

and set

$$B = \tilde{B} \backslash N'.$$

Then, it is obvious that

$$l_B \leq \bar{x}_B \leq u_B, \qquad \bar{z}_B = 0,$$

and the tableau (4.26) can be rewritten (with f column omitted) as

x_B	$x_{N'}$	x_N	rhs
I		$\bar{N}_R$	$\bar{b}_R$
	I'	$\bar{N}_{R'}$	$\bar{b}_{R'}$
	$\bar{z}_{N'}^{\mathrm{T}}$	$\bar{z}_N^{\mathrm{T}}$	$-\bar{f}$

(5.7)

Without loss of generality, assume that B is $m \times k$ pseudo-basis with $0 < k < m$ and that $\bar{b}_{R'}$ is nonzero.

Obviously, the dual search direction in the $z_{N' \cup N}$-space given by (4.34) is still applicable

The $\bar{z}_B = 0$ is fixed. So, the dual solution is updated by the following formula:

$$\hat{z}_{N'} = \bar{z}_{N'} + \beta \Delta z_{N'}, \tag{5.8}$$

$$\hat{z}_N = \bar{z}_N + \beta \Delta x_N, \tag{5.9}$$

where β is the dual stepsize.

Introduce the following index set:

$$J = \{j \in N \mid \Delta z_j < 0,\ \bar{x}_j = l_j\} \cup \{j \in N \mid \Delta z_j > 0,\ \bar{x}_j = u_j\}. \tag{5.10}$$

If J is nonempty, the following column rule is well-defined.

Rule 5.1.1 (Dual Column Rule) Select the column index q such that

$$\beta = -\bar{z}_q / \Delta z_q = \min_{j \in J} -\bar{z}_j / \Delta z_j \geq 0. \tag{5.11}$$

In case of a tie, select the index q corresponding to the biggest $|\delta z_q|$.

The β is the largest possible stepsize to keep signs of components of the new dual solution unchanged.

It can be shown that Theorem 4.3.3 is still valid, that is, the new solution increases the dual objective value by the amount of $\Delta f = \beta \, \|\bar{b}_{R'}\|^2$.

When J defined by (5.10) is empty, we make the following claim.

Theorem 5.1.1 *Assume that $\bar{b}_{R'}$ is nonzero. If J is empty, there is no feasible solution to the variable-bounded problem (3.1).*

Proof Consider the line search scheme (5.8)–(5.9). Obviously, for any given Stepsize $\beta \geq 0$, it determines a new dual solution $(\hat{y}_{R'}, \hat{z}_{N' \cup N})$ (the nonactive part is omitted).

Moreover, components of $\hat{z}_{N'}$ bear certain signs for large enough β, and signs of $\hat{z}_N$ keep unchanged, under the nonempty assumption on J. Therefore, the new solution is dual feasible for large β. Since it holds that

$$\bar{b}_{R'}^T \hat{y}_{R'} = \beta \bar{b}_{R'}^T \bar{b}_{R'} \quad \rightarrow \quad +\infty, \quad \text{as} \quad \beta \rightarrow +\infty,$$

the corresponding dual objective value goes to infinity. Therefore, there is no feasible solution to the primal standard problem. □

5.2 Row Rule and Updating Primal Solution

Assume that the dual solution was updated and column index q is available. What to do next is to select the row index p from R'.

To do so, introduce the following duality gaps:

$$\delta_{j_i} = \begin{cases} (l_{j_i} - \bar{x}_{j_i})\bar{z}_{j_i}, & \text{if } \bar{z}_{j_i} \geq 0, \\ (u_{j_i} - \bar{x}_{j_i})\bar{z}_{j_i}, & \text{if } \bar{z}_{j_i} < 0, \end{cases} \qquad j_i \in N'. \tag{5.12}$$

Then, the following rule applies.

Rule 5.2.1 (Dual Row Rule) Select the row index p such that

$$j_p \in \arg\max_{j_i \in N'} |\delta_{j_i}|. \tag{5.13}$$

Return if δ_{j_p} is nonzero; else, reselect the p such that

$$j_p \in \arg\max_{j_i \in N'} \rho_{j_i} = \begin{cases} l_{j_i} - \bar{x}_{j_i}, & \text{if } \bar{x}_{j_i} < l_{j_i}, \\ u_{j_i} - \bar{x}_{j_i}, & \text{if } \bar{x}_{j_i} > u_{j_i}, \\ 0, & \text{if } l_j \leq \bar{x}_{j_i} \leq u_{j_i}. \end{cases} \tag{5.14}$$

If the δ_{j_p} or ρ_{j_p} is nonzero, update the primal solution as follows: The current primal solution $\bar{x}$ satisfies the system, i.e.,

$$\bar{x}_{B \cup N'} = \bar{b} - \bar{N}\bar{x}_N. \tag{5.15}$$

If some increment α is added to the $\bar{x}_q$ while keeping the other nonbasics unchanged, the the new basics become

$$\hat{x}_{B \cup N'} = \bar{b} - \sum_{q \neq j \in N} \bar{x}_j \bar{a}_j - (\bar{x}_q + \alpha)\bar{a}_q.$$

Subtracting (5.15) from the preceding gives

$$\hat{x}_{B \cup N'} = \bar{x}_{B \cup N'} - \alpha \bar{a}_q. \tag{5.16}$$

Then, the new value of the leaving variable should attain the hopped-for bound, i.e.,

$$\hat{x}_{j_p} = \bar{x}_{j_p} + \rho_{j_p}, \tag{5.17}$$

comparing which and the pth equation of (5.16) leads to

$$\alpha = -\rho_{j_p} / \bar{a}_{p,q}, \tag{5.18}$$

Consequently, (5.16)–(5.18) and

$$\hat{x}_q = \bar{x}_q + \alpha \tag{5.19}$$

together form the update formula of the primal solution.

Accordingly, update the tableau by making the basis change with (p, q), and adjust the index sets by the following tactic conformably.

$$\begin{array}{l} B = B \cup \{q\} \quad R = R \cup \{p\} \\ N' = N' \backslash \{j_p\} \; R' = R' \backslash \{p\} \; N = N \backslash \{q\} \cup \{j_p\} \end{array} \tag{5.20}$$

5.3 Optimality Condition

It is noted that the search direction is not well-defined if, e.g., the set N' is empty or it is nonempty but $\bar{x}_{N'}$ is zero. In this case, it is time to test for optimality. Only the basics $\bar{x}_B$ need to be checked since the components of $\bar{x}_N$ are within bounds, and the associated duality gaps are 0.

Introduce the following index set:

$$T = \{j_i \in B \mid \bar{z}_{j_i} \neq 0\} \cup \{j_i \in B \mid \bar{x}_{j_i} < l_{j_i}\} \cup \{j_i \in B \mid \bar{x}_{j_i} > u_{j_i}\}. \tag{5.21}$$

Using the preceding notation, we make the following claim.

Theorem 5.3.1 *Assume that the dual solution $\bar{z}$ is feasible and the primal solution $\bar{x}$ satisfies (5.4). If the ρ_{j_p} resulting from Rule 5.2.1 is zero and the set T defined by (5.21) is empty, then the $(\bar{x}, \bar{z})$ are primal and dual optimal solutions.*

Proof Without loss of generality, assume that N' is nonempty.

Note that the satisfaction of (5.4) implies that the duality gaps ρ_N and deviations $\delta_{B \cup N'}$ vanish.

The emptiness of T means that

$$\bar{z}_{j_i} = 0, \qquad l_{j_i} \leq \bar{x}_{j_i} \leq u_{j_i}, \qquad \forall \; j_i \in B,$$

and hence duality gaps vanish, with the set B.

In addition, the zero ρ_{j_p} resulting from (5.14) implies that the duality gaps and deviations also vanish with the set N'. Therefore, the same thing is true for the pair of feasible and dual solutions $(\bar{x}, \bar{z})$.

Let us check for optimality with the condition 1.40.

It is clear that the primal solution $\bar{x}$ is feasible, and hence the condition (i) is satisfied.

Furthermore, setting

$$w_j = \begin{cases} \bar{z}_j \text{ if } \bar{z}_j \geq 0 \\ 0 \quad \text{otherwise} \end{cases} \qquad j = 1, \ldots, n,$$

and

$$v_j = \begin{cases} -\bar{z}_j \text{ if } \bar{z}_j < 0 \\ 0 \end{cases} \qquad j = 1, \ldots, n.$$

exhibit optimal conditions (ii) and (iii) being fulfilled. So, the $(\bar{x}, \bar{z})$ are primal/dual feasible and complementary and hence are primal and dual optimal solutions. □

If T is nonempty, adjust sets by

$$\begin{array}{ll} B = B \backslash T & R = R \backslash T' \\ N' = N' \cup T & R' = R' \cup T' \end{array}. \tag{5.22}$$

Then, carry out the next iteration.

5.4 Generalized Dual Face Algorithm: Tableau Form

Based on discussions in previous sections, we formulate the dual face algorithm in tableau form in this section.

Obviously, the active set N' reduces iteration by iteration until N' becomes empty or $\bar{b}_{R'}$ vanishes when reaching a level dual face. Such a series of iterations is termed *dual face contraction*. It repeated until optimality is achieved or infeasibility of the problem is detected.

Initially, take a partition $A = (B, N)$ with normal basis B, and set

$$\tilde{B} = B \cup N', \qquad R = \{1, \ldots, m\}, \qquad N' = R' = \emptyset. \tag{5.23}$$

Set $\bar{z} = c$ and set $\bar{x}$ by (5.4)–(5.5).

Overall steps are summarized into the following algorithm.

Algorithm 5.4.1 (Generalized Dual Face Algorithm: Tableau Form)

Initial: partition (5.23), dual face tableau and $\bar{x}$. This algorithm solves the variable-bounded LP problem (3.1).

1. *Stop if T defined by (5.21) is empty; else adjust sets by (5.22).*
2. *Compute search direction Δz_N by (4.34).*
3. *Stop if J defined by (5.10) is empty.*
4. *Determine stepsize β and index q by (5.11).*
5. *If $\beta > 0$, add $-\beta\bar{b}_{R'}^T$ times of rows corresponding to R' to the bottom line.*
6. *Select row index p by Rule 5.2.1.*
7. *Go to Step 1 if δ_{j_p} and ρ_{j_p} vanish.*

8. *Update $\bar{x}$ by (5.16)–(5.19).*
9. *Premultiply the p-th row by $1/\bar{a}_{p,q}$, and add $-\bar{a}_{i,q}$ times of the row to the i-th row, for $i = 1, \ldots, m,\ i \neq p$.*
10. *Adjust sets by (5.20).*
11. *Go to Step 2 if N' is nonempty and $\bar{b}_{R'}$ is nonzero.*
12. *Go to Step 1.*

We state the following result with the proof omitted.

Theorem 5.4.1 *If each dual face contraction involves a nondegenerate dual face solution, Algorithm 5.4.1 terminates either at:*

(i) Step 1, giving a pair of primal and dual optimal solutions
(ii) Step 3, declaring primal infeasibility

Example 5.4.1 Solve the following variable-bounded problem by Algorithm 5.4.1:

$$A = \begin{bmatrix} 1 & & & & 4 & -6 & 3 & 8 \\ & 1 & & & 2 & -11 & -4 & 5 \\ & & 1 & & -6 & 3 & & 3 \\ & & & 1 & -2 & 0 & 2 & -5 \end{bmatrix} \qquad b = \begin{bmatrix} -3 \\ 6 \\ 1 \\ -7 \end{bmatrix}$$

$c = [6, 0, -2, 0, -4, 7, 0, 2]^T$
$u = [14, 7, 10, 6, 9, 12, 11, 5]^T$
$l = [-8, 0, -6, -7, -3, -5, 0, -7]^T$

Initial: $B = \{1, 2, 3, 4\},\ R = \{1, 2, 3, 4\},\ N = \{5, 6, 7, 8\},\ N', R' = \emptyset$.

According to (5.4)–(5.5), set primal solution $\bar{x} = [-13, -32, 91, -24, 9, -5, 0, -7]^T$.For convenience, we insert three bottom rows in the tableau to exhibit lower and upper bounds and the $\bar{x}$.

x^T	x_1	x_2	x_3	x_4	x_5	x_6	x_7	x_8	b
	1				4	−6	3	8	−3
		1			2	−11	−4	5	6
			1		−6	3		3	1
				1	−2	0	2	−5	−7
$\bar{z}^T$	6		−2		−4	7		2	
u^T	14	7	10	6	9	12	11	5	–
$\bar{x}^T$	−13	−32	91	−24	9	−5	0	−7	–
l^T	−8	0	−6	−7	−3	−5	0	−7	–

Iteration 1:

1. $T = \{1, 2, 3, 4\}$, $B, R = \emptyset$, $N' = \{1, 2, 3, 4\}$, $R' = \{1, 2, 3, 4\}$.
2. $\Delta z_N = -\bar{N}_{R'}^{\mathrm{T}} \bar{b}_{R'} = [-8, 45, 47, -44]^T$.
3. $J = \{7, 8\}$.
4. $\beta = 0, q = 7$.
6. $\max\ |\delta_{N'}| = \max[|(-8 - (-13))6, 0, |(10 - 91)(-2)|, 0] = [30, 0, 162, 0]$,

$p = 1, j_p = 1$.

7. $\delta_{j_1} \neq 0$.
8. $\alpha = -5/3, \bar{x}_{B \cup N'} = [-13, -32, 91, -24]^T - (-5/3)[3, -4, 0, 2]^T$
   $= [-8, -116/3, 91, -62/3]^T, \bar{x}_7 = 0 + (-5/3) = -5/3, \bar{x}_1 = -8$.
9. $p = 1, q = 7$.

x^T	x_1	x_2	x_3	x_4	x_5	x_6	x_7	x_8	$\bar{b}$
	1/3				4/3	−2	1	8/3	−1
	4/3	1			22/3	−19		47/3	2
			1		−6	3		3	1
	−2/3			1	−14/3	4		−31/3	−5
$\bar{z}^T$	6		−2		−4	7		2	
u^T	14	7	10	6	9	12	11	5	–
$\bar{x}^T$	−8	−116/3	91	−62/3	9	−5	−5/3	−7	–
l^T	−8	0	−6	−7	−3	−5	0	−7	–

10. $B = \{7, 2, 3, 4\}, B = \{7\}, R = \{1\}, N' = \{2, 3, 4\}, R' = \{2, 3, 4\}, N = \{1, 5, 6, 8\}$.
11. $N' \neq \emptyset,\ \bar{x}_{N'} \neq 0$.

Iteration 2:

2. $\Delta z_N = -\bar{N}_{R'}^{\mathrm{T}} \bar{b}_{R'} = [-6, -32, 55, -86]^T$.
3. $J = \{1, 8\}$.
4. $\beta = \min[-(6)/(-6), -(2)/(-86)] = 1/43, q = 8$.
5. (Form the bottom line of the tableau below)
6. $\max |\delta_{N'}| = \max[|(7 - (-116/3))(-2/43)|, |(10 - 91)(-87/43)|, |(-7 - (-62/3))(5/43)|] = 7047/43, p = 3, j_p = 3$.
7. $\delta_{j_3} \neq 0$.
8. $\alpha = 27, \bar{x}_{B \cup N'} = [-5/3, -116/3, 91, -62/3]^T - (27)[8/3, 47/3, 3, -31/3]^T$
   $= [-221/3, -1385/3, 10, 775/3]^T, \bar{x}_8 = -7 + 27 = 20, \bar{x}_3 = 10$.
9. $p = 3, q = 8$.

x^T	x_1	x_2	x_3	x_4	x_5	x_6	x_7	x_8	$\bar{b}$
	1/3		−8/9		20/3	−14/3	1		−17/9
	4/3	1	−47/9		116/3	−104/3			−29/9
			1/3		−2	1		1	1/3
	−2/3		31/9	1	−76/3	43/3			−14/9
$\bar{z}^T$	$\frac{252}{43}$	$\frac{-2}{43}$	−87/43	5/43	−204/43	356/43			−30/43
u^T	14	7	10	6	9	12	11	5	–
$\bar{x}^T$	−8	−1385/3	10	775/3	9	−5	−221/3	20	–
l^T	−8	0	−6	−7	−3	−5	0	−7	–

10. $B = \{7, 2, 8, 4\}$, $B = \{7, 8\}$, $R = \{1, 3\}$, $N' = \{2, 4\}$, $R' = \{2, 4\}$, $N = \{1, 3, 5, 6\}$.
11. $N' \neq \emptyset$, $\bar{x}_{N'} \neq 0$.

Iteration 3:

2. $\Delta z_{N'} = -[-29/9, -14/9]^T$, $\Delta z_N = [88/27, -929/81, 2300/27, -2414/27]^T$.
3. $J = \{5, 6\}$.
4. $\beta = \min[-(-204/43)/(2300/27), -(356/43)/(-2414/27)] = 158/2837$, $q = 5$.
5. (Form the bottom line of the tableau below)
6. $\max |\delta_{N'}| = \max[|(7 - (-1385/3))(-2/43)|, |(-7 - 775/3)(5/43)|] = 3980/129$, $p = 4$, $j_p = 4$.
7. $\delta_{j_4} \neq 0$.
8. $\alpha = -199/19$,
   $\bar{x}_{B \cup N'} = [-221/3, -1385/3, 20, 775/3]^T - (-199/19)[20/3, 116/3, -2, -76/3]^T$,
   $= [-73/19, -1077/19, -18/19, -7]^T$, $\bar{x}_5 = 9 - 199/19 = -28/19$, $\bar{x}_4 = -7$.
9. $p = 4, q = 5$.

x^T	x_1	x_2	x_3	x_4	x_5	x_6	x_7	x_8	$\bar{b}$
	3/19		1/57	5/19		−17/19	1		−131/57
	6/19	1	2/57	29/19		−243/19			−319/57
	1/19		7/114	−3/38		−5/38		1	26/57
	1/38		−31/228	−3/76	1	−43/76			7/114
$\bar{z}^T$	$\frac{1583}{262}$	$\frac{136}{1023}$	$\frac{-1331}{500}$	$\frac{655}{3228}$		$\frac{1277}{387}$			$\frac{-687}{487}$
u^T	14	7	10	6	9	12	11	5	−
$\bar{x}^T$	−8	−1077/19	10	−7	−28/19	−5	−73/19	−18/19	−
l^T	−8	0	−6	−7	−3	−5	0	−7	−

10. $B = \{7, 2, 8, 5\}$, $B = \{7, 8, 5\}$, $R = \{1, 3, 4\}$, $N' = \{2\}$, $R' = \{2\}$, $N = \{1, 3, 4, 6\}$.
11. $N' \neq \emptyset$, $\bar{x}_{N'} \neq 0$.

Iteration 4:

2. $\Delta z_{N'} = -[-319/57]$, $\Delta z_N = (319/57)[6/19, 2/57, 29/19, \quad 243/19]^T$
   $= [638/361, 638/3249, 2033/238, -10{,}808/151]^T$.
3. $J = \{3, 6\}$.
4. $\beta = \min[-(-1331/500)/(638/3249) - (1277/387)/(-10{,}808/151)] = 269/5835$, $q = 6$.
5. (Form the bottom line of the tableau below)
6. $\delta_2 = 0$, $\rho_2 = 0 - (-1077/19) > 0$, $p = 2$, $j_p = 2$.

7. $\delta_{j_2} \neq 0$.
8. $\alpha = 359/81, \bar{x}_{B \cup N'} = [-73/19, -1077/19, -18/19, -28/19]^T - (359/81)[-17/19, -243/19, -5/38, -43/76]^T = [10/81, 0, -59/162, 335/324]^T$,
   $\bar{x}_6 = -5 - 4887/268 = -6227/268, \bar{x}_2 = 0$.
9. $p = 2, q = 6$.

x^T	x_1	x_2	x_3	x_4	x_5	x_6	x_7	x_8	$\bar{b}$
	11/81	−17/243	11/729	38/243			1		−1390/729
	−2/81	−19/243	−2/729	−29/243		1			319/729
	4/81	−5/486	89/1458	−23/243				1	749/1458
	1/81	−43/972	−401/2916	−26/243	1				478/1547
$\bar{z}^T$	$\frac{496}{81}$	$\frac{95}{243}$	$\frac{-1934}{729}$	$\frac{145}{243}$					−962/337
u^T	14	7	10	6	9	12	11	5	–
$\bar{x}^T$	−8	0	10	−7	335/324	−46/81	10/81	−59/162	–
l^T	−8	0	−6	−7	−3	−5	0	−7	–

10. $B = \{7, 6, 8, 5\}, B = \{7, 6, 8, 5\}, R = \{1, 2, 3, 4\}, N', R' = \emptyset, N = \{1, 2, 3, 4\}$.
11. $N' = \emptyset$.

Iteration 5:

1. $T = \emptyset$. Optimality achieved.
   The pair of primal and dual optimal face solution is as follows:
   $\bar{x} = [-8, 0, 10, -7, 335/324, -46/81, 10/81, -59/162]^T$,
   $\bar{z} = [496/81, 95/243, -1934/729, 145/243, 0, 0, 0, 0]^T, \quad \bar{f} = c^T \bar{x} = -6224/81$.

5.5 Generalized Dual Face Algorithm

In this section, we revise Algorithm 5.4.1 into an applicable form. The associated formulas are similar to those in Sect. 4.8.

Initially, take a partition $A = (B, N)$ with full basis B and the following sets:

$$\tilde{B} = B, \qquad R = \{1, \ldots, m\}, \qquad N' = R' = \emptyset. \tag{5.24}$$

Note that $\tilde{B} = (B, N'))$ is always a typical basis.

Assume that the LU factorization is available, i.e.,

$$\tilde{B} = LU.$$

Set $\bar{z} = c$ as the initial dual solution.

As for computing primal solution $\bar{x}$, determine subvector $\bar{x}_N$ by (5.4), and compute $\bar{x}_{\tilde{B}}$ by solving the following triangular systems:

$$Lu = b - N\bar{x}_N, \qquad U\bar{x}_{\tilde{B}} = u. \tag{5.25}$$

The initial $\bar{b}$ is determined by solving triangular systems.

$Lu = b, \qquad U\bar{b} = u.$

Each iteration involves the following items:

(a) Search direction $\Delta z_N^T = -\bar{b}_{R'}^T \bar{N}_{R'}$.
Solve the following two triangular systems:

$$U^T u = h, \qquad L^T w = u, \tag{5.26}$$

where

$$h = -[0^T, \bar{b}_{R'}^T]^T, \tag{5.27}$$

and calculate

$$\Delta z_N = N^T w. \tag{5.28}$$

(b) Pivot column $\bar{a}_q$.
Solve the following two triangle systems:

$$Lw = a_q, \qquad U\bar{a}_q = w. \tag{5.29}$$

(c) Right-hand side $\bar{b}$).
Update by

$$\bar{b} = \bar{b} - \alpha'\bar{a}_q, \qquad \bar{b}_p = \alpha', \tag{5.30}$$

where $\alpha' = \bar{b}_p/\bar{a}_{p,q}$.

Therefore, Algorithm 5.4.1 can be converted to the following revised form.

Algorithm 5.5.1 (Generalized Dual Face Algorithm)
Initial: partition (5.24), $\tilde{B} = LU$ and $(\bar{z}, \bar{x})$. This algorithm solves the variable-bounded problem (3.1).

1. *Stop if T defined by (5.21) is empty; else adjust sets by (5.22).*
2. *Solve two triangular systems (5.26).*
3. *Set $\Delta z_{N'} = -\bar{b}_{R'}$ and calculate Δz_N by (5.28).*
4. *Stop if J defined by (5.10) is empty. (infeasible problem)*
5. *Determine Stepsize β by (5.11).*
6. *If $\beta > 0$, update $\bar{z}_{N' \cup N}$ by (5.8)–(5.9).*

7. *Compute $\bar{a}_q$ by solving the two triangular systems (5.29).*
8. *Determine the row index p by Rule (5.2.1).*
9. *Go to Step 1 if δ_{j_p} and ρ_{j_p} vanish.*
10. *Update $\bar{b}$ by (5.30)*
11. *Update $\bar{x}$ by (5.16)–(5.19).*
12. *Adjust sets by (5.20).*
13. *Update $B = LU$.*
14. *Go to Step 2 if N' is nonempty and $\bar{x}_{N'}$ is nonzero.*
15. *Go to Step 1.*

Example 5.5.1 Solve the following variable-bounded problem by Algorithm 5.4.1:

$$A = \begin{bmatrix} -5 & 0 & -2 & 4 & 2 & -1 & 4 \\ & 3 & -4 & -3 & 0 & -6 & -8 & 1 \\ & & -8 & 4 & -6 & 0 & 1 & -6 \end{bmatrix} \qquad b = \begin{bmatrix} 8 \\ \\ -4 \end{bmatrix}$$

$c = [-3, 0, -9, 2, 0, 7, -2]^T$
$u = [9, 12, 10, 6, 11, 5, 7]^T$
$l = [-4, -6, -7, 0, -5, -9, -3]^T$

Initial: $\tilde{B} = B = \{1, 2, 3\}$, $R = \{1, 2, 3\}$, $N = \{4, 5, 6, 7\}$, $N', R' = \emptyset$.

$$B = LU, \text{ where } L = \begin{bmatrix} 1 & & \\ 3/5 & 1/2 & 1 \\ & & 1 \end{bmatrix}, \qquad U = \begin{bmatrix} 5 & & -2 \\ & -8 & 4 \\ & & -19/5 \end{bmatrix}.$$

$\bar{z} = [-3, 0, -9, 2, 0, 7, -2]^T$.
$\bar{x}_N = [0, -5, -9, 7]^T, \bar{x}_B = B^{-1}(b - N\bar{x}_N)$
$= [[170/19, 1529/152, 1211/38]^T$.
$\bar{x} = [170/19, 1529/152, 1211/38, 0, -5, -9, 7]^T$.
$\bar{b} = B^{-1}b = [36/19, 33/38, 14/19]^T$.
Iteration 1:

1. $T = \{1, 3\}, B = \{2\}, R = \{2\}, N' = \{1, 3\}, R' = \{1, 3\}$.
2. $h = -[36/19, 0, 14/19]^T$, $u = [-36/95, 0, 142/361]^T$, $w = [-222/361, 142/361, -71/361]^T$.
3. $\Delta z_{N'} = [-36/19, -14/19]^T$, $\Delta z_N = [-462/361, -1296/361, -985/361, -320/361]^T$.
4. $J = \{4, 5, 6\}$.
5. $\beta = \min[-2/(-462/361), 0, -7/(-985/361)] = 0, q = 5$.
7. $u = [2, 0, -36/5]^T, \bar{a}_5 = [22/19, 18/19, 36/19]^T$.
8. $\max \ |\delta_{N'}| = \max[|(9 - 170/19)(-3)|, |(10 - 1211/38)(-9)|] = [3/19, 7479/38]$,
   $p = 3, j_p = 3$.
9. $\delta_{j3} \neq 0$.
10. $\alpha' = (14/19)/(36/19) = 7/18$,
    $\bar{b} = [36/19, 33/38, 14/19]^T - (7/18)[22/19, 18/19, 36/19]^T$
    $= [13/9, 1/2, 0]^T, \bar{b} = [-16/27, -4/9, 7/18]^T$.
11. $\alpha = 831/38/(36/19) = 277/24, \bar{x}_B = [170/19, 1529/152, 1211/38]^T$

$- (277/24)[22/19, 18/19, 36/19]^T = [-53/12, -7/8, 10]^T$,
$\bar{x}_5 = -5 + 277/24 = 157/24, \bar{x}_3 = 10$.
$\bar{x} = [-53/12, -7/8, 10, 0, 157/24, -9, 7]^T$.

12. $B = \{1, 2, 5\}, B = \{2, 5\}, R = \{2, 3\}, N' = \{1\}, R' = \{1\}, N = \{3, 4, 6, 7\}$.
13. $L = \begin{bmatrix} 1 & & \\ 3/5 & 1/2 & 1 \\ & & 1 \end{bmatrix}, \quad U = \begin{bmatrix} 5 & & 2 \\ & -8 & \\ & & -36/5 \end{bmatrix}$.
14. $N' \neq \emptyset$.(switch to Step 2)

Iteration 2:

2. $h = -[-16/27, 0, 0]^T, \ u = [16/135, 0, 8/243]^T, \ w = [8/81, 8/243, -4/243]^T$.
3. $\Delta z_{N'} = [16/27], \ \Delta z_N = [-88/243, 40/81, -92/243, 128/243]^T$.
4. $J = \{6, 7\}$.
5. $\beta = \min[-7/(-92/243), -(-2)/(128/243)] = 243/64, q = 7$.
6. $\bar{z}_{N'} = -3 + (243/64)(16/27) = -3/4$,
$\bar{z}_N = [-9, 2, 7, -2]^T + (243/64)[-88/243, 40/81, -92/243, 128/243]^T$
$= [-83/8, 31/8, 89/16, 0]^T, \bar{z} = [-3/4, 0, -83/8, 31/8, 0, 89/16, 0]^T$.
7. $u = [4, -6, 8/5]^T, \bar{a}_7 = [8/9, 3/4, -2/9]^T$.
8. $\max|\delta_{N'}| = \max[|(9 - (-53/12))(-3/4)|], p = 1, j_p = 1$.
9. $\delta_{j_1} \neq 0$.
10. $\alpha' = (-16/27)/(8/9) = -2/3$,
$\bar{b} = [-16/27, -4/9, 7/18]^T - (-2/3)[8/9, 3/4, -2/9]^T$
$= [0, 1/18, 13/54]^T, \bar{b} = [-2/3, 1/18, 13/54]^T$.
11. $\alpha = -(161/12)/(8/9) = -483/32, \bar{x}_B = [-53/12, -7/8, 157/24]^T$
$- (-483/32)[8/9, 3/4, -2/9]^T = [9, 1337/128, 51/16]^T$,
$\bar{x}_7 = 7 - 483/32 = -259/32, \bar{x}_1 = 9$.
$\bar{x} = [9, 1337/128, 10, 0, 51/16, -9, -259/32]^T$.
12. $B = \{7, 2, 5\}, B = \{7, 2, 5\}, R = \{1, 2, 3\}, N', R' = \emptyset, N = \{1, 3, 4, 6\}$.
13. $L = \begin{bmatrix} -2/3 & 1 & 1 \\ -1/6 & 1 & \\ 1 & & \end{bmatrix}, \quad U = \begin{bmatrix} -6 & -8 & \\ & -16/3 & -6 \\ & & 8 \end{bmatrix}$.
14. $N' = \emptyset$. (switch to Step 1)

Iteration 3:

1. $T = \{7\}, B = \{2, 5\}, R = \{2, 3\}, N' = \{7\}, R' = \{1\}$.
2. $h = -[-2/3, 0, 0]^T, \ u = [-1/9, 1/6, 1/8]^T, \ w = [1/8, 1/24, -1/48]^T$.
3. $\Delta z_{N'} = [2/3], \ \Delta z_N = [3/4, -11/24, 5/8, -23/48]^T$.
4. $J = \{1, 6\}$.
5. $\beta = \min[-(-3/4)/(3/4), -(89/16)/(-23/48)] = 1, q = 1$.
6. $\bar{z}_{N'} = [0] + [2/3] = [2/3]$,
$\bar{z}_N = [-3/4, -83/8, 31/8, 89/16]^T + [3/4, -11/24, 5/8, -23/48]^T$
$= [0, -65/6, 9/2, 61/12]^T, \bar{z} = [0, 0, -65/6, 9/2, 0, 61/12, 2/3]^T$.
7. $u = [0, 3, 2]^T, \bar{a}_1 = [9/8, -27/32, 1/4]^T$.

8. $\max |\delta_{N'}| = (-3 - (-259/32))(2/3) = 163/48$, $p = 1$, $j_p = 7$.
9. $\delta_{j_1} \neq 0$.
10. $\alpha' = (-2/3)/(9/8) = -16/27$,
    $\bar{b} = [-2/3, 1/18, 13/54]^T - (-16/27)[9/8, -27/32, 1/4]^T = [0, -4/9, 7/18]^T$,
    $\bar{b} = [-16/27, -4/9, 7/18]^T$.
11. $\alpha = -(163/32)/(9/8) = -163/36$, $\bar{x}_B = [-259/32, 1337/128, 51/16]^T - (-163/36)[9/8, -27/32, 1/4]^T = [-3, 53/8, 311/72]^T$, $\bar{x}_1 = 9 - 163/36 = 161/36$,
    $\bar{x}_7 = -3$, $\bar{x} = [161/36, 53/8, 10, 0, 311/72, -9, -3]^T$.
12. $B = \{1, 2, 5\}$, $B = \{1, 2, 5\}$, $R = \{1, 2, 3\}$, N', $R' = \emptyset$, $N = \{3, 4, 6, 7\}$.
13. $L = \begin{bmatrix} 1 & & \\ 3/5 & 1/2 & 1 \\ & & 1 \end{bmatrix}, \quad U = \begin{bmatrix} 5 & & 2 \\ & -8 & \\ & & -36/5 \end{bmatrix}.$
14. $N' = \emptyset$.(switch to Step 1)

Iteration 4:

1. $T = \emptyset$, optimality achieved.

   The primal and dual optimal basic solutions:
   $\bar{x} = [161/36, 53/8, 10, 0, 311/72, -9, -3]^T$,
   $\bar{z} = [0, 0, -65/6, 9/2, 0, 61/12, 2/3]^T$, $\quad \bar{f} = c^T \bar{x} = -1925/12$.

5.6 Practicable Dual Face Algorithm

The dual face method needs to compute the search direction $\Delta z_{N' \cup N}$ in each iteration. Limiting the size of N' is wise to reduce computing costs. In this section, we modify the algorithm by considering duality gaps in conjunction with the N-M pivot rule (Sect. 1.12.3).

The trick is to take a certain percent of the top components of the N' for calculating the search direction (based on duality gaps).

Predetermine be an optimality tolerance ϵ and a proportion μ, like $0.6 - 0, 9$.

Using the lower bound $l_j = 0$ and a sufficiently large upper bound on x_j, like

$$u_j = 10^{20},$$

the duality gaps δ_j defined by (4.62) can be determined.

There are the following two cases:

(a) The N' is empty or $\bar{b}_{R'}$ vanishes.
    Optimality is achieved if

$$\max_{j_i \in R} |\delta_{j_i}| < \epsilon. \tag{5.31}$$

Otherwise, determine the following threshold:

$$\sigma_1 = \mu \max_{j_i \in B} \delta_{j_i}, \tag{5.32}$$

and set

$$T = \{j_i \in B \mid \delta_{j_i} > \sigma_1\}. \tag{5.33}$$

(b) The N' is nonempty and $\bar{b}_{R'}$ is nonzero.
Determine the following threshold:

$$\sigma = \mu \max_{j_i \in N'} \delta_{j_i}, \tag{5.34}$$

and reset

$$N' = \{j_i \in N' \mid \delta_{j_i} > \sigma\}, \qquad B = \tilde{B} \backslash N'. \tag{5.35}$$

Incorporating the preceding techniques into the dual face algorithm gives the following algorithm.

Algorithm 5.6.1 (Practicable Dual Face Algorithm)

Initial: partition $A = (B, N', N)$ with basis B and empty N', feasible dual solution $\bar{z}$, factorization $B = LU$, optimality tolerance ϵ, and proportion μ. This algorithm is the same as Algorithm 4.8.1 except for Steps 2–4 replaced by

2. *If N' is nonempty and $\bar{x}_{N'}$ is nonzero, reset N' by (5.35) and go to Step 5.*
3. *Stop if the condition (5.31) is satisfied. (optimality achieved)*
4. *Set T by (5.33), and adjust $N' = N' \cup T$, $B = B \backslash T$.*

5.7 Tentative Variant

The generalized dual face algorithm involves the right-hand side $\bar{b}$, which is only used to determine the search direction. In this section, we simplify the algorithm by using $x_{B \cup N'}$ instead of $\bar{b}$ in the search direction formula.

Doing so is correct for solving standard problems because the right-hand side of the dual face tableau gives the $x_{B \cup N'}$ itself explicitly.

For the variable-bounded problem, we can handle the subproblem (4.32) with the objective function replaced by

$$\bar{x}_{N'}^T y_{R'}$$

to derive the search direction, which is still in the null of the coefficient matrix of the constraint equations. The substitute problem shares the same primal and dual optimal solutions, except for the dual solution in the y-space, which can be restored after the process ends, if required.

Therefore, we can use the following search direction instead:

$$\Delta z_{N'} = -\bar{x}_{N'}, \qquad \Delta z_N = -\bar{N}_{R'}^{\mathrm{T}}\bar{x}_{N'}. \tag{5.36}$$

It leads to the following tentative algorithm.

Algorithm 5.7.1 (Tentative Algorithm)

Initial: partition (5.24), $\tilde{B} = LU$ and $(\bar{z}, \bar{x})$. This algorithm solves the variable-bounded problem.

1. *Stop if T defined by (5.21) is empty (optimality achieved); else adjust sets by (5.22).*
2. *Solve two triangular systems (5.26) with $\bar{b}_{R'}$ replaced by $\bar{x}_{N'}$.*
3. *Set $\Delta z_{N'} = -\bar{x}_{N'}$ and calculate Δz_N by (5.28).*
4. *Stop if J defined by (5.10) is empty. (infeasible problem)*
5. *Determine Stepsize β by (5.11).*
6. *If $\beta > 0$, update $\bar{z}_{N' \cup N}$ by (5.8)–(5.9).*
7. *Compute $\bar{a}_q$ by solving the two triangular systems (5.29).*
8. *Determine the row index p by (5.13).*
9. *Go to Step 1 if ρ_{j_p} defined by (5.14) vanishes.*
10. *Update $\bar{x}$ by (5.16)–(5.19).*
11. *Adjust sets by (5.20).*
12. *Update $B = LU$.*
13. *Go to Step 2 if N' is nonempty and $\bar{x}_{N'}$ is nonzero.*
14. *Go to Step 1.*

Example 5.7.1 Solve the following variable-bounded problem by Algorithm 5.7.1:

$$A = \begin{bmatrix} -9 & 5 & -2 & 0 & -4 & 5 & 6 & -4 & 2 \\ 4 & 0 & -3 & 4 & 3 & -6 & 0 & 3 & 1 \\ -7 & -6 & 0 & -2 & 0 & 2 & -5 & 8 & -7 \\ 0 & 7 & 3 & 8 & -2 & -4 & -6 & 0 & -6 \end{bmatrix} \qquad b = \begin{bmatrix} 8 \\ -2 \\ 0 \\ -4 \end{bmatrix}$$

$c = [-9, 2, 0, 7, -2 - 3, 0, 3, -7]$
$u = [11, 5, 7, 9, 12, 10, 6, 5, 4]$
$l = [-4, 0, -5, -9, -6, -2, -7 - 3, 0]$

Initial: $\tilde{B} = B = \{1, 2, 3, 4\}$, $R = \{1, 2, 3, 4\}$, $N = \{5, 6, 7, 8, 9\}$, $N', R' = \emptyset$.
$\bar{z} = [-9, 2, 0, 7, -2 - 3, 0, 3, -7]^T$.
$\bar{x}_N = [12, 10, -7, -3, 4]^T, \bar{x}_B = B^{-1}(b - N\bar{x}_N) = [-1633/858, 1239/1685, -2910/809, 2007/337]^T$,
$\bar{x} = [-1633/858, 1239/1685, -2910/809, 2007/337, 12, 10, -7, -3, 4]^T$.

$$B = LU, \text{ where } L = \begin{bmatrix} 1 & & & \\ -4/9 & -20/89 & -63/73 & 1 \\ 7/9 & 1 & & \\ & -63/89 & 1 & \end{bmatrix},$$

$$U = \begin{bmatrix} -9 & 5 & -2 & \\ & -89/9 & 14/9 & -2 \\ & & 365/89 & 586/89 \\ & & & 674/73 \end{bmatrix}.$$

Iteration 1:

1. $\delta_B = [(11 - (-1633/858))(-9), (0 - 1239/1685)(2), (-5 - (-2910/809))$ $(0), (-9 - 2007/337)(7)]$
   $T = \{1, 2, 3, 4\}, B, R = \emptyset, N' = \{1, 2, 3, 4\}, R' = \{1, 2, 3, 4\}.$
2. $h = -[-1633/858, 1239/1685, -2910/809, 2007/337]^T,$
   $u = [-188/889, -129/3961, 1608/2045, -416/343]^T,$
   $w = [-260/703, -416/343, -162/331, -646/2481]^T.$
3. $\Delta z_{N'} = -[-1633/858, 1239/1685, -2910/809, 2007/337]^T$
   , $\Delta z_N = [-1948/1189, 2004/365, 1298/725, -2527/416, 3822/1259]^T.$
4. $J = \{6, 8, 9\}.$
5. $\beta = \min[-(-3)/(2004/365), -3/(-2527/416), -(-7)/(3822/1259)]$
   $= 1248/2527, q = 8.$
6. $\bar{z}_{N'} = [-9, 2, 0, 7]^T + (1248/2527)\Delta z_{N'}$
   $= [-3893/483, 2123/1297, 1192/671, 6145/1514]^T,$
   $\bar{z}_N = [-2 - 3, 0, 3, -7]^T + (1248/2527)\Delta z_N$
   $= [-677/241, -613/2125, 481/544, 0, -3669/667]^T,$
   $\bar{z} = [-3893/483, 2123/1297, 1192/671, 6145/1514, -677/241, -613/2125,$ $481/544, 0, -3669/667]^T.$
7. $u = [-4, 100/9, 700/89, 767/73]^T, \bar{a}_8 = [-539/1685, -801/598, 153/1685,$ $767/674]^T.$
8. $\max|\delta_{N'}| = \max[|(11 - (-1633/858))(-3893/483)|, |(0 - 1239/1685)$ $(2123/1297)|,$
   $|(-5 - (-2910/809))(1192/671)|, |(-9 - 2007/337)(6145/1514)|] =$ $124{,}177/1194,$
   $p = 1, j_p = 1, \rho_{j_1} = 11 - (-1633/858) = 11{,}071/858 \neq 0.$
10. $\alpha = -(11{,}071/858)/(-539/1685) = 3106/77,$
   $\bar{x}_B = [-1633/858, 1239/1685, -2910/809, 2007/337]^T$
   $- (3106/77)\bar{a}_8 - [11, 4217/77, -559/77, -3076/77]^T, \bar{x}_8 = -3 +$ $3106/77 = 2875/77.$
   $\bar{x} = [11, 4217/77, -559/77, -3076/77, 12, 10, -7, 2875/77, 4]^T.$
11. $B = \{8, 2, 3, 4\}, B = \{8\}, R = \{1\}, N' = \{2, 3, 4\}, R' = \{2, 3, 4\}, N =$ $\{1, 5, 6, 7, 9\}.$

12. $L = \begin{bmatrix} -1/2 & 2/7 & 80/111 & 1 \\ 3/8 & 9/28 & 1 & \\ 1 & & & \\ & 1 & & \end{bmatrix}, \quad U = \begin{bmatrix} 8 & -6 & & -2 \\ & 7 & 3 & 8 \\ & & -111/28 & 61/28 \\ & & & -539/111 \end{bmatrix}.$
13. $N' \neq \emptyset$.(switch to Step 2)

Iteration 2:

2. $h = -[0, 4217/77, -559/77, -3076/77]^T$,
   $u = [0, -2042/261, -2969/383, -6001/244]^T$,
   $w = [-6001/244, 6054/607, -5597/349, -1525/381]^T$.
3. $\Delta z_{N'} = [-4217/77, 559/77, 3076/77]^T$
   $\Delta z_N = [49{,}676/133, 16{,}629/122, -11{,}336/57, -477/11, 6309/65]^T$.
4. $J = \{1, 5, 7, 9\}$.
5. $\beta = \min[-(-3893/483)/(49{,}676/133), -(-677/241)/(16{,}629/122), -(481/544)/(-477/11), -(-3669/667)/(6309/65)] = 231/11{,}329, q = 7$.
6. $\bar{z}_{N'} = [2123/1297, 1192/671, 6145/1514]^T + (231/11{,}329)\Delta z_{N'}$
   $= [1535/2951, 739/384, 731/150]^T$,
   $\bar{z}_N = [-3893/483, -677/241, -613/2125, 481/544, -3669/667]^T + (231/11{,}329)\Delta z_N$
   $= [-247/556, -522/17{,}467, -2819/649, 0, -1708/485]^T$, $\bar{z} = [-247/556, 2123/1297, 1192/671, 6145/1514, -522/17{,}467, -2819/649, 0, 0, -1708/485]^T$.
7. $u = [-5, -6, 213/56, 183/74]^T, \bar{a}_7 = [-302/539, 138/539, -668/539, -549/1078]^T$.
8. $\max |\delta_{N'}| = \max[|(0 - (4217/77))(2123/1297)|, |(-5 - (-559/77))(1192/671)|, |(-9 - (-3076/77))(6145/1514)|] = 19{,}721/157$,
   $p = 4, j_p = 4, \rho_{j_4} = -9 - (-3076/77) = 2383/77 \neq 0$.
10. $\alpha = -(2383/77)/(-549/1078) = 8143/134$,
    $\bar{x}_B = [2875/77, 4217/77, -559/77, -3076/77]^T - (8143/134)\bar{a}_7 = [7210/101, 7175/183, 19{,}327/284, -9]^T$, $\bar{x}_7 = -7 + 8143/134 = 7205/134$.
    $\bar{x} = [11, 7175/183, 19{,}327/284, -9, 12, 10, 7205/134, 7210/101, 4]^T$.
11. $B = \{8, 2, 3, 7\}, B = \{8, 7\}, R = \{1, 4\}, N' = \{2, 3\}, R' = \{2, 3\}, N = \{1, 4, 5, 6, 9\}$.
12. $L = \begin{bmatrix} -1/2 & 2/7 & 80/111 & 1 \\ 3/8 & 9/28 & 1 & \\ 1 & & & \\ & 1 & & \end{bmatrix}, \quad U = \begin{bmatrix} 8 & -6 & & -2 \\ & 7 & 3 & 8 \\ & & -111/28 & 61/28 \\ & & & -539/111 \end{bmatrix}.$
13. $N' \neq \emptyset$.(switch to Step 2)

Iteration 3:

2. $h = -[0, 4217/77, -559/77, 0]^T$, $u = [0, -1025/183, 6089/471, -3749/112]^T$,
   $w = [-3749/112, 704/19, -7811/255, -1502/189]^T$.
3. $\Delta z_{N'} = [-7175/183, -19{,}327/284]^T$,
   $\Delta z_N = [5975/9, 4231/29, 19{,}049/73, -13{,}413/32, 5573/24]^T$.
4. $J = \{1, 5, 9\}$.
5. $\beta = \min[-(-247/556)/(5975/9), -(-522/17{,}467)/(19{,}049/73), -(-1708/485)/(5573/24)]^T$
   $= 3/26{,}195, q = 5$.
6. $\bar{z}_{N'} = [2123/1297, 1192/671]^T + (3/26{,}195)\Delta z_{N'} = [2189/1341, 5237/2961]^T$,
   $\bar{z}_N = [-247/556, 6145/1514, , -522/17{,}467, -2819/649, -1708/485]^T + (3/26{,}195)\Delta z_N$
   $= [-278/755, 3509/861, 0, -628/143, -5655/1618]^T$,
   $\bar{z} = [-278/755, 2189/1341, 5237/2961, 3509/861, 0, -628/143, 0, 0, -5655/1618]^T$.
7. $u = [0, -2, 51/14, -224/37]^T, \bar{a}_5 = [-415/183, -60/61, -598/183, -448/183]^T$.
8. $\max|\delta_{N'}| = \max[|(0-(7175/183))(2189/1341)|, |(-5-19{,}327/284)(5237/2961)|]$
   $= 9432/73, p = 3, j_p = 3, \rho_{j_3} = -5 - 19{,}327/284 = -20{,}747/284 \neq 0$.
10. $\alpha = -(-20{,}747/284)/(-598/183) = -7355/329$,
   $\bar{x}_B = [7210/101, 7175/183, 19{,}327/284, 7205/134]^T$
   $-(-7355/329)\bar{a}_5 = [7117/344, 6698/389, -5, -1842/1919]^T$,
   $\bar{x}_5 = 12 - 7355/329 = -3407/329$.
   $\bar{x} = [11, 6698/389, -5, -9, -3407/329, 10, -1842/1919, 7117/344, 4]^T$
11. $B = \{8, 2, 5, 7\}, B = \{8, 5, 7\}, R = \{1, 3, 4\}, N' = \{2\}, R' = \{2\}, N = \{1, 3, 4, 6, 9\}$.
12. $$L = \begin{bmatrix} -1/2 & 2/7 & -16/17 & 1 \\ 3/8 & 9/28 & 1 & \\ 1 & & & \\ & 1 & & \end{bmatrix}, \quad U = \begin{bmatrix} 8 & -6 & & -5 \\ & 7 & -2 & -6 \\ & & 51/14 & 213/56 \\ & & & 299/34 \end{bmatrix}.$$
13. $N' \neq \emptyset$.(switch to Step 2)

Iteration 4:

2. $h = -[0, 6698/389, 0, 0]^T$, $u = [0, -1621/659, -1576/1167, -1104/1009]^T$,
   $w = [-1104/1009, -1640/689, 652/1887, -2145/1552]^T$,
3. $\Delta z_{N'} = [-6698/389]^T$
   $\Delta z_N = [-1609/769, 1871/361, -9337/439, 4479/298, 1304/999]^T$.
4. $J = \{4, 6, 9\}$.
5. $\beta = \min[-(3509/861)/(-9337/439), -(-628/143)/(4479/298), -(-5655/1618)/(1304/999)]^T = 663/3460, q = 4$.

6. $\bar{z}_{N'} = [2189/1341]^T + (663/3460)\Delta z_{N'} = -1567/940$,
$\bar{z}_N = [-278/755, 5237/2961, 3509/861, -628/143, -5655/1618]^T + (663/3460)\Delta z_N$
$= [-653/849, 1113/403, 0, -1244/823, -1762/543]^T$,
$\bar{z} = [-653/849, -1567/940, 1113/403, 0, 0, -1244/823, 0, 0, -1762/543]^T$.
7. $u = [-2, 8, 61/28, -21/17]^T, \bar{a}_4 = [176/299, 1108/897, 668/897, -42/299]^T$.
8. $\max|\delta_{N'}| = \max[|(5 - 6698/389)(-1567/940)|]$,
$p = 2, j_p = 2, \rho_{j_2} = 5 - 6698/389 = -4753/389 \neq 0$.
10. $\alpha = -(-4753/389)/(1108/897) = 2740/277$,
$\bar{x}_B = [7117/344, 6698/389, -3407/329, -1842/1919]^T$
$-(2740/277)\bar{a}_4 = [3895/262, 5, -4909/277, 180/419]^T$,
$\bar{x}_4 = -9 + 2740/277 = 1375/1542$.
$\bar{x} = [11, 5, -5, 1375/1542, -4909/277, 10, 180/419, 3895/262, 4]^T$
11. $B = \{8, 4, 5, 7\}, B = \{8, 4, 5, 7\}, R = \{1, 2, 3, 4\}, N', R' = \emptyset, N = \{1, 2, 3, 6, 9\}$.
12. $L = \begin{bmatrix} -1/2 & -1/8 & 1 & 0 \\ 3/8 & 19/32 & -67/68 & 1 \\ 1 & 0 & 0 & 0 \\ 0 & 1 & 0 & 0 \end{bmatrix}, \quad U = \begin{bmatrix} 8 & -2 & 0 & -5 \\ 0 & 8 & -2 & -6 \\ 0 & 0 & -17/4 & 11/4 \\ 0 & 0 & 0 & 277/34 \end{bmatrix}$.
13. $N' = \emptyset$.(switch to Step 1)

Iteration 5:

1. $\delta_B = 0, \rho_B = [5 - 3895/262 < 0, 0, -6 - (-4909/277) > 0, 0]$.
$T = \{8, 5\}, B = \{8, 4, 5, 7\}, B = \{4, 7\}, R = \{2, 4\}, N' = \{8, 5\}, R' = \{1, 3\}$.
2. $h = -[3895/262, 0, -4909/277, 0]^T$,
$u = [-3895/2096, -859/1849, -2189/554, -127/853]^T$,
$w = [12{,}586/199, -17{,}795/4943, 2899/485, -6910/291, 6633/277]^T$.
3. $\Delta z_{N'} = -[3895/262, -4909/277]^T$
$\Delta z_N = [12{,}586/199, -17{,}795/4943, 2899/485, -6910/291, 6633/277]^T$.
4. $J = \{1, 9\}$.
5. $\beta = \min[-(-653/849)/(12{,}586/199), -(-1762/543)/(6633/277)] = 74/6085, q = 1$.
6. $\bar{z}_{N'} = [0, 0]^T + \beta[-3895/262, 4909/277] = [-32/177, 1786/8287]^T$,
$\bar{z}_N = [-653/849, -1567/940, 1113/403, -1244/823, -1762/543]^T + \beta\Delta z_N$
$= [0, -491/287, 3305/1166, -1136/631, -1149/389]^T$,
$\bar{z} = [0, -491/287, 3305/1166, 0, 1786/8287, -1136/631, 0, -32/177, -1149/389]^T$.
7. $u = [-7, 0, -25/2, -387/68]^T, \bar{a}_1 = [-713/554, 109/1108, 1379/554, -387/554]^T$.
8. $\max|\delta_{N'}| = \max[|(5 - (3895/262))(-32/177)|, |(-6 - (-4909/277))(1786/8287)|]$

$= 22{,}853/9046$, $p = 3$, $j_p = 5$, $\rho_{j_3} = -6 - (-4909/277) = 3247/277 \neq 0$.

10. $\alpha = -(3247/277)/(1379/554) = -3579/760$,
$\bar{x}_B = [3895/262, 1375/1542, -4909/277180/419]^T$
$-(-3579/760)\bar{a}_1 = [2809/31, 668/493, -6, -1124/393]^T$,
$\bar{x}_1 = 11 - 3579/760 = 4781/760$.
$\bar{x} = [4781/760, 5, -5, 668/493, -6, 10, -1124/393, 2809/319, 4]^T$.
11. $B = \{8, 4, 1, 7\}$, $B = \{4, 1, 7\}$, $R = \{2, 3, 4\}$, $N' = \{8\}$, $R' = \{1\}$, $N = \{2, 3, 5, 6, 9\}$.
12. $L = \begin{bmatrix} -1/2 & -1/8 & 1 & \\ 3/8 & 19/32 & -53/100 & 1 \\ 1 & & & \\ & 1 & & \end{bmatrix}$, $\quad U = \begin{bmatrix} 8 & -2 & -7 & -5 \\ & 8 & & -6 \\ & & -25/2 & 11/4 \\ & & & 1379/200 \end{bmatrix}$.
13. $N' \neq \emptyset$.(switch to Step 2)

Iteration 6:

2. $h = -[2809/319, 0, 0, 0]^T$,
$u = [-1716/1559, -429/1559, 1361/2208, -1897/1478]^T$,
$w = [-533/8347, -1897/1478, -566/869, 1306/2727]^T$,
3. $\Delta z_{N'} = -2809/319$,
$\Delta z_N = [3769/543, 1814/335, -947/208, 841/202, 634/2309]^T$.
4. $J = \{2, 5, 6, 9\}$.
5. $\beta = \min[-(-491/287)/(3769/543), -(1786/8287)/(-947/208), (-1136/631)/(841/202), -(-1149/389)/(634/2309)] = 439/9274$, $q = 5$.
6. $\bar{z}_{N'} = -32/177 + \beta[-2809/319] = -854/1429$,
$\bar{z}_N = [-491/287, 3305/1166, 1786/8287, -1136/631, -1149/389]^T + \beta\Delta z_N$
$= [-1027/743, 2621/848, 0, -396/247, -1935/658]^T$,
$\bar{z} = [0, -1027/743, 2621/848, 0, 0, -396/247, 0, -854/1429, -1935/658]^T$.
7. $u = [0, -2, -17/4, 387/200]^T$, $\bar{a}_5 = [713/1379, -109/2758, 554/1379, 387/1379]^T$.
8. $\max|\delta_{N'}| = \max[|(5 - 2809/319)(-32/177)|]$,
$p = 1$, $j_p = 8$, $\rho_{j_1} = 5 - 2809/319 = -1214/319 \neq 0$.
10. $\alpha = -(-1214/319)/(713/1379) = 2083/283$,
$\bar{x}_B = [2809/319, 668/493, 4781/760, -1124/393]^T$
$-(2083/283)\bar{a}_5 = [5, 1134/689, 2327/698, -3512/713]^T$, $\bar{x}_5 = -6 + 2083/283 = 4781/760$.
$\bar{x} = [2327/698, 5, -5, 1134/689, 2963/2178, 10, 5, 4]^T$.
11. $B = \{5, 4, 1, 7\}$, $B = \{5, 4, 1, 7\}$, $R = \{1, 2, 3, 4\}$, N', $R' = \emptyset$, $N = \{2, 3, 6, 8, 9\}$.
12. $L = \begin{bmatrix} 1 & & & \\ -3/4 & 1/2 & 40/47 & 1 \\ & -1/4 & 1 & \\ 1/2 & 1 & & \end{bmatrix}$, $\quad U = \begin{bmatrix} -4 & & -9 & 6 \\ & 8 & 9/2 & -9 \\ & & -47/8 & -29/4 \\ & & & 713/47 \end{bmatrix}$.
13. $N' = \emptyset$.(switch to Step 1)

Iteration 7:

1. $T = \emptyset$. Optimality achieved.

The primal and dual optimal solutions:

$\bar{x} = [2327/698, 5, -5, 1134/689, 2963/2178, 10, 5, 4]^T$, $\bar{f} = c^T\bar{x} = -2656/49$.

$\bar{z} = [0, -1027/743, 2621/848, 0, 0, -396/247, 0, -854/1429, -1935/658]^T$.

Chapter 6
Reduced Face Method

The term "reduced algorithm" first appeared in the simplex context in Pan (p.357, [34]). It might be the first simplex-type method that searches along the *objective-edge direction.* Contrary to the simplex algorithms, it determines the pivot row and the pivot column. This chapter transforms it into a face algorithm with the same framework. This method deserves particular attention because of its latent capacity to solve integer.

In the next section, we derive the search direction and update the formula of the face solution. In Sect. 6.2, we present the optimality condition. In Sect. 6.3, we formulate the reduced face algorithm in tableau form. Section 6.4 gives a double-phase reduced face algorithm. Section 6.5 presents the reduced face algorithm in revised form. In the final section, we deal with the generalized reduced face algorithms, including the tableau form, revised form, and double-phase version.

The standard LP problem (2.1) can be converted to the so-called reduced LP problem. Regarding its objective function as an additional variable, move it to the constraint, converting the standard problem to the following problem, equivalently:

$$\begin{array}{ll} \min & f \\ \text{s.t.} & \begin{bmatrix} A \\ c^T \end{bmatrix} x = \begin{bmatrix} 0 \\ f \end{bmatrix} + \begin{bmatrix} b \\ 0 \end{bmatrix}, \qquad x \geq 0, \end{array}$$

Resetting

$$A := \begin{bmatrix} A \\ c^T \end{bmatrix}, \qquad b := \begin{bmatrix} b \\ 0 \end{bmatrix},$$

The preceding problem can be expressed as the following so-called reduced LP problem:

P.-Q. Pan, *Face Method*, https://doi.org/10.1007/978-3-031-93594-7_6

$$\begin{aligned} &\min\ f \\ &\text{s.t.}\ \ Ax = e_{m+1} f + b, \qquad x \geq 0, \end{aligned} \tag{6.1}$$

where $A \in \mathcal{R}^{(m+1)\times n}$, $m < n$, $b \in \mathcal{R}^{m+1}$, and Rank $(A) = m + 1$.

It is noted that the objective function only involves a single variable f, and the right-hand side of the equality constraint includes f's term, whose coefficient is the unit vector e_{m+1}.

The rank assumption on A is reasonable. It is straightforward to show that it is so if and only if the objective function of the standard problem is not constant over the feasible region. It is also evident that (x, f) is a feasible/optimal solution to the reduced problem if and only if so is x to the standard problem. Geometrically, iterates produced in the solution process will be face (or feasible) solutions situated on the faces of the underlying polyhedron.

In this chapter, we will not distinguish between a face solution and a feasible solution since any feasible solution can be regarded as a face solution, even an interior solution is on the maximum face—the feasible region itself.

6.1 Search Direction and Updating Face Solution

Let (B, N) be a partition:

$$B = \{j_1, \ldots, j_{m+1}\}, \qquad N = A\backslash B.$$

Assume that the equality constraint of (6.1) is transformed into the following so-called *canonical form*, equivalently:

$$x_B + \bar{N} x_N = f\bar{d} + \bar{b}, \tag{6.2}$$

where

$$\bar{B} \triangleq I, \qquad \bar{N} = B^{-1}N, \qquad \bar{d} = B^{-1}e_{m+1} \neq 0, \qquad \bar{b} = B^{-1}b.$$

Assume that $(\bar{x}, \bar{f})$ is the current face solution, satisfying the canonical system (6.2) and the nonnegativity condition, i.e.,

$$\bar{x}_B + \bar{N}\bar{x}_N = \bar{f}\bar{d} + \bar{b}, \qquad \bar{x}_B, \bar{x}_N \geq 0. \tag{6.3}$$

There is an available search direction $\bar{d}$, which is just the current objective column. To improve the face solution $\bar{x}$, we fix $x_N = \bar{x}_N$ and reduce $\bar{f}$ along the direction by some amount α, so that the according $\bar{x}_B$ becomes $\hat{x}_B$ such that

$$\hat{x}_B + \bar{N}\bar{x}_N = (\bar{f} - \alpha)\bar{d} + \bar{b}. \tag{6.4}$$

Subtracting the first equation of (6.3) from (6.4) leads to

$$\hat{x}_B - \bar{x}_B = -\alpha \bar{d}.$$

Consequently, the new solution $(\hat{x}, \hat{f})$ can be expressed as

$$\hat{x}_B = \bar{x}_B - \alpha \bar{d}, \qquad \hat{f} = \bar{f} - \alpha, \qquad \hat{x}_N = \bar{x}_N, \tag{6.5}$$

where α is termed *stepsize* to be determined.

Then, the validity of the following claim is obvious.

Proposition 6.1.1 *If $(\bar{x}, \bar{f})$ is a solution to the equality constraint, so is the $(\hat{x}, \hat{f})$, defined by (6.5), for any real number α.*

To fulfill the nonnegativity of the new solution, ensuring $\hat{x}_B \geq 0$, the stepsize α has to be limited, in general. To see this, consider the following row index set:

$$R = \{i \mid \bar{d}_i > 0, i = 1, \ldots, m+1\}, \tag{6.6}$$

which includes indices blocking the α from infinitely increasing.

If R is nonempty, the following rule determines the maximum possible stepsize.

Rule 6.1.1 (Row Rule) Determine the row index p and stepsize α by

$$\alpha = \bar{x}_{j_p}/\bar{d}_p = \min_{i \in R} \bar{x}_{j_i}/\bar{d}_i, \tag{6.7}$$

In the case of a tire, select one corresponding to the largest $\bar{d}_p$.

The preceding rule determines α and p and the leaving index j_p.

We can make the following claim.

Theorem 6.1.1 *Let the face solution be $(\bar{x}, \bar{f})$. If R is nonempty, the new solution $(\hat{x}, \hat{f})$ given by (6.5) with (6.7) is a face solution.*

Proof The non-emptiness of R ensures that the formula (6.7) is well-defined. Note that $\bar{x}_B \geq 0$ and $\alpha \geq 0$.

According to Proposition 6.1.1, $(\hat{x}, \hat{f})$ satisfies the equality constraint. So, it only needs to show $\hat{x} \geq 0$.

It holds that

$$\hat{x}_{j_i} = \bar{x}_{j_i} - \alpha \bar{d}_i \geq 0. \qquad \forall\, i \in R. \tag{6.8}$$

On the other hand, by (6.7), it is known that

$$\alpha \leq \bar{x}_{j_i}/\bar{d}_i, \qquad \forall\, i \in R,$$

from which it follows that

$$\bar{x}_{j_i} - \alpha \bar{d}_i \geq 0, \qquad \forall\, i \in R.$$

The preceding rule together with (6.8) leads to

$$\hat{x}_B \geq 0,$$

which with $\hat{x}_N = \bar{x}_N$ together ensues $\hat{x} \geq 0$. Therefore, the new solution is a face solution. □

Notably, the stepsize α determined by (6.7) equals the decrement in objective value. Note that if $\bar{x}_{j_p}$ is zero, α vanishes, and hence the so-called new solution is not really "new" but the same as the old.

A face solution is said to be *degenerate* if $\bar{x}_{j_i}$ equals 0 for some $i \in R$. Zero stepsize occurs if and only if the face solution is degenerate.

On the other hand, Rule 6.1.1 is not well-defined if the row index set R is empty. In this case, we are still done, as the following says.

Lemma 6.1.2 *Assume that $(\bar{x}, \bar{f})$ is a face solution. If set R is empty, the problem is lower unbounded.*

Proof The assumption implies $\bar{d} \leq 0$ and $\bar{x} \geq 0$. Note that $\hat{x}_N = \bar{x}_N \geq 0$.

Then, from the update formula (6.5, it follows that

$$\hat{x}_B = \bar{x}_B - \alpha \bar{d} \geq 0, \qquad \forall\, \alpha \geq 0.$$

Considering Proposition 6.1.1, we assert that the new solution $(\hat{x}, \hat{f})$ is feasible for any $\alpha \geq 0$. Accordingly, we have

$$\hat{f} = \bar{f} - \alpha \to -\infty \quad \text{as} \quad \alpha \to \infty,$$

which reveals the lower unboundedness of the problem. □

6.2 Optimality Condition

Now, assume that p and α were determined, and the new solution given by the update formula (6.5) is a face solution. Then, it is time to check further to see if the optimality is achieved.

To do so, extract the pth row from the canonical form, i.e.,

$$v_N^T \triangleq e_p^T \bar{N}, \tag{6.9}$$

Moreover, form the index set below:

$$R' = \{j \in N \mid v_j < 0\} \cup \{j \in N \mid v_j > 0,\ \bar{x}_j > 0\}. \tag{6.10}$$

The set B' can serve as a criterion for optimality, as the following says.

Theorem 6.2.1 *Under the same assumption of Theorem 6.1.1, if the set B' defined by (6.10) is empty, $(\hat{x}, \hat{f})$, given by (6.5) with (6.7), is an optimal face solution.*

Proof According to Theorem 6.1.1, $\hat{x}$ is a face solution to the original standard LP problem. We show that it fulfills the optimal condition.

The pth row of the canonical form (6.2) represents the following equation:

$$x_{j_p}/\bar{d}_p + \sum_{j\in B, j\neq j_p} (v_j/\bar{d}_p)x_j + \sum_{j\in N}(v_j/\bar{d}_p)x_j = f + \bar{b}_p/\bar{d}_p, \tag{6.11}$$

where

$$v_j = 0, \qquad j \in B, j \neq j_p. \tag{6.12}$$

Since (6.11) exhibits the objective function, the coefficients of its left-hand side give components of the dual solution. Note that the dual component corresponding to j_p is $1//\bar{d}_p$. We show that it matches the primal solution $\hat{x}$ determined by the first two expressions of (6.5) with (6.7).

According to Theorem 6.1.1, the new $\hat{x}$ is a face solution. It is obvious that

$$v_j/\bar{d}_p = 0, \qquad \hat{x}_j \geq 0, \qquad \forall\, j \in B, j \neq j_p. \tag{6.13}$$

In addition, it holds by (6.7) that

$$\alpha = \bar{x}_{j_p}/\bar{d}_p, \qquad \bar{d}_p > 0.$$

from which it follows that

$$\hat{x}_{j_p} = \bar{x}_{j_p} - \alpha\bar{d}_p = 0, \tag{6.14}$$

Therefore, $\hat{x}_{j_p}$ and the dual component $1/\bar{d}_p$ exhibit complementary. In addition, emptiness of B' and feasibility of $\hat{x}$ guarantee that the following two expressions hold:

$$v_j/\bar{d}_p \geq 0, \qquad \hat{x}_j \geq 0, \forall\, j \in N. \tag{6.15}$$

and

$$v_j/\bar{d}_p = 0 \qquad \text{or} \qquad \hat{x}_j = 0.\ j \in N \tag{6.16}$$

Therefore, $\hat{x}$ is an optimal solution to the problem. □

Corollary 6.2.2 *Under the same assumption of Theorem 6.1.1, $(\hat{x}, \hat{f})$, given by (6.5) with (6.7), is an optimal face solution if the following condition is satisfied:*

$$v_j \geq 0 \qquad \text{and} \qquad v_j > 0, \;\; \bar{x}_j = 0, \;\; \forall\, j \in N,$$

where $v_N^T = e_p^T \bar{N}$.

In the other case, if B' is nonempty, select the column index using the following rule.

Rule 6.2.1 (Column Rule) Select the column index q such that

$$q = \arg\max_{j \in B'} |v_j| \tag{6.17}$$

In the case of a tire, select the index corresponding to the largest $\bar{x}_q$.

The preceding rule would be far from being the best, merely for convenience of illustrations. From a practical point of view, the following column rule should be recommended.

Taking into account the size of $\bar{x}_j$, we set an upper bound on it by a sufficiently large number, like

$$u_j = 10^{20},$$

and introduce the following so-called *duality gaps*:

$$\delta_j = \begin{cases} (u_j - \bar{x}_j)v_j, & \text{if } v_j < 0, \\ \bar{x}_j v_j, & \text{if } v_j > 0, \end{cases} \qquad j \in B', \tag{6.18}$$

which reflect the potential improvements of objective value with respect to indices in B'. Therefore, it is logical to favor the following rule.

Rule 6.2.2 (Column Rule) Select a column index q to enter B such that

$$q \in \arg\max_{j \in B'} \delta_j, \tag{6.19}$$

where δ_j is defined by (6.18).

Executing the preceding rule will remove from B' index q, corresponding to the maximum dual gap δ_p.

Then, the basis change is performed with the determined (p, q), and the next iteration is carried out.

6.3 Reduced Face Algorithm: Tableau Form

The previous discussions show that the canonical system (6.2) includes all items required in an iteration. If the optimal condition is not satisfied, the solution process has to transfer to a new canonical system by basis change.

It can be done using the following tableau, contracted from the canonical system (6.2):

$$\begin{array}{cc|c} \hline x_B & x_N & f \\ \hline I & \bar{N} & \bar{d} \\ \hline \bar{x}_B^T & \bar{x}_N^T & \bar{f} \\ \hline \end{array} \tag{6.20}$$

which is termed *reduced face tableau* with the rightmost *objective column*, associated with objective variable f. The bottom line is added for exhibiting the current face solution (with the $\bar{b}$ column omitted) for reference.

Notably, the reduced face tableau includes the objective column instead of the objective row.

As discussed previously, in each iteration, the row index p and stepsize α are determined by updating the face solution. If B' is empty, optimality is achieved; otherwise, a column index q is selected. Then, according to elementary transformations, they are performed to update the tableau.

The overall steps are summarized using the following algorithm.

Algorithm 6.3.1 (Reduced Face Algorithm: Tableau Form)

Initial: reduced tableau in the form (6.20); face solution $(\bar{x}, \bar{f})$. This algorithm solves the standard problem.

1. *Stop if row index set R defined by (6.6) is empty.*
2. *Determine stepsize α and row index p as well as j_p by Rule 6.1.1.*
3. *If $\alpha \neq 0$, update $\bar{f} = \bar{f} - \alpha$ and $\bar{x}_B = \bar{x}_B - \alpha\bar{d}$.*
4. *Stop if the index set B', defined by (6.10), is empty.*
5. *Select column index q by Rule 6.2.1.*
6. *Adjust $B = B\backslash j_p \cup q$, $N = N\backslash q \cup j_p$.*
7. *Convert $\bar{a}_{pq}$ to 1, and eliminate the other nonzeros in the pivot column by elementary transformations.*
8. *Go to Step 1.*

Theorem 6.3.1 *Under the assumption of nondegeneracy, Algorithm 6.5.1 terminates either at*

(i) Step 1, detecting lower unboundedness of the problem or at
(ii) Step 4, producing an optimal face solution $(\bar{x}, \bar{f})$.

Proof According to Theorem 6.1.1, each iteration yields a face solution.

Under nondegeneracy, stepsize α is positive in each iteration. Therefore, the objective value decreases strictly monotonically. Since there are only finitely many faces, the corresponding faces never appear repeatedly, and the algorithm must terminate.

The meanings of the exits are obvious by Lemma 6.1.2 and Theorem 6.2.1. □

6.4 Double-Phase Reduced Face Algorithm

The reduced face algorithm presented in the previous section needs a face solution to get itself started. This section presents a phase I procedure using an auxiliary reduced problem to gain a double-phase algorithm for general-purpose use.

Let $\bar{x}$ be any n-dimensional nonnegative vector. To construct the auxiliary problem, introduce the following vector:

$$h = A\bar{x} - b. \tag{6.21}$$

If h equals 0, then arx is already a face solution. Let us assume that this is not the case and proceed accordingly.

Using the normalized vector $d = h/\|h\|$, we construct the auxiliary reduced problem below:

$$\begin{array}{ll} \min & f' \\ \text{s.t.} & Ax = f'd + b, \qquad x \leq 0, \;\; f' \geq 0, \end{array} \tag{6.22}$$

It is evident that the preceding problem has the following face solution:

$$(\bar{x}, \bar{f}' = \|h\|).$$

Let $A = (B, N)$ be a partition with the basis B. Assume that Gauss-Jordan elimination transforms the equality constraint of the auxiliary problem to the canonical form below:

$$x_B + \bar{N}x_N = f'\bar{d} + \bar{b},$$

which can be put into the following reduced face tableau:

x_B	x_N	f'
I	$\bar{N}$	$\bar{d}$
$\bar{x}_B$	$\bar{x}_N$	$\bar{f}'$

(6.23)

where $(\bar{x}, \bar{f}' = \|h\|)$ is inserted as the bottom line for reference (with the $\bar{b}$ column omitted).

From the preceding tableau, Algorithm 6.3.1 can get itself started. Once the objective value vanishes, a reduced tableau to the original problem is obtained to carry over to phase II. If it finishes with the end $\bar{f}'$ value greater than zero, it can be asserted that there is no feasible solution to the original problem.

We suggest the following slight variant of Algorithm 6.3.1 as the phase I solver.

Subalgorithm 6.4.1 (Reduced Face Phase I: Tableau Form)

Initial: tableau in the form (6.23); solution $(\bar{x},\ \bar{f}')$. This algorithm finds a face solution:

1. *Determine stepsize α, row index p with j_p by Rule 6.1.1.*
2. *If $\bar{f}' - \alpha \leq 0$, reset $\alpha = \bar{f}'$.*
3. *If $\alpha \neq 0$, update $\bar{f} = \bar{f} - \alpha$ and $\bar{x}_B = \bar{x}_B - \alpha \bar{d}$.*
4. *Return if $\bar{f}' = 0$. (feasibility achieved)*
5. *Return if index set B', defined by (6.10), is empty. (there is no feasible solution).*
6. *Select column index q by Rule 6.2.1.*
7. *Adjust $B = B \backslash j_p \cup q,\ N = N \backslash q \cup j_p$.*
8. *Convert $\bar{a}_{pq}$ to 1, and eliminate the other nonzeros in the pivot column by elementary transformations.*
9. *Go to Step 1.*

Theorem 6.4.1 *Assuming nondegeneracy, Subalgorithm 6.4.1 terminates either at*

(i) Step 4, achieving feasibility, or at
(ii) Step 5, detecting no feasible solution to the problem.

Proof The validity is from the proof of Theorem 6.3.1. □

Once Subalgorithm 6.4.1 terminates at Step 4 with the auxiliary function value $\bar{f}'$ vanishing, the following steps apply.

Tactic 6.4.1 (Transition from Phase I to Phase II)

(i) Insert the costs of the original problem as the $(m+1)$th row.
(ii) Use the f column e_{m+1} to replace the f' column.
(iii) Eliminate basic costs (if any) by elementary transformations.
(iv) Take $p = m+1$ as the pivot row index and determine the pivot column index q by

$$q \in \arg\max_{j \in B'} |v_j|,$$

where B' is defined by (6.10) and $v_{B'}^T$ is from the pth row.

(v) Make the according basis change.

Starting from the resulting reduced face tableau and the face solution $(\bar{x}, \bar{f} = c^T \bar{x})$, Algorithm 6.3.1 is called to solve the original problem. A combination of Algorithm 6.3.1 and Subalgorithm 6.4.1 forms the double reduced face algorithm.

Example 6.4.1 Solve the following standard problem by the double-phase reduced face algorithm :

$$A = \begin{bmatrix} -4 & -1 & 5 & -2 & 1 & & & \\ 3 & 1 & 3 & 5 & & 1 & & \\ -6 & -7 & -1 & 4 & & & 1 & \\ 1 & -3 & -2 & -1 & & & & 1 \end{bmatrix} \qquad b = \begin{bmatrix} -3 \\ 12 \\ -9 \\ -7 \end{bmatrix}$$

$c = [6, 4, 0, 0, -8, 0, -6, 0]^T$

Although it seems favorable to set components of initial $\bar{x}$ differing from each other, we take the following for simplicity:

$\bar{x} = [1, 1, 1, 1, 1, 1, 1, 1]^T$.

Set the objective value and associated column as follows:

$\bar{f}' = \|A\bar{x} - b\| = 3476/929$.

$\bar{d} = (A\bar{x} - b)/\bar{f} = [929/1738, 929/3476, 0, 809/1009]^T$,

Put the auxiliary problem into the following tableau:

x_1	x_2	x_3	x_4	x_5	x_6	x_7	x_8	f'
−4	−1	5	−2	1				929/1738
3	1	3	5		1			929/3476
−6	−7	−1	4			1		
1	−3	−2	−1				1	809/1009
1	1	1	1	1	1	1	1	3476/929

Note that the end column exhibits $\bar{d}$.

Phase I: Call Subalgorithm 6.4.1.

Iteration 1:

1. $\alpha = \min[\bar{x}(5)/\bar{d}(1), \bar{x}(6)/\bar{d}(2), \bar{x}(8)/\bar{d}(4)]$
$= \min[1738/929, 3476/929, 1009/809] = 1009/809, \ p = 4, \ jp = 8.$
2. $\bar{f}' - \alpha = 2018/809 > 0$.
3. $\bar{x}_B = \bar{x}_B - \alpha\bar{d} = [1/3, 2/3, 1, 0]^T, \ \bar{f}' = 2018/809$.
4. $\bar{f}' > 0$.
5. $B' = \{1, 2, 3, 4\}$.
6. $q = \arg\max[|1|, |-3|, |-2|, |-1|] = 2$.
7. $B = \{5, 6, 7, 2\}, \ N = \{1, 3, 4, 8\}$.
8. Multiply row 4 by $1/(-3)$ and add $1, -1, 7$ times of row 4 to rows 1, 2, and 3:

x_1	x_2	x_3	x_4	x_5	x_6	x_7	x_8	f'
−13/3		17/3	−5/3	1			−1/3	929/3476
10/3		7/3	14/3		1		1/3	929/1738
−25/3		11/3	19/3			1	−7/3	−1738/929
−1/3	1	2/3	1/3				−1/3	−929/3476
1	1	1	1	1/3	2/3	1	0	2018/809

Iteration 2:

1. $\alpha = \min[\bar{x}(5)/\bar{d}(1), \bar{x}(6)/\bar{d}(2)]$
   $= [1009/809, 1009/809] = 1009/809, \ p = 1, jp = 5.$
2. $\bar{f}' - \alpha = 1009/809 > 0.$
3. $\bar{x}_B = \bar{x}_B - \alpha\bar{d} = [0, 0, 10/3, 4/3]^T, \ \bar{f}' = 1009/809.$
4. $\bar{f}' > 0.$
5. $B' = \{1, 3, 4, 8\}.$
6. $q = \arg\min[| - 13/3|, |17/3|, | - 5/3|, |1|, | - 1/3|] = 3.$
7. $B = \{3, 6, 7, 2\}, \ N = \{1, 4, 5, 8\}.$
8. Multiply row 1 by 17/3, and add $-7/3, -11/3, -2/3$ times of row 1 to rows 2, 3, and 4:

x_1	x_2	x_3	x_4	x_5	x_6	x_7	x_8	f'
−13/17		1	−5/17	3/17			−1/17	587/12,446
87/17			91/17	−7/17	1		8/17	1835/4323
−94/17			126/17	−11/17		1	−36/17	−934/457
3/17	1		9/17	−2/17			−5/17	−1590/5323
1	4/3	1	1	0	0	10/3	0	1009/809

Iteration 3:

1. $\alpha = \min[\bar{x}(3)/\bar{d}(1), \bar{x}(6)/\bar{d}(2)]$
   $= \min[12{,}446/587, 0] = 0, \ p = 2, jp = 6.$
5. $B' = \{1, 4, 5\}.$
6. $q = \arg\max[|87/17, |91/17|, | - 7/17|] = 4.$
7. $B = \{3, 4, 7, 2\}, \ N = \{1, 5, 6, 8\}.$
8. Multiply row 2 by $1/(91/17)$, and add $5/17, -126/17, -9/17$ times of row 2 to rows 1, 3, and 4:

x_1	x_2	x_3	x_4	x_5	x_6	x_7	x_8	f'
−44/91		1		2/13	5/91		−3/91	326/4625
87/91			1	−1/13	17/91		8/91	167/2106
−164/13				−1/13	−18/13	1	−36/13	−1671/635
−30/91	1			−1/13	−9/91		−31/91	−448/1315
1	160/91	83/91	82/91	0	0	86/13	0	0

Iteration 4:

1. $\alpha = \min[\bar{x}(3)/\bar{d}(1), \bar{x}(4)/\bar{d}(2)]$
   $= \min[4625/326, 2106/167] = 2106/167, \ p = 2, jp = 4.$
2. $\bar{f}' - \alpha = -12{,}284/1081 < 0, \alpha = \bar{f}'.$

3. $\bar{x}_B = \bar{x}_B - \alpha \bar{d} = [83/91, 82/91, 86/13, 160/91]^T$,
   $\bar{f}' = 0$.
4. Feasibility achieved. The face solution:

$$\bar{x} = [1, 160/91, 83/91, 82/91, 0, 0, 86/13, 0]^T$$

The original costs and corresponding objective value are

$$c^T = [6, 4, 0, 0, -8, 0, -6, 0], \qquad \bar{f} = c^T \bar{x} = -2426/91.$$

We execute Tactic 6.4.1 as follows:

(i) Insert the costs as the fifth row.
(ii) Use e_5 to replace the end column.
(iii) Eliminate the basic costs by adding -4 times of the fourth row to the fifth row, and six times of the third row to the fifth row. The tableau becomes

x_1	x_2	x_3	x_4	x_5	x_6	x_7	x_8	f
$-44/91$		1		$2/13$	$5/91$		$-3/91$	
$87/91$			1	$-1/13$	$17/91$		$8/91$	
$-164/13$				$-1/13$	$-18/13$	1	$-36/13$	
$-30/91$	1			$-1/13$	$-9/91$		$-31/91$	
$-6222/91$				$-106/13$	$-720/91$		$1388/91$	1
1	$160/91$	$83/91$	$82/91$	0	0	$86/13$	0	$-2426/91$

(iv) Set row index $p = 5$; the formula (6.10) gives $B' = \{1, 5, 6\}$. Thus, we have $q = \arg\max[|-6222/91|, |-106/13|, |-720/91|] = 5$.
(v) The according basis change leads to the following reduced face tableau:

x_1	x_2	x_3	x_4	x_5	x_6	x_7	x_8	f
$-94/53$		1			$-5/53$		$-17/53$	$1/53$
$594/371$			1		$97/371$		$86/371$	$-1/106$
$-4441/371$					$-486/371$	1	$-974/371$	$-1/106$
$117/371$	1				$-9/371$		$-73/371$	$-1/106$
$3111/371$				1	$360/371$		$694/371$	$-13/106$
1	$160/91$	$83/91$	$82/91$	0	0	$86/13$	0	$-2426/91$

Note that the end column exhibits the objective column $\bar{d}$.

Phase II: Call Algorithm 6.3.1

Iteration 5: $B = \{3, 4, 7, 2, 5\}, \ N = \{1, 6, 8\}$

1. $I = \{1\}$.
2. $\alpha = \min[\bar{x}(3)/\bar{d}(1)] = 2106/167, \ p = 1, \ jp = 3$.
3. $\bar{x}_B = \bar{x}_B - \alpha \bar{d} = [0, 19/14, 99/14, 31/14, 83/14]^T$,
   $\bar{f} = \bar{f} - \alpha = -75$.
4. $B' = \{1, 6, 8\}$.

5. $q = \arg\max[| -94/53|, | -5/53|, | -17/53|] = 1$.
6. $B = \{1, 4, 7, 2, 5\}$, $N = \{3, 6, 8\}$.
7. Multiply row 1 by $1/(-94/53)$, and add $-594/371, 4441/371, -117/371, -3111/371$ times of row 1 to rows 2, 3, 4, and 5:

x_1	x_2	x_3	x_4	x_5	x_6	x_7	x_8	f
1		−53/94			5/94		17/94	−1/94
		297/329	1		58/329		−19/329	5/658
		−2207/327			−443/658	1	−303/658	−45/329
	1	117/658			−27/658		−167/658	−2/329
		1877/397		1	345/658		233/658	−11/329
1	31/14	0	19/14	83/14	0	99/14	0	−75

Iteration 6:

1. $I = \{2\}$.
2. $\alpha = \min[\bar{x}(4)/\bar{d}(2)] = 893/5$, $p = 2$, $jp = 4$.
3. $\bar{x}_B = \bar{x}_B - \alpha\bar{d} = [29/10, 0, 63/2, 33/10, 119/10]^T$, $\bar{f} = \bar{f} - \alpha = -1268/5$.
4. $B' = \{8\}$.
5. $q = 8$.
6. $B = \{1, 8, 7, 2, 5\}$, $N = \{3, 4, 6\}$.
7. Multiply row 2 by $1/(-19/329)$, and add $-17/94, 303/658, 167/658, -233/658$ times of row 2 to rows 1, 3, 4, and 5:

x_1	x_2	x_3	x_4	x_5	x_6	x_7	x_8	f
1		43/19	119/38		23/38			1/76
		−297/19	−329/19		−58/19		1	−5/38
		−265/19	−303/38		−79/38	1		−15/76
	1	−72/19	−167/38		−31/38			−3/76
		195/19	233/38	1	61/38			1/76
29/10	33/10	0	0	119/10	0	63/2	0	−1268/5

Iteration 7:

1. $I = \{1, 5\}$.
2. $\alpha = \min[\bar{x}(1)/\bar{d}(1), \bar{x}(5)/\bar{d}(5)]$
   $= \min[1102/5, 4522/5] = 1102/5$, $p = 1$, $jp = 1$.
3. $\bar{x}_B = \bar{x}_B - \alpha\bar{d} = [0, 29, 75, 12, 9]^T$, $f = f - \alpha = -474$.
4. $B' = \emptyset$.
   Optimality is achieved. Optimal solution:

$\bar{x} = [0, 12, 0, 0, 9, 0, 75, 29]^T, \qquad \bar{f} = -474.$

All the nonbasic coefficients of the fifth row are positive, so there exists a unique optimal vertex solution to the problem.

6.5 Reduced Face Algorithm

It is possible to compute necessary quantities from the original data. It concerns the solution of linear systems.

Let (B, N) be the current partition. The reduced face tableau (6.20) is theoretically

equivalent to the following revised tableau:

$$\begin{array}{cc|c} x_B & x_N & f \\ \hline I & B^{-1}N & B^{-1}e_{m+1} \\ \hline \bar{x}_B & \bar{x}_N & \bar{f} \end{array} \tag{6.24}$$

It is favorable to use LU factorization to handle linear systems. To do so, compute the initial LU factorization of B, i.e.,

$$B = LU.$$

Subsequently, the LU factors are updated iteration by iteration.

Thereby, the objective column $\bar{d} = B^{-1}e_{m+1}$ can be obtained by solving the two triangular systems below:

$$Lw = e_{m+1}, \qquad U\bar{d} = w.$$

The solution to the first system is readily available, and hence, only the following system needs to be solved:

$$U\bar{d} = (1/\mu)e_{m+1}, \tag{6.25}$$

where $\mu = 1$ is the last diagonal entry of L.

To obtain the nonbasic entries of the pth row v_N^T, defined by (6.9), solve the following two triangular systems:

$$U^T w = e_p, \qquad L^T u = w. \tag{6.26}$$

and then compute

$$v_N^T = u^T N. \tag{6.27}$$

Based on equivalence between tableaus (6.20) and (6.24), it is straightforward to convert Algorithm 6.3.1 to the following revised version.

Algorithm 6.5.1 (Reduced Face Algorithm)

Initial: (B, N), face solution $(\bar{x}_B, \bar{f})$, factorization $B = LU$. This algorithm solves the standard problem.

1. *Solve the triangular system (6.25) for $\bar{d}$.*
2. *Stop if $R = \{i \mid \bar{d}_i > 0, i = 1, \ldots, m+1\}$ is empty (unbounded problem).*
3. *Determine stepsize α and row index p and column index j_p by (6.7).*
4. *If $\alpha > 0$, update $\bar{x}_B = \bar{x}_B - \alpha\bar{d}$ and $\bar{f} = \bar{f} - \alpha$.*
5. *Solve two triangular systems (6.26), and compute $v_N^T = u^T N$.*
6. *Stop if the index set B', defined by (6.10), is empty (optimality achieved).*
7. *Determine column index $q = \arg\max_{j \in B'} |v_j|$.*
8. *Update (B, N) by exchanging j_p and q.*
9. *Update LU factors.*
10. *Go to Step 1.*

Note: Practically, it is recommended to use Rule 6.2.2 in Step 7 instead.
This algorithm solves three triangular systems per iteration (Steps 1 and 5).

Example 6.5.1 Solve the following standard problem by Algorithm 6.5.1 :

$$A = \begin{bmatrix} -5 & 1 & -2 & 4 & 2 & -11 & 2 \\ 3 & -4 & -3 & 0 & -6 & 8 & -6 \\ -1 & -8 & 4 & -6 & 0 & 1 & 0 \end{bmatrix} \qquad b = \begin{bmatrix} -4 \\ -24 \\ -18 \end{bmatrix}$$

$c = [-3, -1, 0, 2, 0, 5, -4]^T$

Initial face solution: $\bar{x} = [1, 2, 1, 1, 2, 1, 2]^T$, $\bar{f} = -6$

Still use A to denote the coefficient matrix of the corresponding reduce problem, i.e.,

$$A = \begin{bmatrix} -5 & 1 & -2 & 4 & 2 & -11 & 2 \\ 3 & -4 & -3 & 0 & -6 & 8 & -6 \\ -1 & -8 & 4 & -6 & 0 & 1 & 0 \\ -3 & -1 & 0 & 2 & 0 & 5 & -4 \end{bmatrix}$$

Initially, it might as well take

$$B = \{1, 2, 3, 4\}, \qquad N = \{5, 6, 7\}.$$

permutated LU factors:

$$L = \begin{bmatrix} 1 & 0 & 0 & 0 \\ -3/5 & 17/41 & 1 & 0 \\ 1/5 & 1 & 0 & 0 \\ 3/5 & 8/41 & -14/247 & 1 \end{bmatrix} \qquad U = \begin{bmatrix} -5 & 1 & -2 & 4 \\ 0 & -41/5 & 22/5 & -34/5 \\ 0 & 0 & -247/41 & 214/41 \\ 0 & 0 & 0 & 302/247 \end{bmatrix}.$$

Iteration 1:

1. $w = [0, 0, 0, 1]^T$, $\bar{d} = [47/151, -45/151, 107/151, 247/302]^T$.
2. $R = \{1, 3, 4\}$.
3. $\alpha = \min[\bar{x}(1)/\bar{d}(1), \bar{x}(3)/\bar{d}(3), \bar{x}(4)/\bar{d}(4)]$
   $= \min[151/47, 151/107, 302/247] = 302/247,\ p = 4,\ jp = 4$.
4. $\bar{x}_B = [1, 2, 1, 1]^T - \alpha[47/151, -45/151, 107/151, 247/302]^T$
   $= [153/247, 584/247, 33/247, 0]^T,\ \bar{f} = \bar{f} - \alpha = -1784/247$.
   $\bar{x} = [153/247, 584/247, 33/247, 0, 2, 1, 2]^T$.
5. $w = [0, 0, 0, 494/151]^T$, $u = [-258/151, 28/151, -108/151, 494/151]^T$,
   $v_N^T = [-684/151, 5424/151, -2660/151]$.
6. $B' = \{5, 6, 7\}$.
7. $q = \arg\max[|-684/151|, |5424/151|, |-2660/151|] = 6$.
8. $B = \{1, 2, 3, 6\}$, $N = \{4, 5, 7\}$.
9. Permuted LU factors:

$$L = \begin{bmatrix} 1 & 0 & 0 & 0 \\ -3/5 & 17/41 & 1 & 0 \\ 1/5 & 1 & 0 & 0 \\ 3/5 & 8/41 & -14/247 & 1 \end{bmatrix} \quad U = \begin{bmatrix} -5 & 1 & -2 & 4 \\ 0 & -41/5 & 22/5 & -34/5 \\ 0 & 0 & -247/41 & 214/41 \\ 0 & 0 & 0 & 302/247 \end{bmatrix}.$$

Iteration 2:

1. $w = [0, 0, 0, 1]^T$, $\bar{d} = [-175/904, 49/1356, 1/904, 247/2712]^T$.
2. $R = \{2, 3, 4\}$.
3. $\alpha = \min[\bar{x}(2)/\bar{d}(2), \bar{x}(3)/\bar{d}(3), \bar{x}(4)/\bar{d}(4)]$
   $= \min[5169/79, 29{,}832/247, 2712/247] = 2712/247, p = 4, jp = 6$.
4. $\bar{x}_B = [153/247, 584/247, 33/247, 1]^T - \alpha[-175/904, 49/1356, 1/904, 247/2712]^T = [678/247, 486/247, 30/247, 0]^T,\ \bar{f} = \bar{f} - \alpha = -4496/247$.
   $\bar{x} = [678/247, 486/247, 30/247, 0, 2, 0, 2]^T$.
5. $w = [0, 0, 0, 247/2712]^T$, $u = [-43/904, 7/1356, -9/452, 247/2712]^T$,
   $v_N^T = [151/1356, -57/452, -665/1356]$.
6. $B' = \{5, 7\}$.
7. $q = \arg\max[|-57/452|, |-665/1356|] = 7$.
8. $B = \{1, 2, 3, 7\}$, $N = \{4, 5, 6\}$.
9. Permuted LU factors:

$$L = \begin{bmatrix} 1 & 0 & 0 & 0 \\ -3/5 & 17/41 & 1 & 0 \\ 1/5 & 1 & 0 & 0 \\ 3/5 & 8/41 & -14/247 & 1 \end{bmatrix} \quad U = \begin{bmatrix} -5 & 1 & -2 & 2 \\ 0 & -41/5 & 22/5 & -2/5 \\ 0 & 0 & -247/41 & -190/41 \\ 0 & 0 & 0 & -70/13 \end{bmatrix}.$$

Iteration 3:

1. $w = [0, 0, 0, 1]^T$, $\bar{d} = [-4/35, 3/35, 1/7, -13/70]^T$.
2. $R = \{2, 3\}$.
3. $\alpha = \min[\bar{x}(2)/\bar{d}(2), \bar{x}(3)/\bar{d}(3)]$
   $= \min[5670/247, 210/247] = 210/247, p = 3, jp = 4$.
4. $\bar{x}_B = [678/247, 486/247, 30/247, 2]^T - \alpha[-4/35, 3/35, 1/7, -13/70]^T$
   $= [54/19, 36/19, 0, 41/19]^T,\ \bar{f} = \bar{f} - \alpha = -362/19$.

$\bar{x} = [54/19, 36/19, 0, 0, 2, 0, 41/19]^T$.

5. $w = [0, 0, -41/247, 1/7]^T$, $u = [-25/133, -3/19, 5/133, 1/7]^T$, $v_N^T = [-92/133, 4/7, 207/133]$.
6. $B' = \{4, 5\}$.
7. $q = \arg\max[|-92/133|, |4/7|] = 4$.
8. $B = \{1, 2, 4, 7\}$, $N = \{3, 5, 6\}$.
9. Permuted LU factors:

$$L = \begin{bmatrix} 1 & 0 & 0 & 0 \\ -3/5 & 17/41 & 1 & 0 \\ 1/5 & 1 & 0 & 0 \\ 3/5 & 8/41 & 19/107 & 1 \end{bmatrix} \quad U = \begin{bmatrix} -5 & 1 & 4 & 2 \\ 0 & -41/5 & -34/5 & -2/5 \\ 0 & 0 & 214/41 & -190/41 \\ 0 & 0 & 0 & -460/107 \end{bmatrix}.$$

Iteration 4:

1. $w = [0, 0, 0, 1]^T$, $\bar{d} = [-51/230, 21/115, -19/92, -107/460]^T$.
2. $R = \{2\}$.
3. $\alpha = \min[\bar{x}(2)/\bar{d}(2)] = 1380/133$, $p = 2$, $jp = 2$.
4. $\bar{x}_B = [54/19, 36/19, 0, 41/19]^T - \alpha[-51/230, 21/115, -19/92, -107/460]^T$ $= [36/7, 0, 15/7, 32/7]^T$, $\bar{f} = \bar{f} - \alpha = -206/7$.
   $\bar{x} = [36/7, 0, 0, 15/7, 2, 0, 32/7]^T$.
5. $w = [0, -5/41, -17/107, 21/115]^T$, $u = [-24/115, -22/115, -9/115, 21, /115]^T$ $v_N^T = [78/115, 84/115, 8/5]$.
6. $B' = \{5\}$.
7. $q = 5$.
8. $B = \{1, 5, 4, 7\}$, $N = \{2, 3, 6\}$.
9. Permuted LU factors:

$$L = \begin{bmatrix} 1 & 0 & 0 & 0 \\ -3/5 & 1 & 0 & 0 \\ 1/5 & 1/12 & 1 & 0 \\ 3/5 & 1/4 & 1/7 & 1 \end{bmatrix} \quad U = \begin{bmatrix} -5 & 2 & 4 & 2 \\ 0 & -24/5 & 12/5 & -24/5 \\ 0 & 0 & -7 & 0 \\ 0 & 0 & 0 & -4 \end{bmatrix}.$$

Iteration 5:

1. $w = [0, 0, 0, 1]^T$, $\bar{d} = [0, 1/4, 0, -1/4]^T$.
2. $R = \{2\}$.
3. $\alpha = \min[\bar{x}(5)/\bar{d}(2)] = 8$, $p = 2$, $jp = 5$.
4. $\bar{x}_B = [36/7, 2, 15/7, 32/7]^T - \alpha[0, 1/4, 0, -1/4]^T$ $= [36/7, 0, 15/7, 46/7]^T$, $\bar{f} = \bar{f} - \alpha = -262/7$.
   $\bar{x} = [36/7, 0, 0, 15/7, 0, 0, 46/7]^T$.
5. $w = [0, -5/24, -1/14, 1/4]^T$, $u = [-2/7, -11/42, -3/28, 1/4]^T$, $v_N^T = [115/84, 13/14, 46/2]$.
6. $B' = \emptyset$. Optimality is achieved.

Optimal face solution:
$\bar{x} = [36/7, 0, 0, 15/7, 0, 0, 46/7]^T$, $\qquad \bar{f} = -262/7$.

6.6 Generalized Reduced Face Algorithm

Although we always consider the standard LP problem, the LP problems from practice are diverse, which can be transformed into the so-called *variable-bounded* LP problem. In this section, we generalize the previous reduced algorithm to deal with such problems. At first, we formulate the tableau form of the algorithm. Then, we present a double-phase algorithm for general-purpose use. Finally, we formulate the reduced face algorithm in revised form.

Replacing the sign restriction with the general bound restriction, the reduced problem (6.1) can be converted to the *reduced variable-bounded problem* below:

$$\begin{array}{ll} \min & f \\ \text{s.t.} & Ax = e_{m+1}f + b, \qquad l \le x \le u, \end{array} \tag{6.28}$$

where $A \in \mathcal{R}^{(m+1)\times n}$, $b \in \mathcal{R}^{m+1}$, Rank $A = m+1$, $m < n$; l/u are the given lower/upper bounds. f is termed *objective variable.*

6.6.1 Generalized Reduced Face Algorithm: Tableau Form

Assume that $A = (B, N)$ is the current basis and nonbasis:

$$B = \{j_1, \ldots, j_m\}, \quad N = A\backslash B.$$

It is noticeable that the constraints of (6.28) are equivalent to the following canonical form:

$$l_B \le x_B = f\bar{d} + \bar{b} - \bar{N}x_N \le u_B, \quad l_N \le x_N \le u_N, \tag{6.29}$$

where

$$\bar{b} = B^{-1}b, \quad \bar{N} = B^{-1}N, \quad \bar{d} = B^{-1}e_{m+1}. \tag{6.30}$$

Assume that $(\bar{x}, \bar{f})$ fulfills the condition (6.29) and hence is a feasible solution.

If each component of $\bar{x}_N$ is on one of its bounds, it is a vertex (basic) solution. It is termed *face solution* if some component of $\bar{x}$ is on one of its bounds. It is an interior point if no $\bar{x}$'s component is on any bounds. The latter can be viewed as the feasible region—the maximum face. So, we will not distinguish between feasible and face solutions.

Consider the typical line search scheme below:

$$\hat{x}_B = \bar{x}_B - \alpha\bar{d}, \quad \hat{f} = \bar{f} - \alpha, \quad \hat{x}_N = \bar{x}_N, \tag{6.31}$$

where the following rule determines the stepsize α.

Using the following items:

$$\alpha_i = \begin{cases} (\bar{x}_{j_i} - u_{j_i})/\bar{d}_i, & \text{if } \bar{d}_i < 0, \\ (\bar{x}_{j_i} - l_{j_i})/\bar{d}_i, & \text{if } \bar{d}_i > 0, \end{cases} \quad i = 1, \ldots, m+1, \tag{6.32}$$

we determine α by the following rule.

Rule 6.6.1 (Row Rule) Determine the row index p and the stepsize α such that

$$\alpha = \alpha_p = \min\{\alpha_i \mid i = 1, \ldots, m+1\}, \tag{6.33}$$

Face solution is said to be *degenerate* if the according stepsize α vanishes.

One should consider the problem unbounded if the stepsize α becomes too large.

Lemma 6.6.1 *If $(\bar{x}, \bar{f})$ is a face solution, then the solution $(\hat{x}, \hat{f})$, defined by (6.31)–(6.32), is a face solution, associated with a non-increasing or strictly non-decreasing objective value if $(\bar{x}, \bar{f})$ is nondegenerate.*

Proof Note that $\hat{x}_N = \bar{x}_N$.

From (6.31), $\bar{d} = B^{-1}e_{m+1}$ and $(\bar{x}, \bar{f})$, satisfying the equality constraint, and it follows that

$$\begin{aligned} A\hat{x} &= B\hat{x}_B + N\hat{x}_N \\ &= B(\bar{x}_B - \alpha\bar{d}) + N\bar{x}_N \\ &= B\bar{x}_B + N\bar{x}_N - \alpha e_{m+1} \\ &= \bar{f}e_{m+1} + b - \alpha e_{m+1} \\ &= \hat{f}e_{m+1} + b. \end{aligned}$$

So, the new solution $(\hat{x}, \hat{f})$ satisfies the equality constraint.

In addition, by (6.33)–(6.32), it holds that

$$l \leq \hat{x} \leq u.$$

Therefore, $(\hat{x}, \hat{f})$ is a face solution. Obviously, it holds that $\alpha \geq 0$, and hence that

$$\hat{f} = \bar{f} - \alpha \leq \bar{f},$$

which means that the objective value decreases by α. If $(\bar{x}, \bar{f})$ is nondegenerate, then $\alpha > 0$, and the objective strictly decreases. □

α, determined by (6.32), is the largest possible stepsize to maintain the new solution within the feasible region.

It deserves attention that the decrement of the objective value is equal to the stepsize α.

Note that the j_p-indexed component of the new solution is equal to the associated upper/lower bound as follows:

$$\hat{x}_{j_p} = \begin{cases} u_{j_p}, & \text{if } \bar{d}_p < 0, \\ l_{j_p}, & \text{if } \bar{d}_p > 0. \end{cases} \tag{6.34}$$

Denote the nonbasic part of the pth row as follows:

$$v_N^T \triangleq e_p^T B^{-1} N. \tag{6.35}$$

Introduce the following sets:

$$N1 = \{j \in N \mid v_j \bar{d}_p > 0\}, \qquad N2 = \{j \in N \mid v_j \bar{d}_p < 0\} \tag{6.36}$$

Then, the following is valid for optimality condition.

Theorem 6.6.2 *The new solution $(\hat{x}, \hat{f})$, defined by (6.31), is an optimal face solution if the following set is empty:*

$$B' = \{\, j \in N1 \mid \bar{x}_j - l_j > 0\} \cup \{\, j \in N2 \mid u_j - \bar{x}_j > 0\}, \tag{6.37}$$

Proof Note that the emptiness of B' is equivalent to the following condition:

$$\bar{x}_{N1} = l_{N1}, \qquad \bar{x}_{N2} = u_{N2}, \tag{6.38}$$

By (6.33)–(6.32), it is known that $\bar{d}_p \neq 0$. Extracting the pth equation from the canonical system, we have

$$f = -\bar{b}_p/\bar{d}_p + x_{j_p}/\bar{d}_p + (1/d_p) v_N^{\mathrm{T}} x_N. \tag{6.39}$$

Substituting $(\hat{x}, \hat{f})$ into (6.39) gives

$$\hat{f} = -\bar{b}_p/\bar{d}_p + \hat{x}_{j_p}/\bar{d}_p + (1/\bar{d}_p) v_N^{\mathrm{T}} \bar{x}_N, \tag{6.40}$$

Assume that (x', f') is any face solution. It satisfies

$$l_j \le x'_j \le u_j, \qquad j = 1, \ldots, n. \tag{6.41}$$

and the pth equation, i.e.,

$$f' = -\bar{b}_p/\bar{d}_p + x'_{j_p}/\bar{d}_p + (1/\bar{d}_p) v_N^{\mathrm{T}} x'_N. \tag{6.42}$$

Only two cases arise:

(i) $\bar{d}_p > 0$.

In this case, by (6.36), it holds that

$$v_{N1}^T > 0, \qquad v_{N2}^T < 0\} \tag{6.43}$$

Therefore, from $x'_{N1} \geq l_{N1}$ and $x'_{N2} \leq u_{N2}$, it follows that

$$v_N^T x'_N = v_{N1}^T x'_{N1} + v_{N2}^T x'_{N2} \geq v_{N1}^T l_{N1} + v_{N2}^T u_{N2},$$

which (6.38) together give

$$v_N^T x'_N \geq v_{N1}^T \bar{x}_{N1} + v_{N2}^T \bar{x}_{N2} = v_N^T \bar{x}_N. \tag{6.44}$$

In addition, by (6.34), it holds that

$$\hat{x}_{j_p} = l_{j_p},$$

and hence

$$x'_{j_p}/\bar{d}_p \geq l_{j_p}/\bar{d}_p = \hat{x}_{j_p}/\bar{d}_p, \tag{6.45}$$

Consequently, (6.42), (6.45), (6.44), and (6.40) together lead to

$$f' \geq \hat{f}.$$

Therefore, $(\hat{x}, \hat{f})$ is an optimal face solution.
(ii) $\bar{d}_p < 0$.
The statement's validity can be shown analogously and hence is omitted. □

In the case of $B' \neq \emptyset$, the following rule is well-defined.

Rule 6.6.2 (Column Rule) Select the column index

$$q \in \arg\max_{j \in B'} |v_j|. \tag{6.46}$$

Considering the size of $\bar{x}_j$, we introduce the following so-called *duality gaps* in present context:

$$\delta_j = \begin{cases} (\bar{x}_j - l_j)|v_j|, & \text{if } j \in N1,, \\ (u_j - \bar{x}_j)|v_j|, & \text{if } j \in N2 \end{cases} \qquad j \in N, \tag{6.47}$$

which reflects the potential improvement of the objective value associated with index $j \in N$ (Theorem 6.6.2).

For practice, we recommend the following rule.

Rule 6.6.3 (Column Rule) Select a column index q to enter B such that

$$q \in \arg\max_{j \in B'} \delta_j, \tag{6.48}$$

where δ_j is defined by (6.47).

Favorably, the preceding rule removes from B' index q, associated with the maximum duality gap.

Then, conduct the according basis change so that j_p leaves and q enters the basis, and an iteration is complete. The new basis corresponds to the new solution $(\hat{x}, \hat{x})$, given by (6.31)–(6.32).

The canonical form of the system can be put into the following so-called *reduced face tableau*, where the right-hand side is omitted. The associated face solution $(\bar{x}, \bar{f})$ is inserted as the bottom line for convenience.

x_B	x_N	f
I	$\bar{N}$	$\bar{d}$
$\bar{x}_B$	$\bar{x}_N$	$\bar{f}$

(6.49)

The overall steps can be put into the following algorithm.

Algorithm 6.6.1 (Generalized Reduced Face Algorithm: Tableau Form) *Initial: reduced face tableau of form (6.49); face solution* $(\bar{x}, \bar{f})$. *This algorithm solves the reduced variable-bounded problem:*

1. *Determine row index* p *and stepsize* α *by (6.33)–(6.32).*
2. *If* $\alpha \neq 0$, *update* $\bar{x}_B = \bar{x}_B - \alpha\bar{d}$ *and* $\bar{f} = \bar{f} - \alpha$.
3. *Stop if index set* B', *defined by (6.37), is empty (optimality achieved).*
4. *Determine column index* $q \in \arg\max_{j \in B'} |v_j|$, *where* $v_{B'}^T$ *is from the pth row.*
5. *Adjust* (B, N) *by exchanging* j_p *and* q.
6. *Convert* v_q *to 1, and eliminate the other nonzeros in the column by elementary transformations.*
7. *Go to Step 1.*

Theorem 6.6.3 *Assuming nondegeneracy, the sequence of face solutions produced by Algorithm 6.6.2 terminates at Step 4, offering an optimal face solution.*

Proof The validity is from Lemma 6.6.1, Theorem 6.6.2, and the discussions preceding the algorithm. □

Example 6.6.1 Solve the following variable-bounded problem by Algorithm 6.6.1:

$$A = \begin{bmatrix} 5 & 4 & & 3 & 5 & 1 & & \\ 6 & 2 & 8 & -5 & 4 & & 1 & \\ -2 & -8 & -4 & 1 & -3 & & & 1 \end{bmatrix} \qquad b = \begin{bmatrix} 78 \\ 44 \\ -63 \end{bmatrix}$$

$$c = [3, -2, -8, 0, 5, 0, 0, 0]^T$$

$u = [5, 7, 2, 4, 10, 8, 3, 6]^T$
$l = [0, 0, 0, 0, 0, 0, 0, 0]^T$
Initial face solution: $\bar{x} = [2, 5, 13, 7, 4, 1, 3]^T$.

Put the problem into the following tableau with $\bar{x}$ inserted as the bottom line:

x_1	x_2	x_3	x_4	x_5	x_6	x_7	x_8	f
5	4	0	3	5	1	0	0	0
6	2	8	−5	4	0	1	0	0
−2	−8	−4	1	−3	0	0	1	0
3	−2	−8	0	5	0	0	0	1
2	5	1	3	7	4	1	3	23

Convert the preceding tableau to the reduced face tableau by making the basis change with $p = 4$, $q = \arg\max[|3|, |-2|, |-8|, |0|, |5|] = 3$:

x_1	x_2	x_3	x_4	x_5	x_6	x_7	x_8	f
5	4	0	3	5	1	0	0	0
9	0	0	−5	9	0	1	0	1
−7/2	−7	0	1	−11/2	0	0	1	−1/2
−3/8	1/4	1	0	−5/8	0	0	0	−1/8
2	5	1	3	7	4	1	3	23

Iteration 1: $u^T = [5, 7, 2, 4, 10, 8, 3, 6]$, $B = \{6, 7, 8, 3\}$, $\bar{x}_B = [4, 1, 3, 1]^T$

1. $\alpha = \min[(1-0)/1, (3-6)/(-1/2), (1-2)/(-1/8)] = 1$, $p = 2$, $j_p = 7$.
2. $\bar{x}_B = \bar{x}_B - \alpha\bar{d} = [4, 0, 7/2, 9/8]^T$, $\bar{f} = 23 - 1 = 22$.
3. $B' = \{1, 4, 5\}$
4. $q = \arg\max[|9|, |-5|, |9|] = 1$.
5. $B = \{6, 1, 8, 3\}$, $N = \{2, 4, 5, 7\}$.
6.

x_1	x_2	x_3	x_4	x_5	x_6	x_7	x_8	f
0	4	0	52/9	0	1	−5/9	0	−5/9
1	0	0	−5/9	1	0	1/9	0	1/9
0	−7	0	−17/18	−2	0	7/18	1	−1/9
0	1/4	1	−5/24	−1/4	0	1/24	0	−1/12
2	5	9/8	3	7	4	0	7/2	22

Iteration 2:

1. $\alpha = \min[(4-8)/(-5/9), (2-0)/(1/9), (7/2-6)/(-1/9), (9/8-2)/(-1/12)]$ $36/5,\ p = 1, j_p = 6.$
2. $\bar{x}_B = \bar{x}_B - \alpha \bar{d} = [8, 6/5, 43/10, 69/40]^T$, $\bar{f} = 22 - 36/5 = 74/5.$
3. $B' = \{2, 4\}$
4. $q = \arg\max\{4, 52/9\} = 4.$
5. $B = \{4, 1, 8, 3\},\ N = \{2, 5, 6, 7\}.$
6.

x_1	x_2	x_3	x_4	x_5	x_6	x_7	x_8	f
0	9/13	0	1	0	9/52	−5/52	0	−5/52
1	5/13	0	0	1	5/52	3/52	0	3/52
0	−165/26	0	0	−2	17/104	31/104	1	−21/104
0	41/104	1	0	−1/4	15/416	9/416	0	−43/416
6/5	5	69/40	3	7	8	0	43/10	74/5

Iteration 3:

1. $\alpha = \min[(3-4)/(-5/52), (6/5-0)/(3/52), (43/10-6)/(-21/104),$ $(69/40-2)/(-43/416)] = 572/215,\ p = 4, j_p = 3.$
2. $\bar{x}_B = \bar{x}_B - \alpha \bar{d} = [140/43, 45/43, 208/43, 2]^T$, $\bar{f} = 74/5 - 572/215 = 522/43.$
3. $B' = \{2, 5, 7\}$
4. $q = \arg\max\{41/104, |-1/4|, 9/416\} = 2.$
5. $B = \{4, 1, 8, 2\},\ N = \{3, 5, 6, 7\}.$
6.

x_1	x_2	x_3	x_4	x_5	x_6	x_7	x_8	f
0	0	−72/41	1	18/41	9/82	−11/82	0	7/82
1	0	−40/41	0	51/41	5/82	3/82	0	13/82
0	0	660/41	0	−247/41	61/82	53/82	1	−153/82
0	1	104/41	0	−26/41	15/164	9/164	0	−43/164
45/43	5	2	140/43	7	8	0	208/43	522/43

Iteration 4:

1. $\alpha = \min[(140/43-0)/(7/82), (45/43-0)/(13/82), (208/43-6)/(-153/82),$ $(5-7)/(-43/164)] = 1381/2216,\ p = 3, j_p = 8.$
2. $\bar{x}_B = \bar{x}_B - \alpha \bar{d} = [490/153, 145/153, 6, 790/153]^T,$

$\bar{f} = \bar{f} - \alpha = 1762/153.$

3. $B' = \{5, 7\}$
4. $q = \arg\max\{|-247/41|, 53/82\} = 5.$
5. $B = \{4, 1, 5, 2\},\ N = \{3, 6, 7, 8\}.$
6.

x_1	x_2	x_3	x_4	x_5	x_6	x_7	x_8	f
0	0	−144/247	1	0	81/494	−43/494	18/247	−25/494
1	0	580/247	0	0	53/247	42/247	51/247	−56/247
0	0	−660/247	0	1	−61/494	−53/494	−41/247	153/494
0	1	16/19	0	0	1/76	−1/76	−2/19	−5/76
145/153	790/153	2	490/153	7	8	0	6	1762/153

Iteration 5:

1. $\alpha = \min[(490/153 - 4)/(-25/494), (145/153 - 5)/(-56/247), (7 - 0)/(153/494), (790/153 - 7)/(-5/76)]$
   $= 1381/2216,\ p = 1,\ j_p = 4.$
2. $\bar{x}_B = \bar{x}_B - \alpha\bar{d} = [4, 113/25, 53/25, 31/5]^T,$
   $\bar{f} = \bar{f} - \alpha = -106/25.$
3. $B' = \{3\}$
4. $q = 3.$
5. $B = \{3, 1, 5, 2\},\ N = \{4, 6, 7, 8\}.$
6.

x_1	x_2	x_3	x_4	x_5	x_6	x_7	x_8	f
0	0	1	−247/144	0	−9/32	43/288	−1/8	25/288
1	0	0	145/36	0	7/8	−13/72	1/2	−31/72
0	0	0	−55/12	1	−7/8	7/24	−1/2	13/24
0	1	0	13/9	0	1/4	−5/36	0	−5/36
113/25	31/5	2	4	53/25	8	0	6	−106/25

Iteration 6:

1. $\alpha = \min[(2 - 0)/(25/288), (113/25 - 5)/(-31/72), (53/25 - 0)/(13/24), (31/5 - 7)/(-5/36)]$
   $= 864/775,\ p = 2,\ j_p = 1.$
2. $\bar{x}_B = \bar{x}_B - \alpha\bar{d} = [59/31, 5, 47/31, 197/31]^T,$
   $\bar{f} = \bar{f} - \alpha = -166/31.$
3. $B' = \emptyset$. Optimal face solution:
   $\bar{x} = [5, 197/31, 59/31, 4, 47/31, 8, 0, 6]^T, \qquad \bar{f} = -166/31.$

The maximum optimal face is zero-dimensional—vertex.

6.6.2 Double-Phase Generalized Reduced Face Algorithm

The generalized reduced face Algorithm 6.6.1 needs a face solution to get itself started. This section presents a generalized reduced phase I procedure, a variant of the single-artificial-variable Algorithm 1.6.2. It turns out that phase I is a good match for Algorithm 6.6.1.

Take any vector $\bar{x} \in R^n$ such that

$$l \leq \bar{x} \leq u. \tag{6.50}$$

Let (B, N) be a partition. Assume that the corresponding canonical system of the equality constraint is

$$x_B + \bar{N} x_N = \bar{b},$$

where $\bar{N} = B^{-1}N,\ \bar{b} = B^{-1}b$. Consider the following vector:

$$h \triangleq \bar{x}_B + \bar{N}\bar{x}_N - \bar{b}. \tag{6.51}$$

If h is equal to zero, then $(\bar{x}_B, \bar{x}_N)$ is a face solution to the problem.

In the other case, we construct and solve the following auxiliary problem.

Using the normalized vector

$$\bar{d} = h/\|h\|, \tag{6.52}$$

we construct the following auxiliary problem:

$$\begin{array}{ll} \min & f' \\ \text{s.t.} & x_B + \bar{N}x_N = f'\bar{d} + \bar{b}, \qquad l \leq x \leq u,\ f' \geq 0. \end{array} \tag{6.53}$$

The preceding problem is a typical generalized reduced problem. It is obvious that

$$(\bar{x}, \bar{f}' = \|h\|)$$

is a face solution to it.

Put the problem into the following reduced face tableau:

x_B	x_N	f'
I	$\bar{N}$	$\bar{d}$
$\bar{x}_B$	$\bar{x}_N$	$\bar{f}'$

(6.54)

where $(\bar{x} = h, \bar{f}' = \|h\|)$ is inserted as the bottom line for reference (with $\bar{b}$ omitted).

From the preceding tableau, Algorithm 6.6.1 can get itself started. Once the objective value vanishes, a reduced tableau to the original problem can be easily obtained to carry on phase II. If it ends with an optimal value greater than zero, it can be asserted that there is no feasible solution to the original problem.

We suggest the following slight variant of Algorithm 6.6.1 as the phase I solver.

Subalgorithm 6.6.1 (Generalized Reduced Phase I: Tableau Form)
Initial : tableau in the form (6.54). $(\bar{x}, \bar{f}')$. *This algorithm finds a reduced face tableau for the variable-bounded problem:*

1. *Determine row index* p *and stepsize* α *by (6.33)–(6.32).*
2. *If* $\bar{f}' - \alpha \leq 0$, *reset* $\alpha = \bar{f}'$.
3. *If* $\alpha \neq 0$, *update* $\bar{x}_B = \bar{x}_B - \alpha\bar{d}$ *and* $\bar{f}' = \bar{f}' - \alpha$.
4. *Return if* $\bar{f}' = 0$.
5. *Return if index set* B', *defined by (6.37), is empty.*
6. *Determine column index* $q \in \arg\max_{j \in B'} |v_j|$, *where* $v_{B'}^T$ *is from the pth row.*
7. *Adjust* (B, N) *by exchanging* j_p *and* q.
8. *Convert* v_q *to* 1, *and eliminate nonzeros in column* q *by elementary transformations.*
9. *Go to Step 1.*

Theorem 6.6.4 *Assuming nondegeneracy, Subalgorithm 6.6.1 terminates either at*
(i) Step 4, giving a face solution, or at
(ii) Step 5, detecting the nonexistence of a feasible solution.

Proof The validity is obvious from the proof of Theorem 6.6.3. □

Once Subalgorithm 6.6.1 terminates at Step 4 with the auxiliary function value $\bar{f}'$ vanishing, the following steps are relevant:

Tactic 6.6.1 (Transition from Phase I to Phase II)

(i) *Use the costs of the original problem to replace the* $(m + 1)$*th row.*
(ii) *Use the* f *column* e_{m+1} *to replace the* f' *column.*
(iii) *Eliminate basic costs by elementary transformations.*
(iv) *Take* $p = m + 1$ *as the row index and determine the column index* q *by*

$$q \in \arg\max_{j \in B'} |v_j|,$$

where B' *is defined by (6.37) and* $v_{B'}^T$ *is from the pth row.*
(v) *Make the according basis change.*

Starting from the resulting reduced face tableau and the face solution $(\bar{x}, \bar{f} = c^T\bar{x})$, Algorithm 6.6.1 is called to solve the original problem. It is the pair of algorithms forming the double generalized reduced face algorithm.

Example 6.6.2 Solve the following problem by the double-phase generalized reduced face algorithm:

$$A = \begin{bmatrix} 1 & -3 & & -5 & 1 & & & \\ & & -2 & 6 & & 1 & & \\ -2 & 4 & & -3 & & & 1 & \\ & 6 & -3 & & & & & 1 \end{bmatrix} \qquad b = \begin{bmatrix} 7 \\ 0 \\ -15 \\ 2 \end{bmatrix}$$

$c = [-4, 0, 0, 6, 0, 3, 0, -7]^T$
$u = [6, 9, 12, 11, 5, 14, 7, 10]^T$
$l = [-7, -3, -5, -4, -7, -8, -7, -6]^T$

Take $B = \{5, 6, 7, 8\},\ N = \{1, 2, 3, 4\}$. It might as well set

$$\bar{x} = [0, 0, 0, 0, 0, 0, 0, 0]^{\mathrm{T}},$$

so that by (6.51)–(6.52), it is simple:

$$\bar{d} = -\bar{b}/\|b\| = [-1050/2501, 0, 2250/2501, -300/2501]^T,$$

corresponding to the auxiliary objective value $\bar{f}' = 2501/150$.

The initial reduced tableau is then

x^T	x_1	x_2	x_3	x_4	x_5	x_6	x_7	x_8	f'
	1	−3		−5	1				−1050/2501
			−2	6		1			0
	−2	4		−3			1		2250/2501
		6	−3					1	−300/2501
u^T	6	9	12	11	5	14	7	10	–
$\bar{x}^T$	0	0	0	0	0	0	0	0	2501/150
l^T	−7	−3	−5	−4	−7	−8	−7	−6	0

Phase I: Call Subalgorithm 6.6.1.

Iteration 1: $B = \{5, 6, 7, 8\},\ N = \{1, 2, 3, 4\},\ \bar{f}' = 2501/150$

1. $\alpha = \min[(0 - u(5))/\bar{d}(1), (0 - l(7))/\bar{d}(3), (0 - u(8))/\bar{d}(4)]$;
   $= \min[2501/210, 4723/607, 2501/30] = 4723/607,\ p = 3$.
2. $2501/150 - 4723/607 > 0$.
3. $\bar{x}_B = [49/15, 0, -7, 14/15]^{\mathrm{T}},\ \bar{f}' = 3059/344$.
4. $B' = \{1, 2, 4\}$.
5. $q = \arg\max\{|-2|, |4|, |-3|\} = 2$.
6. $B = \{5, 6, 2, 8\},\ N = \{1, 3, 4, 7\}$.
7. Multiply row 1 by 1/4, then add 3, −6 times of row 3 to rows 1 and 4:

x^T	x_1	x_2	x_3	x_4	x_5	x_6	x_7	x_8	f'
	$-1/2$			$-29/4$	1		$3/4$		$1288/5053$
			-2	6		1			0
	$-1/2$	1		$-3/4$			$1/4$		$688/3059$
	3		-3	$9/2$			$-3/2$	1	$-1249/850$
u^T	6	9	12	11	5	14	7	10	–
$\bar{x}^T$	0	0	0	0	$49/15$	0	-7	$14/15$	$3059/344$
l^T	-7	-3	-5	-4	-7	-8	-7	-6	0

Iteration 2:

1. $\alpha = \min[(x(5) - l(5))/\bar{d}(1), (x(2) - l(2))/\bar{d}(3),$ $(x(8) - u(8))/\bar{d}(4)] = \min[9143/227, 5002/375, 2283/370]$ $= 2283/370,\ p = 4.$
2. $3059/344 - 2283/370 > 0.$
3. $\bar{x}_B = [83/49, 0, -68/49, 10]^{\mathrm{T}},\ \bar{f}' = 3302/1213.$
5. $B' = \{1, 3, 4\} \neq \emptyset.$
6. $q = \arg\max[|3|, |-3|, |9/2|] = 4.$
7. $B = \{5, 6, 2, 4\},\ N = \{1, 3, 7, 8\}.$
8. Multiply row 4 by $1/(9/2)$, and then add $29/4, -6, 3/4$ times of row 4 to rows 1, 2, and 3:

x^T	x_1	x_2	x_3	x_4	x_5	x_6	x_7	x_8	f'
	$13/3$		$-29/6$		1		$-5/3$	$29/18$	$-2216/1049$
	-4		2			1	2	$-4/3$	$1201/613$
		1	$-1/2$				0	$1/6$	$-50/2501$
	$2/3$		$-2/3$	1			$-1/3$	$2/9$	$-1217/3727$
u^T	6	9	12	11	5	14	7	10	–
$\bar{x}^T$	0	$-68/49$	0	0	$83/49$	0	-7	10	$3302/1213$
l^T	-7	-3	-5	-4	-7	-8	-7	-6	0

Iteration 3:

1. $\alpha = \min[(x(5) - u(5))/\bar{d}(1), (x(6) - l(6))/\bar{d}(2),$ $(x(2) - u(2))/\bar{d}(3), (x(4) - u(4))/\bar{d}(4)]$ $= \min[1504/961, 5002/1225, 21{,}823/42, 3874/115] = 1504/961,\ \ p = 1.$
2. $3302/1213 - 1504/961 > 0.$
3. $\bar{x}_B = [5, -972/317, -430/317, 162/317]^{\mathrm{T}},\ \ \bar{f}' = 5324/4601.$
5. $B' = \{1, 3\} \neq \emptyset.$
6. $q = \arg\max[|13/3|, |-29/6|] = 3.$
7. $B = \{3, 6, 2, 4\},\ N = \{1, 5, 7, 8\}.$
8. Multiply row 1 by $1/(-29/6)$, and then add $-2, 1/2, 2/3$ times of row 1 to rows 2, 3, and 4:

x^T	x_1	x_2	x_3	x_4	x_5	x_6	x_7	x_8	f'
	−26/29		1		−6/29		10/29	−1/3	441/1009
	−84/29				12/29	1	38/29	−2/3	982/905
	−13/29	1			−3/29		5/29	0	245/1234
	2/29			1	−4/29		−3/29	0	−131/3726
u^T	6	9	12	11	5	14	7	10	–
$\bar{x}^T$	0	−430/317	0	162/317	5	−972/317	−7	10	5324/4601
l^T	−7	−3	−5	−4	−7	−8	−7	−6	0

Iteration 4:

1. $\alpha = \min[(x(3) - l(3))/\bar{d}(1), (x(6) - l(6))/\bar{d}(2),$
   $(x(2) - l(2))/\bar{d}(3), (x(4) - u(4))/\bar{d}(4)],$
   $= [4759/416, 4897/1077, 1697/205, 69{,}512/233] = 4897/1077, \ p = 2.$
2. $4897/1077 > 5324/4601$; set $\alpha = 5324/4601$
3. $\bar{x}_B = [-44/87, -376/87, -46/29, 16/29]^{\mathrm{T}}, \ \bar{f}' = 0.$
4. The feasibility is achieved. The feasible face solution to the original problem is

$$\bar{x} = [0, -46/29, -44/87, 16/29, 5, -376/87, -7, 10]^T, \ \bar{f} = c^T\bar{x} = -2310/29.$$

To transit to phase I, execute Tactic 6.6.1:

(i) Insert its costs

$$c^T = [-4, 0, 0, 6, 0, 3, 0, -7]$$

as the fifth row. The objective value is $\bar{f} = c^T\bar{x} = -2310/29$.

(ii) Replace the f' column by the f column e_5

(iii) Eliminate basic costs by adding -6 times of the fourth row to the fifth row, and -3 times of the second row to the fifth row.
The tableau becomes

x^T	x_1	x_2	x_3	x_4	x_5	x_6	x_7	x_8	f
	−26/29		1		−6/29		10/29	−1/3	
	−64/29				12/29	1	38/29	−2/3	
	−13/29	1			−3/29		5/29	0	
	2/29			1	−4/29		−3/29	0	
	64/29				−12/29		−96/29	−5	1
u^T	6	9	12	11	5	14	7	10	–
$\bar{x}^T$	0	−46/29	−44/87	16/29	5	−376/87	−7	10	−2310/29
l^T	−7	−3	−5	−4	−7	−8	−7	−6	–

(iv) Set $p = 5$. By (6.37), we have $B' = \{1, 7\}$, and hence $q = \arg\max[|64/29|, |-96/29|] = 7$.

(v) The according basis change adds index 7 to the basis, resulting in $B = \{3, 6, 2, 4, 7\},\ N = \{1, 5, 8\}$:

x^T	x_1	x_2	x_3	x_4	x_5	x_6	x_7	x_8	f
	$-2/3$		1		$-1/4$			$-41/48$	$5/48$
	$-4/3$				$1/4$	1		$-127/48$	$19/48$
	$-1/3$	1			$-1/8$			$-25/96$	$5/96$
				1	$-1/8$			$5/32$	$-1/32$
	$-2/3$				$1/8$		1	$145/96$	$-29/96$
u^T	6	9	12	11	5	14	7	10	–
$\bar{x}^T$	0	$-46/29$	$-44/87$	$16/29$	5	$-376/87$	-7	10	$-2310/29$
l^T	-7	-3	-5	-4	-7	-8	-7	-6	–

Note that the end column exhibits the current $\bar{d}$.

Phase II: Call Algorithm 6.6.1.

Iteration 5: $B = \{3, 6, 2, 4, 7\},\ N = \{1, 5, 8\}$.

1. $\alpha = \min[(\bar{x}(3) - l(3))/\bar{d}(1), (\bar{x}(6) - l(6))/\bar{d}(2),$
   $(\bar{x}(2) - l(2))/\bar{d}(3),$
   $(\bar{x}(4) - u(4))/\bar{d}(4), (\bar{x}(7) - u(7))/\bar{d}(5)]$
   $= [6256/145, 5120/551, 3936/145, 9696/29, 1344/29] = 5120/551,\ p = 2,\ jp = 6.$
2. $\bar{x}_B = [-28/19, -8, -118/57, 16/19, -239/57]^{\mathrm{T}}$,
   $\bar{f} = \bar{f} - \alpha = -1690/19.$
3. $B' = \{1, 5\}$.
4. $q = \arg\max[|-4/3|, |1/4|] = 1$.
5. $B = \{3, 1, 2, 4, 7\},\ N = \{5, 6, 8\}$.
6.

x^T	x_1	x_2	x_3	x_4	x_5	x_6	x_7	x_8	f
			1		$-3/8$	$-1/2$		$15/32$	$-3/32$
	1				$-3/16$	$-3/4$		$127/64$	$-19/64$
		1			$-3/16$	$-1/4$		$77/192$	$-3/64$
				1	$-1/8$			$5/32$	$-1/32$
						$-1/2$	1	$17/6$	$-1/2$
u^T	6	9	12	11	5	14	7	10	–
$\bar{x}^T$	0	$-118/57$	$-28/19$	$16/19$	5	-8	$-239/57$	10	$-1690/19$
l^T	-7	-3	-5	-4	-7	-8	-7	-6	–

Iteration 6:

1. $\alpha = \min[(\bar{x}(3) - u(3))/\bar{d}(1), (\bar{x}(1) - u(1))/\bar{d}(2),$
   $(\bar{x}(2) - u(2))/\bar{d}(3), (\bar{x}(4) - u(4))/\bar{d}(4),$

$(\bar{x}(7) - u(7))/\bar{d}(5)]$
$= [8192/57, 384/19, 12{,}989/55, 6176/19, 1276/57] = 384/19, \; p = 2.$

2. $\bar{x}_B = \bar{x}_B - \alpha\bar{d} = [8/19, 6, -64/57, 28/19, 337/57]^T$,
$\bar{f} = \bar{f} - \alpha = -2074/19.$
3. $B' = \{5, 6, 8\}$.
4. $q = \arg\max[|-3/16|, |-3/4|, |127/64|] = 8$.
5. $B = \{3, 8, 2, 4, 7\}, \; N = \{1, 5, 6\}$.
6.

x^T	x_1	x_2	x_3	x_4	x_5	x_6	x_7	x_8	f
	−30/127		1		−42/127	−41/127			−3/127
	64/127				−12/127	−48/127		1	−19/127
	−77/381	1			−19/127	−25/254			5/381
	−10/127			1	−14/127	15/254			−1/127
	−544/381				34/127	145/254	1		−29/381
u^T	6	9	12	11	5	14	7	10	–
$\bar{x}^T$	6	−64/57	8/19	28/19	5	−8	337/57	10	−2074/19
l^T	−7	−3	−5	−4	−7	−8	−7	−6	–

Iteration 7:

1. $\alpha = \min[(\bar{x}(3) - u(3))/\bar{d}(1), (\bar{x}(8) - u(1))/\bar{d}(2),$
$(\bar{x}(2) - l(2))/\bar{d}(3), (\bar{x}(4) - u(4))/\bar{d}(4),$
$(\bar{x}(7) - u(7))/\bar{d}(5)]$
$= [27{,}940/57, 0, 13{,}589/95, 22{,}987/19, 7874/551] = 0, \; p = 2, \; j_p = 8.$
3. $B' = \{5\}$.
4. $q = 5$.
5. $B = \{3, 5, 2, 4, 7\}, \; N = \{1, 6, 8\}$.
6.

x^T	x_1	x_2	x_3	x_4	x_5	x_6	x_7	x_8	f
	−2		1			1		−7/2	1/2
	−16/3				1	4		−127/12	19/12
	−1	1				1/2		−19/12	1/4
	−2/3			1		1/2		−7/6	1/6
						−1/2	1	17/6	−1/2
u^T	6	9	12	11	5	14	7	10	–
$\bar{x}^T$	6	−5/3	−2/3	10/9	14/9	−8	7	10	−2074/19
l^T	−7	−3	−5	−4	−7	−8	−7	−6	–

Iteration 8:

1. $\alpha = \min[(\bar{x}(3) - l(3))/\bar{d}(1), (\bar{x}(5) - l(5))/\bar{d}(2), (\bar{x}(2) - l(2))/\bar{d}(3),$

$(\bar{x}(4) - l(4))/\bar{d}(4), (\bar{x}(7) - u(7))/\bar{d}(5)],$
$= [206/19, 144/19, 428/57, 624/19, 124/57] = 124/57, \ p = 5.$

2. $\bar{x}_B = \bar{x}_B - \alpha d = [-2/3, 14/9, -5/3, 10/9, 7]^T,$
   $\bar{f} = \bar{f} - \alpha = -334/3.$
3. $B' = \emptyset$. Optimality achieved. Optimal face solution:

$$\bar{x} = [6, -5/3, -2/3, 10/9, 14/9, -8, 7, 10]^{\mathrm{T}}, \qquad \bar{f} = -334/3.$$

From the fifth row of the end tableau, we see that the optimal set is a one-dimensional face.

6.6.3 Generalized Reduced Face Algorithm

Let us turn to the revised version of the algorithm.

The required quantities in the reduced face tableau can be expressed through the LU factorization of B. At the moment, we use the inverse of B instead.

It is obvious that

$$\bar{N} = B^{-1}N, \quad \bar{d} = B^{-1}e_{m+1}. \tag{6.55}$$

Thus, $\bar{d}$ can be obtained by solving the following system:

$$B\bar{d} = e_{m+1}. \tag{6.56}$$

The pth row can be expressed by

$$v_N^T = y^T N, \tag{6.57}$$

where $y^T = e_p^T B^{-1}$ can be obtained by solving the following system:

$$B^T y = e_p. \tag{6.58}$$

Alternatively, $\bar{d}$ can be updated in each iteration. If the entering index q is determined, the corresponding column $\bar{a}_q = B^{-1}a_q$ can be obtained by solving the following system:

$$B\bar{a}_q = a_q. \tag{6.59}$$

If $\bar{a}_q$ and v_N^T are available, the update formula of $\bar{d}$ is

$$\hat{d} = \bar{d} + (\bar{d}_p/v_q)(\bar{a}_q - e_p). \tag{6.60}$$

Then, it is straightforward to transform Algorithm 6.6.1 into the following form.

Algorithm 6.6.2 (Generalized Reduced Face Algorithm)

Initial: (B, N); *face solution* $(\bar{x}, \bar{f})$. *This algorithm solves the variable-bounded problem.*

1. *Solve system* $B\bar{d} = e_{m+1}$ *for* $\bar{d}$.
2. *Determine the row index* p *and stepsize* α *by (6.33)–(6.32).*
3. *If* $\alpha \neq 0$, *update* $\bar{x}_B = \bar{x}_B - \alpha\bar{d}$ *and* $\bar{f} = \bar{f} - \alpha$.
4. *Solve system* $B^T y = e_p$, *and compute* $v_N^T = y^T N$.
5. *Stop if* B' *defined by (6.37) is empty. (optimality achieved)*
6. *Determine column index* q *by (6.46).*
7. *Adjust* (B, N) *by exchanging* j_p *and* q.
8. *Go to Step 1.*

Note: The algorithm solves three triangle systems per iteration (Steps 1 and 4). It is recommended to use Rule 6.6.3 in Step 6 instead.

Example 6.6.3 Solve the following variable-bounded problem by Algorithm 6.6.2:

$$A = \begin{bmatrix} -2 & 1 & -3 & -1 & & 1 & & \\ 3 & & 1 & & -2 & & 1 & \\ & -6 & & -4 & 6 & & & 1 \end{bmatrix} \qquad b = \begin{bmatrix} 3 \\ 2 \\ -5 \end{bmatrix}$$

$c = [-1, 0, 3, 0, -5, 0, 0, 0]^T$

$u = [10, 12, 15, 27, 14, 18, 13, 21]^T$

$l = [0, 0, 0, 0, 0, 0, 0, 0]^T$

Initial strict interior point (five-dimensional face solution):

$\bar{x} = [6, 5, 7, 14, 5, 1, 4, 10]^{\mathrm{T}}, \ \bar{f} = -10.$

The initial reduced tableau is

x_1	x_2	x_3	x_4	x_5	x_6	x_7	x_8	f
−2	1	−3	−1	0	1			
3	0	1	0	−2		1		
0	−6	0	−4	6			1	
−1	0	3	0	−5				1
6	5	7	14	5	1	4	10	−10

Convert the preceding tableau to a reduced face tableau by making the basis change with $p = 4,\ q = \arg\max[|-1|, |3|, |-5|] = 5$: $B = \{6, 7, 8, 5\},\ N = \{1, 2, 3, 4\}$. The LU factors of B are as follows:

Iteration 1:

$\bar{x}_B = [1, 4, 10, 5]^{\mathrm{T}},\ u_B = [18, 13, 21, 14]^T;\ \bar{x}_N = [6, 5, 7, 14]^T,\ u_N = [10, 12, 15, 27]^T$

1. $\bar{d} = [0, -2/5, 6/5, -1/5]^T$.

2. $\alpha = \min[(4-13)/(-2/5), (10-0)/(6/5), (5-14)/(-1/5)] = 25/3$, $p = 3$, $j_p = 8$.
3. $\bar{x}_B = \bar{x}_B + \alpha\bar{d} = [1, 22/3, 0, 20/3]^T$, $\bar{f} = \bar{f} - \alpha = -55/3$,
4. $y^T = [0, 0, 1, 6/5]$, $v_N^T = [-6/5, -6, 18/5, -4]$.
5. $B' = \{1, 2, 3, 4\}$
6. $q = \arg\max[|-6/5|, |-6|, |18/5|, |-4|] = 2$.
7. $B = \{6, 7, 2, 5\}, /, N = \{1, 3, 4, 8\}$.

Iteration 2:

$\bar{x}_B = [1, 22/3, 5, 20/3]^{\mathrm{T}}, u_B = [18, 13, 12, 14]^T; \bar{x}_N = [6, 5, 14, 0]^T, u_N = [10, 15, 27, 21]^T$

1. $\bar{d} = [1/5, -2/5, -1/5, -1/5]^T$.
2. $\alpha = \min[(1-0)/(1/5), (22/3-13)/(-2/5), (5-12)/(-1/5), (20/3-14)/(-1/5)]$
   $= 5, p = 1, jp = 6$.
3. $\bar{x}_B = \bar{x}_B + \alpha\bar{d} = [0, 28/3, 6, 23/3]^T$, $\bar{f} = \bar{f} - \alpha = -70/3$,
4. $y^T = [1, 0, 1/6, 1/5], v_N^T = [-11/5, -12/5, -5/3, 1/6]$.
5. $B' = \{1, 3, 4\}$
6. $q = \arg\max[|-11/5|, |-12/5|, |-5/3|] = 3$.
7. $B = \{3, 7, 2, 5\}, /, N = \{1, 4, 6, 8\}$.

Iteration 3:

$\bar{x}_B = [7, 28/3, 6, 23/3]^{\mathrm{T}},\ u_B = [15, 13, 12, 14]^T;$
$\bar{x}_N = [6, 14, 0, 0]^T,\ u_N = [10, 27, 18, 21]^T$

1. $\bar{d} = [-1/12, -5/12, -1/4, -1/4]^T$.
2. $\alpha = \min[(7-15)/(-1/12), (28/3-13)/(-5/12), (6-12)/(-1/4),$
   $(23/3-14)/(-1/4)] = 44/5, p = 2, jp = 7$.
3. $\bar{x}_B = \bar{x}_B + \alpha\bar{d} = [116/15, 13, 41/5, 148/15]^T$,
   $\bar{f} = \bar{f} - \alpha = -482/15$.
4. $y^T = [-1/12, 1, -1/72, -5/12], v_N^T = [43/12, 5/36, -1/12, -1/72]$.
5. $B' = \{1, 4\}$
6. $q = \arg\max[|43/12|, |5/36|] = 1$.
7. $B = \{3, 1, 2, 5\}, /, N = \{4, 6, 7, 8\}$.

Iteration 4:

$\bar{x}_B = [116/15, 6, 41/5, 148/15]^{\mathrm{T}},\ u_B = [15, 10, 12, 14]^T;$
$\bar{x}_N = [14, 0, 13, 0]^T,\ u_N = [27, 18, 13, 21]^T$

1. $\bar{d} = [1/43, -5/43, -7/43, -7/43]^T$.
2. $\alpha = \min[(116/15-0)/(1/43), (6-10)/(-5/43), (41/5-12)/(-7/43),$
   $(148/15-14)/(-7/43)] = 817/35, p = 3, jp = 2$.
3. $\bar{x}_B = \bar{x}_B + \alpha\bar{d} = [151/21, 61/7, 12, 41/3]^T$,
   $\bar{f} = \bar{f} - \alpha = -1165/21$,
4. $y^T = [-10/43, -9/43, -53/258, -7/43],\ v_N^T = .$
   $= [136/129, -10/43, -9/43, -53/258]$.
5. $B' = \{4, 7\}$.

6. $q = \arg\max[|-10/43|, |-53/258|] = 4$.
7. $B = \{3, 1, 4, 5\}, /, N = \{2, 6, 7, 8\}$.

Iteration 5:

$\bar{x}_B = [151/21, 61/7, 14, 41/3]^{\mathrm{T}}, u_B = [15, 10, 27, 14]^T$;
$\bar{x}_N = [12, 0, 13, 0]^T,\ u_N = [12, 18, 13, 21]^T$.

1. $\bar{d} = [1/8, -15/136, -21/136, -7/68]^T$.
2. $\alpha = \min[(151/21-0)/(1/8), (61/7-10)/(-15/136), (14-27)/(-21/136), (41/3-14)/(-7/68)] = [1208/21, 408/35, 1768/21, 68/21]. = 68/21, p = 4, jp = 5$.
3. $\bar{x}_B = \bar{x}_B + \alpha\bar{d} = [95/14, 127/14, 29/2, 14]^T, \bar{f} = \bar{f} - \alpha = -411/7$, ($\bar{x} = [127/14, 12, 95/14, 29/2, 14, 0, 13, 0]^T$).
4. $y^T = [-5/34, -9/68, 5/136, -7/68],\ v_N^T = [-25/68, -5/34, -9/68, 5/136]$.
5. $B' = \{2, 7, 8\}$.
6. $q = \arg\max[|-25/68|, |-9/68|, |5/136|] = 2$.
7. $B = \{3, 1, 4, 2\}, /, N = \{5, 6, 7, 8\}$.

Iteration 6:

$\bar{x}_B = [95/14, 127/14, 29/2, 12], u_B = [15, 10, 27, 12]$,
$\bar{x}_N = [14, 0, 13, 0], u_N = [14, 18, 13, 21]$

1. $\bar{d} = [3/10, -1/10, -21/50, 7/25]^T$.
2. $\alpha = \min[(95/14-0)/(3/10), (127/14-10)/(-1/10), (29/2-27)/(-21/50), (12-0)/(7/25)] = [475/21, 65/7, 625/21, 300/7]. = 65/7, p = 2, jp = 1$.
3. $\bar{x}_B = \bar{x}_B + \alpha\bar{d} = [4, 10, 92/5, 47/5]^T, \bar{f} = \bar{f} - \alpha = -68$, ($\bar{x} = [10, 47/5, 4, 92/5, 14, 0, 13, 0]^T$).
4. $y^T = [0, 3/10, 0, -1/10],\ v_N^T = [-1/10, 0, 3/10, 0]$.
5. $B' = \{5\}$.
6. $q = 5$.
7. $B = \{3, 5, 4, 2\}, /, N = \{1, 6, 7, 8\}$.

Iteration 7:

$\bar{x}_B = [4, 14, 92/5, 47/5],\ u_B = [15, 14, 27, 12];\ \bar{x}_N = [10, 0, 13, 0],\ u_N = [10, 18, 13, 21]$

1. $\bar{d} = [2, 1, -3, 3]^T$.
2. $\alpha = \min[(4-0)/(2), (14-0)/(1), (92/5-27)/(-3), (47/5-0)/(3)] = [2, 14, 43/15, 47/15]. = 2, p = 1, jp = 3$.
3. $\bar{x}_B = \bar{x}_B + \alpha\bar{d} = [4, 14, 92/5, 47/5]^T, \bar{f} = \bar{f} - \alpha = -70$, ($\bar{x} = [10, 17/5, 0, 122/5, 12, 0, 13, 0]^T$).
4. $y^T = [0, -5, 0, 2],\ v_N^T = [-17, 0, -5, 0]$.
5. $B' = \emptyset$. Optimality achieved. The optimal face solution is as follows:

$$\bar{x} = (10, 17/5, 0, 122/5, 12, 0, 13, 0)^{\mathrm{T}}, \qquad \bar{f} = -70.$$

The maximum optimal face achieved is two-dimensional.

Chapter 7
Dual Reduced Face Method

This chapter derives the dual version of the reduced face method. In the next section, we present the dual reduced face tableau. In Sect. 7.2, we derive the search direction and update of the dual face solution. In Sect. 7.3, we address row pivoting operation. Section 7.4 formulates the dual reduced face algorithm in tableau form. To get started from an arbitrary solution, Sect. 7.5 presents a double-phase dual reduced face algorithm. Then, we formulate the revised dual reduced face algorithm in Sect. 7.6. The last section generalizes the dual reduced face algorithm for solving variable-bounded problems.

Consider the following problem:

$$\begin{array}{ll} \min\ f = c^{\mathrm{T}}x & \\ \text{s.t.}\ \ Ax = e_r, & x \geq 0, \end{array} \tag{7.1}$$

where the right-hand side is the unit vector e_r with $1 \leq r \leq m$, while the other items are the same as the standard LP problem. Clearly, the dual objective function of the preceding problem involves only a single variable. So, its dual problem can be regarded as a "reduced" LP problem.

We will develop the dual method by handling this unique LP problem, but it is suitable for solving general standard LP problems.

7.1 Dual Reduced Face Tableau

Assume a dual feasible (face) solution $(\bar{y}, \bar{z})$ is available.

At the moment, assume that $r = m$ in the problem (7.1). Let $A = (B', N)$ be a partition with the following pseudo-basis:

$$B' = \{j_1, \ldots, j_{m-1}\}$$

P.-Q. Pan, *Face Method*, https://doi.org/10.1007/978-3-031-93594-7_7

such that the $m \times m$ matrix $[B', e_m]$ is nonsingular.

Put the system of the equality constraint to the following tableau (with the f column omitted):

x_N	$x_{B'}$	rhs
N	B'	e_m
$\bar{z}_N^{\mathrm{T}}$	$\bar{z}_B^{\mathrm{T}}$	μ

where the bottom line is added as the primal objective row.

Assume that Gauss-Jordan elimination transforms the preceding tableau to the following "canonical form":

x_N	$x_{B'}$	rhs
$\bar{N}$	I'	e_m
$\bar{z}_N^{\mathrm{T}}$	$\bar{z}_{B'}^{\mathrm{T}}$	$-\bar{y}_m$

(7.2)

which is called the *dual reduced face tableau*. It includes the unit matrix, i.e.,

$$[I', e_m] \triangleq [e_1, \ldots, e_{m-1}, e_m].$$

The mth row is called the *dual objective row*, corresponding to the dual objective variable y_m.

It is noted that the mth row of I' vanishes, i.e.,

$$e_m^T I' = 0.$$

Therefore, the mth equation of the tableau can be expressed as

$$\sum_{j \in N} \bar{a}_{m,j} x_j = 1. \tag{7.3}$$

The dual reduced face tableau (7.2) also represents the following dual problem:

$$\begin{array}{ll} \max & y_m, \\ \text{s.t.} & \begin{bmatrix} \bar{N}^{\mathrm{T}} \\ (I')^T \end{bmatrix} y + \begin{bmatrix} z_N \\ z_{B'} \end{bmatrix} = \begin{bmatrix} \bar{z}_N \\ \bar{z}_{B'} \end{bmatrix}, \quad z \geq 0. \end{array} \tag{7.4}$$

The preceding problem has the following dual feasible (face) solution:

$$(\bar{y} = 0, \bar{z})$$

$\bar{z}$ is a dual face solution to the original standard problem, although $\bar{y} = 0$ may not. Fortunately, we can only focus on the former, ignoring the latter, as it can be calculated at the end of the solution process if needed (see p. 130 in Pan [36]).

7.2 Search Direction

The dual reduced face tableau has an apparent dual search direction.

To show this, introduce the following row index set:

$$R' = \{1, \ldots, m - 1\}.$$

Set Δy by

$$\Delta y_m = 1, \qquad \Delta y_i = 0, \qquad i \in R', \tag{7.5}$$

and extract the dual objective row, i.e.,

$$\Delta z^T = [-\bar{a}_{m,N}, -\bar{a}_{m,B'}] = [-\bar{a}_{m,N}, 0]. \tag{7.6}$$

Using the preceding notation, we make the following claim:

Theorem 7.2.1 $(\Delta y, \Delta z)$ *is an ascent search direction.*

Proof Obviously, Δy is the steepest uphill of the objective function of the program (7.4).

In addition, it is verified that the direction defined by (7.7)–(7.8) satisfies the following condition:

$$\begin{bmatrix} \bar{N}^{\mathrm{T}} \\ (I')^T \end{bmatrix} \Delta y + \Delta z = 0.$$

Therefore, it is eligible to be the search direction. □

Thereby, the according line search of the dual solution is as follows:

$$\hat{y}_m = \bar{y}_m + \beta, \qquad \hat{z}_N^T = \bar{z}_N^T - \beta \bar{a}_{m,N}, \tag{7.7}$$

where β, termed *dual stepsize*, is to be determined.

Introduce the following index set:

$$J = \{j \in N \mid \bar{a}_{m,j} > 0\}. \tag{7.8}$$

Assume that J is nonempty. Then, the following column rule is well-defined.

Rule 7.2.1 (Dual Column Rule) Determine β and column index q such that

$$\beta = \bar{z}_q / \bar{a}_{m,q} = \min_{j \in J} \bar{z}_j / \bar{a}_{m,j}. \tag{7.9}$$

In case of a tie, select that corresponding to the maximum $\bar{a}_{m,q}$.

Once the stepsize β is available, the new dual face solution can be computed by the update formula (7.7). It is known that the stepsize β is the largest possible to keep the new dual solution feasible. Alternatively, the update can be actualized with the tableau by adding $-\beta$ times of the dual objective row to the bottom row.

It is noted that if $\bar{z}_j = 0$ holds for some $j \in J$, the stepsize β vanishes. In this case, the dual face solution is said to be dual *degenerate*.

Lemma 7.2.2 *The dual problem is upper unbounded if the index set defined by (7.8) is empty.*

Proof Note that $\bar{z}_N \geq 0$ and $\bar{a}_{m,N} \neq 0$.

Emptiness of J implies that $\Delta z_N = -\beta \bar{a}_{m,N}$ is nonnegative for any positive β, so is $\hat{z}_N$, corresponding to infinite dual objective y_r. □

It is then evident that there is no feasible solution to the original problem if J is empty.

7.3 Row Pivoting Operation

Assume that index q was determined and the dual face solution was updated.

On the other hand, the tableau also gives a primal basic solution, as the following statement says.

Proposition 7.3.1 *Let (7.2) be a dual reduced face tableau. If, for some $q \in N$, it holds that $\bar{a}_{m,q} \neq 0$, then the tableau gives the following (primal) basic solution:*

$$\bar{x}_q = 1/\bar{a}_{m,q}, \qquad \bar{x}_{B'} = -\bar{x}_q \bar{a}_{R,q}, \qquad \bar{x}_j = 0, \; q \neq j \in N. \tag{7.10}$$

Proof It is obvious that the solution (7.10) satisfies the system represented by tableau (7.2) and hence the equality constraint of the primal problem. Besides, if one enters column q into B', then he can convert (7.2) to a normal simplex tableau, corresponding to the basic solution (7.10). □

It is noted that for the column index q determined by (7.9), it holds that $\bar{a}_{m,q} > 0$ and hence so is $\bar{x}_q$. Besides, the primal basic solution calculation involves the pivot column $\bar{a}_q$ only.

Then, the following row rule applies.

Rule 7.3.1 (Dual Row Rule) Select row index p such that

$$p \in \arg\min_{i \in R} \bar{x}_{j_i}. \tag{7.11}$$

Assume that $\bar{x}_{j_p} < 0$. Then, the corresponding basic index j_p should leave the basis. If, otherwise, $\bar{x}_{j_p} \geq 0$, then the basic solution given by (7.8) is feasible.

Although the dual solution $\bar{z}$ is also feasible, optimality can still not be declared because it is unknown whether or not the complementarity condition is fulfilled.

In this case, therefore, the following rule must be performed.

Rule 7.3.2 (Dual Row Rule) Select the row index p such that

$$p \in \arg\max_{i \in R} \bar{x}_{j_i} \bar{z}_{j_i}. \tag{7.12}$$

Theorem 7.3.1 *Assume that the dual solution is feasible. If $\bar{x}_{j_p}$ is nonnegative for p selected by (7.11), and $\bar{x}_{j_p}\bar{z}_{j_p}$ vanishes for p selected by (7.12), then $(\bar{x}, \bar{z})$ are primal and dual optimal solutions.*

Proof The dual feasibility assumption and nonnegativeness of $\bar{x}_{j_p}$ for p selected by (7.11) indicate that $\bar{x}$ and $\bar{z}$ are primal and dual feasible, respectively. In addition, that $\bar{x}_{j_p}\bar{z}_{j_p}$ vanishes for p selected by (7.12) implies that

$$\bar{x}_B^T \bar{z}_B = 0,$$

and hence

$$\bar{x}^T \bar{z} = 0.$$

Therefore, complementarity is ensured; hence, the tableau gives primal and dual optimal solutions. □

If $\bar{x}_{j_p}\bar{z}_{j_p} > 0$, j_p is then moved from the pseudo-basis B' to N, and q from N to B'. This can actualized by elementary transformations. It is noted that the pseudo-basis change does not affect the right-hand side at all.

7.4 Dual Reduced Face Algorithm: Tableau Form

Starting from a dual reduced face tableau, the method generates a series of such tableaus, corresponding to a series of dual face solutions $\bar{z}$ until achieving optimality, unless dual upper unboundedness or primal infeasibility is detected.

It is unnecessary to compute the primal basic solution all the time in each iteration. It is cheaper to use the pivot column $\bar{a}_q$ instead in the row index selection and calculate the optimal solution at the end.

We previously took the mth row as the dual objective row for exposition. It can be the rth row for any

$$r \in \{1, \ldots, m\}$$

without any essential adaptation.

The initial dual reduced face tableau can be obtained from a normal simplex tableau as follows. Assume that $\bar{b}$ is the right-hand side of the simplex tableau and the reduced costs are nonnegative. Determine the row index r such that

$$r = \arg\min \{\bar{b}_i \mid i = 1, \ldots, m\}.$$

The tableau is already optimal if $\bar{b}_r \geq 0$. Otherwise, enter $\bar{b}$ to the basis to replace the rth basic column with elementary eliminations, resulting in a reduced face tableau with the rth dual objective row.

The overall steps can be put into the following model.

Algorithm 7.4.1 (Dual Reduced Face Algorithm: Tableau Form)
Initial : dual reduced face tableau in (7.2) with the rth dual objective row. This algorithm solves the standard problem:

1. *Stop if $J = \{j \in N \mid \bar{a}_{r,j} > 0\} = \emptyset$.*
2. *Determine β and column index q such that $\beta = \bar{z}_q/\bar{a}_{r,q} = \min_{j \in J} \bar{z}_j/\bar{a}_{r,j}$.*
3. *If $\beta > 0$, add $-\beta$ times of dual objective row to the bottom row.*
4. *Select row index $p \in \arg\max \{\bar{a}_{i,q} \mid i = 1, \ldots, m; i \neq r\}$.*
5. *If $\bar{a}_{p,q} \leq 0$, then*
   (1) Reselect row index $p \in \arg\max \{-\bar{a}_{i,q}\bar{z}_{j_i} \mid r \neq i = 1, \ldots, m\}$
   (2) Stop if $-\bar{a}_{p,q}\bar{z}_{j_p}$ vanishes
6. *Convert $\bar{a}_{pq}$ to 1, and eliminate the other nonzeros in the column by elementary transformations.*
7. *Go to Step 1.*

Theorem 7.4.1 *Under dual nondegeneracy assumption, Algorithm 7.4.1 terminates either at*

(i) Step 1, detecting infeasibility of the problem, or at
(ii) Step 5(2), producing a pair of optimal primal/dual face solutions.

Proof That the solution process does not terminate means that it performs infinitely many iterations, corresponding to infinitely many dual reduced face tableaus, among which a finite number of tableaus repeat infinitely times since there are only finitely many different tableaus. The repetition contradicts that $\bar{f}$ increases strictly in each iteration under the dual nondegeneracy assumption. Therefore, the algorithm terminates.

Meanings of the exits come from Lemma 7.2.2 and Theorem 7.3.1. □

Example 7.4.1 Solve the following standard problem by Algorithm 7.4.1: $A = \begin{bmatrix} 3 & -8 & -1 & 2 & 5 & -3 & -4 \\ -4 & -1 & 2 & -4 & -3 & 6 & -2 \\ -8 & -4 & 1 & -3 & & -2 & -6 \end{bmatrix}$ $\quad b = \begin{bmatrix} -2 \\ -4 \\ -7 \end{bmatrix}$
$c = [3, 0, 1, 2, 1, 3, 2]^T$

The corresponding dual reduced problem in the form (6.1) can be put into the following tableau:

x_1	x_2	x_3	x_4	x_5	x_6	x_7	x_8(rhs)
3	− 8	− 1	2	5	− 3	− 4	− 2
− 4	− 1	2	− 4	− 3	6	− 2	− 4
− 8	− 4	1	− 3		− 2	− 6	− 7
3		1	2	1	3	2	

which can be converted to the following dual reduced face tableau ($r = 3$) by elementary transformations:

x_1	x_2	x_3	x_4	x_5	x_6	x_7	x_8(rhs)
31/45	− 8/15	1/9	− 4/45	1/45	1		
− 137/45	107/30	4/9	− 52/45	− 199/90		1	
32/9	− 7/3	− 5/9	13/9	17/9			1
3		1	2	1	3	2	

Iteration 1:

1. $J = \{1, 4, 5\}$.
2. $\beta = \min[\bar{z}_1/w_1, \bar{z}_4/w_4, \bar{z}_5/w_5] = \min[27/32, 18/13, 9/17] = 9/17,\ q = 5.$
3. $\bar{z}_N = \bar{z}_N - \beta * w_N = [19/17, 21/17, 22/17, 21/17, 0].$
4. $p = 1,\ j_p = 6$.
5. $\bar{a}_{p,q} > 0.$
6.

x_1	x_2	x_3	x_4	x_5	x_6	x_7	x_8(rhs)
31	− 24	5	− 4	1	45		
131/2	− 99/2	23/2	− 10		199/2	1	
− 55	43	− 10	9		− 85		1
19/17	21/17	22/17	21/17		3	2	− 9/17

Iteration 2:

1. $J = \{2, 4\}$.
2. $\beta = \min[\bar{z}_2/w_2, \bar{z}_4/w_4] = \min[21/731, 7/51] = 21/731,\ q = 2.$
3. $\bar{z}_N = \bar{z}_N - \beta * w_N = [116/43, 0, 68/43, 42/43, 234/43].$
4. $p = 1, j_p = 5.$
5. $\bar{a}_{p,q} \leq 0.$

   (1) $\max[-(-24)0, -(99/2)2] = [0, 99/43] = 99,\ p = 2, j_p = 7.$
   (2) $\bar{x}_{j_p}\bar{z}_{j_p} \neq 0.$

6.

x_1	x_2	x_3	x_4	x_5	x_6	x_7	x_8(rhs)
– 25/33		– 19/33	28/33	1	– 107/33	– 16/33	
– 131/99	1	– 23/99	20/99		– 199/99	– 2/99	
188/99		– 1/99	31/99		142/99	86/99	1
116/43		68/43	42/43		234/43	2	– 24/43

Iteration 3:

1. $J = \{1, 4, 6, 7\}$.
2. $\beta = \min[\bar{z}_1/w_1, \bar{z}_4/w_4, \bar{z}_6/w_6, \bar{z}_7/w_7] = \min[1753/1234, 2249/721, 755/199, 99/43] = 1753/1234, \ q = 1$.
3. $\bar{z}_N = \bar{z}_N - \beta * w_N = [0, 75/47, 25/47, 160/47, 36/47]$.
4. $p = 1, \ j_p = 5$.
5. $\bar{a}_{p,q} \leq 0$.

   (1) $\max[-(-25/33))3, -(-131/99)2] = [25/11, 262/99] = 262/99, \ p = 2, \ j_p = 7$.
   (2) $\bar{x}_{j_p}\bar{z}_{j_p} \neq 0$.

7.

x_1	x_2	x_3	x_4	x_5	x_6	x_7	x_8(rhs)
	– 75/131	– 58/131	96/131	1	– 274/131	– 62/131	
1	– 99/131	23/131	– 20/131		199/131	2/131	
	188/131	– 45/131	79/131		– 190/131	110/131	1
		75/47	25/47		160/47	36/47	– 93/47

Iteration 4:

1. $J = \{2, 4, 7\}$.
2. $\beta = \min[\bar{z}_2/w_2, \bar{z}_4/w_4, \bar{z}_7/w_7] = 0, \ q = 2$.
4. $p = 1, \ j_p = 5$.
5. $\bar{a}_{p,q} \leq 0$.

   (1) $\max[-(-75/131)0, -(-99/131)0] = 0$.
   (2) Optimality achieved.
   $\bar{x}_2 = 1/w_2 = 131/188, \ \bar{x}_5 = -(131/188)(-75/131) = 75/188, \ \bar{x}_1 = -(131/188)(-99/131) = 99/188$.

Primal and dual optimal solutions are as follows:

$$\bar{x} = [99/188, 131/188, 0, 0, 75/188, 0, 0]^T,$$
$$\bar{z} = [0, 0, 75/47, 25/47, 0, 160/47, 36/47]^T, \qquad \bar{f} = 93/47.$$

7.5 Double-Phase Dual Reduced Face Algorithm

The algorithm described in the previous section has to start from a dual feasible solution. It is available if the costs c of the primal problem are nonnegative. However, a phase I procedure is needed to form a double-phase algorithm. This section presents a phase I approach that solves a unique auxiliary problem.

Let $\bar{z}$ be any nonnegative n-vector. Construct the auxiliary dual problem below:

$$\begin{array}{ll} \max & y_{m+1} \leq 0 \\ \text{s.t.} & A^T y + (\bar{z} - c) y_{m+1} + z = c, \ z \geq 0. \end{array} \tag{7.13}$$

It is obvious that $\bar{z} \geq 0$ with $(\bar{y} = 0, \bar{y}_{m+1} = -1)$ is a dual feasible solution to the preceding problem. If the optimal objective value reaches 0, the corresponding $\bar{z}$ is a dual face solution to the original dual problem; otherwise, there is no dual feasible solution to it, that is, there is no feasible solution to the original problem.

It seems logical to make $\bar{z} \geq 0$ as close to c as possible. Therefore, we set

$$\bar{z}_j = \begin{cases} c_j & \text{if} \quad c_j \geq 0, \\ 0 & \text{otherwise}, \end{cases} \qquad j = 1, \ldots, n. \tag{7.14}$$

Let $A = (N, B)$ be a partition with basis B. Put the canonical system of the associated problem (7.13) into the following tableau:

x_N	x_B	$rhs(x_{n+1})$
$\bar{N}$	I	
$(\bar{z}_N - c_N)^T$	$(\bar{z}_B - c_B)^T$	1
$\bar{z}_N^T$	$\bar{z}_B^T$	$-\bar{y}_{m+1}$

The preceding tableau can be converted to the following dual reduced face tableau by eliminating $(\bar{z}_B - c_B)^T$ by elementary transformations, i.e.,

x_N	x_B	$rhs(x_{n+1})$
$\bar{N}$	I	
w_N^T		1
$\bar{z}_N^T$	$\bar{z}_B^T$	$-\bar{y}_{m+1}$

(7.15)

where $r = m + 1$ is the dual objective row.

Since our present goal is to raise the dual objective y_{m+1} to 0, we define the following index set instead of (7.8):

$$J = \{j \in N \mid w_j > 0\} \cup \{n + 1\} \tag{7.16}$$

Overall steps, slightly modified from Algorithm 7.4.1, are put into the following model.

Subalgorithm 7.5.1 (Dual Reduced Phase I: Tableau Form)
Initial : tableau in the form (7.15) with $\bar{y}_{m+1} = -1$. *This algorithm provides a dual face solution:*

1. *Eliminate* $(\bar{z}_B - c_B)^T$ *by elementary transformations.*
2. *Determine* β *and* q *by* $\beta = \bar{z}_q/w_q = \min_{j \in J} \bar{z}_j/w_j$, *where* J *is defined by (7.16).*
3. *If* $\beta > 0$, *add* $-\beta$ *times of the* $(m+1)$*th row to the bottom row.*
4. *Return if the entry* $-\bar{y}_{m+1}$ *vanishes.*
5. *Select row index* $p \in \arg\max \{\bar{a}_{i,q} \mid i = 1, \ldots, m\}$.
6. *If* $\bar{a}_{p,q} \leq 0$, *then:*

    (1) *Reselect row index* $p \in \arg\max \{-\bar{a}_{i,q}\bar{z}_{j_i} \mid i = 1, \ldots, m\}$
    (2) *Stop if* $-\bar{a}_{p,q}\bar{z}_{j_p}$ *vanishes*

7. *Convert* $\bar{a}_{pq}$ *to* 1, *and eliminate nonzeros in the column by elementary transformations.*
8. *Go to Step 2.*

The validity of the following claim is apparent.

Theorem 7.5.1 *Under dual nondegeneracy assumption, Algorithm 7.5.1 returns either from*

(i) *Step 4, with a dual face solution, or from*
(ii) *Step 6(2), indicating the nonexistence of a dual face solution.*

In the following example, the right-hand side $\bar{b}$ of the original problem is placed at the end to turn to phase II once phase I is finished.

Example 7.5.1 Solve the following standard problem using the double-phase dual reduced face algorithm:

$$A = \begin{bmatrix} 6 & -2 & 3 & 2 & 1 & & \\ -2 & 4 & -1 & -8 & & 1 & \\ 4 & 5 & 2 & 1 & & & 1 \end{bmatrix} \qquad b = \begin{bmatrix} 2 \\ -4 \\ 6 \end{bmatrix}$$

$$c = [-3, -4, -5, -8, -7, 0, 2]^T$$

According to (7.14), set $\bar{z} = [0, 0, 0, 0, 0, 0, 2]^T$,
and $\bar{z} - c = [3, 4, 5, 8, 7, 0, 0]^T$.
Construct an auxiliary initial tableau below:

x_1	x_2	x_3	x_4	x_5	x_6	x_7	$rhs(x_8)$	rhs
6	− 2	3	2	1				2
− 2	4	− 1	− 8		1			− 4
4	5	2	1			1		6
3	4	5	8	7			1	
						2	1	

Dual-phase I: call Subalgorithm 7.5.1 with the fourth dual objective row.
Iteration 1:

1. Add − 7 times of the first row to the fourth row, so that the latter becomes $[-39, 18, -16, -6, 0, 0, 0, 1, -14]$.
2. $J = \{2, 8\}$, $\beta = \min\{0/18, 1/1)\} = 0$, $q = 2$.
4. $-\bar{y}_{m+1} = 1$.
5. $\max[-2, 4, 5] = 5 > 0$, $p = 3$, $j_p = 7$.
7. Multiply row 3 by 1/5, and add 2, −4, −18 times of row 3 to rows 1, 2, and 4:

x_1	x_2	x_3	x_4	x_5	x_6	x_7	rhs	rhs
38/5		19/5	12/5	1		2/5		22/5
− 26/5		− 13/5	− 44/5		1	− 4/5		− 44/5
4/5	1	2/5	1/5			1/5		6/5
− 267/5		− 116/5	− 48/5			− 18/5	1	− 178/5
						2	1	

Iteration 2:

2. $J = \{8\}$, $\beta = \min[1/1] = 1$, $q = 8$.
3. Add − 1 times of row 4 to row 5, so that the latter becomes $[267/5, 0, 116/5, 48/5, 0, 0, 28/5, 0, 178/5]$.
4. $-\bar{y}_{m+1} = 0$. Dual feasibility is achieved, and phase I is finished.

Then, deleting row 4 and column 8 gives the following tableau:

x_1	x_2	x_3	x_4	x_5	x_6	x_7	rhs
38/5		19/5	12/5	1		2/5	22/5
− 26/5		− 13/5	− 44/5		1	− 4/5	− 44/5
4/5	1	2/5	1/5			1/5	6/5
267/5		116/5	48/5			28/5	178/5

$\min[22/5, -44/5, 6/5] = -44/5$, $r = 2$. To transform the preceding tableau to dual reduced face tableau,

multiply row 2 by $1/(-44/5)$, and add − 22/5, −6/5 times of row 2 to rows 1 and 3:

x_1	x_2	x_3	x_4	x_5	x_6	x_7	rhs
5		5/2	− 2	1	1/2	0	
13/22		13/44	1		− 5/44	1/11	1
1/11	1	1/22	− 1		3/22	1/11	
267/5		116/5	48/5			28/5	178/5

Phase II: call Algorithm 7.4.1 with the second dual objective row.
Iteration 3:

1. $J = \{1, 3, 4, 7\}$.
2. $\beta = \min[(267/5)/(13/22), (116/5)/(13/44), (48/5)/(1), (28/5)/(1/11)] = 48/5,\ q = 4$.
3. Add $-\beta$ times of row 2 to row 4:

x_1	x_2	x_3	x_4	x_5	x_6	x_7	rhs
5		5/2	− 2	1	1/2	0	
13/22		13/44	1		− 5/44	1/11	1
1/11	1	1/22	− 1		3/22	1/11	
525/11		224/11			12/11	52/11	26

4. $\max[-2, -1] = -1 \le 0,\ p = 3,\ j_p = 2$.

3.(1) $\max[-(-2)0, -(-1)0] = 0$.

(2) Optimality is achieved.

$\bar{x}_4 = 1,\ \bar{x}_5 = -(-2) = 2, \bar{x}_2 = -(-1) = 1.$

Primal and dual optimal solutions:

$$\bar{x} = [0, 1, 0, 1, 2, 0, 0]^T,$$
$$\bar{z} = [525/11, 0, 224/11, 0, 0, 12/11, 52/11]^T, \qquad \bar{f} = -26.$$

Let us turn to the revision of Subalgorithm 7.5.1.
Let the following quantities be available:

$$\bar{z} \ge 0, \qquad w = \bar{z} - c \tag{7.17}$$

It is straightforward to have the revised form of the dual reduced face tableau (7.15), i.e.,

x_N	x_B	$rhs(x_{n+1})$
$B^{-1}N$	I	
w_N^T		1
$\bar{z}_N^T$	$\bar{z}_B^T$	$-\bar{y}_{m+1}(\bar{z}_{n+1})$

(7.18)

where

$$w_N^T = w_N^T - w_B^T B^{-1} N. \tag{7.19}$$

Then, Subalgorithm 7.5.1 can be converted into the following version.

Subalgorithm 7.5.2 (Dual Reduced Face Phase I)

Initial : N, B, $B = LU$, $\bar{z}$, *and* w, $\bar{y}_{m+1} = -1$. *This procedure provides dual face solution:*

1. *Solve two triangular systems* $U^T u = w_B$, $L^T v = u$, *and compute* $w_N = w_N - N^T v$.
2. *Determine* β *and column index* q *by* $\beta = \bar{z}_q / w_q = \min_{j \in J} \bar{z}_j / w_j$,
   where $J = \{j \in N \mid w_j > 0\} \cup \{n+1\}$.
3. *If* $\beta > 0$, *update* $\bar{z}_N = \bar{z}_N - \beta w_N$, $\bar{y}_{m+1} = \bar{y}_{m+1} + \beta$.
4. *Return if* $\bar{y}_{m+1}$ *vanishes (dual feasibility achieved).*
5. *Solve two triangular systems* $Lu = a_q$, $U\bar{a}_q = u$ *for* $\bar{a}_q$.
6. *Select row index* $p \in \arg\max \{\bar{a}_{i\,q} \mid i = 1, \ldots, m\}$.
7. *If* $\bar{a}_{p,q} \leq 0$, *then:*
   (1) *Reselect row index* $p \in \arg\max \{-\bar{a}_{i\,q} \bar{z}_{j_i} \mid i = 1, \ldots, m\}$
   (2) *Return if* $-\bar{a}_{p\,q} \bar{z}_{j_p}$ *vanishes (dual infeasibility problem).*
8. *Adjust* (N, B) *by exchanging* q *and* j_p *and update LU factors.*
9. *Go to Step 1.*

7.6 Dual Reduced Face Algorithm

In this section, we formulate the revised version of Algorithm 7.4.1.

Assume that B is the basis matrix, including b as the rth column, and that its LU factorization is available, i.e.,

$$B = LU.$$

The key quantity is the rth dual objective row of the corresponding dual reduced tableau. Denote it by w^T. Only the following part associated with the set N is useful, i.e.,

$$t w_N^T - e_r^T B^{-1} N,$$

To gain it, solve the following two triangular systems for v:

$$U^T u = e_r, \qquad L^T v = u. \tag{7.20}$$

Then calculate by

$$w_N = N^T v. \tag{7.21}$$

The pivot column $\bar{a}_q = B^{-1}a_q$ can be obtained by solving the following two triangular systems:

$$Lu = a_q, \qquad U\bar{a}_q = u. \tag{7.22}$$

It is straightforward to transform Algorithm 7.4.1 into the revised version below.

Algorithm 7.6.1 (Dual Reduced Face Algorithm)

Initial : partition $B.N$, with b as the rth column of B, factorization $B = LU$, and dual face solution $\bar{z} \geq 0$. This algorithm solves the standard problem:

1. *Solve two triangular systems (7.20), and compute $w_N = N^T v$.*
2. *Stop if $J = \{j \in N \mid w_j > 0\}$ is empty (infeasible problem).*
3. *Determine β and column index q such that $\beta = \bar{z}_q/w_q = \min_{j\in J} \bar{z}_j/w_j$.*
4. *If $\beta > 0$, update $\bar{z}_N = \bar{z}_N - \beta w_N,\ \bar{f} = \bar{f} + \beta$.*
5. *Solve two triangular systems (7.22) for $\bar{a}_q$.*
6. *Select row index $p \in \arg\max \{\bar{a}_{i\,q} \mid r \neq i = 1, \ldots, m\}$.*
7. *If $\bar{a}_{p,q} \leq 0$, then:*
   (1) *Reselect row index $p \in \arg\max \{-\bar{a}_{i\,q}\bar{z}_{j_i} \mid r \neq i = 1, \ldots, m\}$*
   (2) *Stop if $-\bar{a}_{p\,q}\bar{z}_{j_p}$ vanishes (optimality achieved).*
8. *Adjust (N, B) by exchanging q and j_p and update LU factors.*
9. *Go to Step 1.*

Example 7.6.1 Solve the following standard problem by Algorithm 7.6.1:

$$A = \begin{bmatrix} 5 & 8 & 3 & 14 & -4 & -5 & 9 & 0 \\ -9 & 0 & 4 & -3 & -7 & -8 & -12 & 4 \\ 0 & 3 & -8 & 2 & -5 & 2 & -6 & 5 \\ 2 & -9 & 0 & -4 & -2 & 22 & 5 & 3 \end{bmatrix} \qquad b = \begin{bmatrix} -6 \\ 13 \\ -9 \\ -8 \end{bmatrix}$$

$c = [2, 1, 7, 0, 1, 2, 0, 1]^T$

Initial: $a_9 = [-6, 13, -9, -8]^T,\ N = \{1, 2, 3, 4, 5\},\ B = \{6, 7, 8, 9\},\ r = 4$
$\bar{z} = [2, 1, 7, 0, 1, 2, 0, 1]^T,\ \bar{f} = 0.$

$$L = \begin{bmatrix} -5/22 & -223/224 & 1 & 0 \\ -4/11 & 1 & 0 & 0 \\ 1/11 & 71/112 & 6/23 & 1 \\ 1 & 0 & 0 & 0 \end{bmatrix} \quad U = \begin{bmatrix} 22 & 5 & 3 & -8 \\ 0 & -112/11 & 56/11 & 111/11 \\ 0 & 0 & 23/4 & 499/224 \\ 0 & 0 & 0 & -4926/323 \end{bmatrix}.$$

Iteration 1: $\bar{z}_N = [2, 1, 7, 0, 1]^T$

1. $u = [0, 0, 0, -323/4926]^T,\ v = [336/19{,}643, 106/1809, -323/4926, 83/2664]^T,$

   $w = N^T v = [-367/967, -621/1825, 679/838, -645/3358, -395/1854]^T.$

2. $J = \{3\}$.
3. $\beta = \min[\bar{z}_3/w_3] = 838/97, \ q = 3$.
4. $\bar{z}_N = \bar{z}_N - \beta w = [2935/556, 1241/315, 0, 609/367, 1889/665]^T$,
   $\bar{f} = \bar{f} + \beta = 838/97$.
   $\bar{z} = [2935/556, 1241/315, 0, 609/367, 1889/665, 2, 0, 1]^T$.
5. $u = [0, 4, 391/56, -173/14]^T$, $\bar{a}_3 = [-107/4520, 1300/1511, 1211/1345, 679/838]^T$.
6. $\max[-107/4520, 1300/1511, 1211/1345] = 1211/1345 > 0, \ p = 3, j_p = 8$.
8. Adjust $N = \{1, 2, 4, 5, 8\}, \ B = \{6, 7, 3, 9\}$

$$L = \begin{bmatrix} -5/22 & -223/224 & -391/590 & 1 \\ -4/11 & 1 & 0 & 0 \\ 1/11 & 71/112 & 1 & 0 \\ 1 & 0 & 0 & 0 \end{bmatrix}$$

$$U = \begin{bmatrix} 22 & 5 & 0 & -8 \\ 0 & -112/11 & 4 & 111/11 \\ 0 & 0 & -295/28 & -1643/112 \\ 0 & 0 & 0 & -1896/253 \end{bmatrix}.$$

Iteration 2: $\bar{z}_N = [2935/556, 1241/315, 609/367, 1889/665, 1]^T$

1. $u = [0, 0, 0, -253/1896]^T$, $v = [-253/1896, -297/3868, -159/1798, -12/239]^T$,
   $w = N^T v = [-403/5264, -725/823, -3013/1867, 3013/1867, -1133/1259]^T$.
2. $J = \{5\}$.
3. $\beta = \min[\bar{z}_5/w_5] = 1211/688, \ q = 5$.
4. $\bar{z}_N = \bar{z}_N - \beta w = [720/133, 1691/308, 9/2, 0, 1261/488]^T$,

   $\bar{f} = \bar{f} + \beta = 3151/303$,
   $\bar{z} = [720/133, 1691/308, 0, 9/2, 0, 2, 0, 1261/488]^T$,

5. $u = [-2, -85/11, 9/112, -2443/202]^T, \bar{a}_5 = [159/986, 1960/1331, -2054/911, 3013/1867]^T$.
6. $\max[159/986, 1960/1331, -2054/911] = 1960/1331 > 0, \ p = 2, j_p = 7$.
8. Adjust $N = \{1, 2, 4, 7, 8\}, \ B = \{6, 5, 3, 9\}$

$$L = \begin{bmatrix} -5/22 & 49/85 & -59/892 & 1 \\ -4/11 & 1 & 0 & 0 \\ 1/11 & 53/85 & 1 & 0 \\ 1 & 0 & 0 & 0 \end{bmatrix}$$

$$U = \begin{bmatrix} 22 & -2 & 0 & -8 \\ 0 & -85/11 & 4 & 111/11 \\ 0 & 0 & -892/85 & -1238/85 \\ 0 & 0 & 0 & -6511/446 \end{bmatrix}.$$

Iteration 3: $z_N = [720/133, 1691/308, 9/2, 0, 1261/488]^T$

1. $u = [0, 0, 0, -446/6511]^T$, $v = [-446/6511, 551/13{,}022, -59/13{,}022, 3/13{,}022]^T$,

$$w = N^T v = [-699/967, -819/1453, -617/563, -617/563, 489/3320]^T.$$

2. $J = \{8\}$.
3. $\beta = \min[\bar{z}_8/w_8] = 9807/559, \ q = 8$.
4. $\bar{z}_N = \bar{z}_N - \beta w = [10{,}079/557, 1907/124, 5813/245, 3480/181, 0]^T$,

$$\bar{f} = \bar{f} + \beta = 2459/88,$$
$$\bar{z} = [10{,}079/557, 1907/124, 0, 5813/245, 0, 2, 3480/181, 0]^T,$$

5. $u = [3, 56/11, 132/85, -959/446]^T, \bar{a}_8 = [309/2360, -2216/3415, -887/2517, 489/3320]^T$.
6. $\max[309/2360, -2216/3415, -887/2517] = 309/2360 > 0, \ p = 1, j_p = 6$.
8. Adjust $N = \{1, 2, 4, 6, 7\}, \ B = \{8, 5, 3, 9\}$

$$L = \begin{bmatrix} 0 & 1 & 0 & 0 \\ 4/5 & 3/4 & 1 & 0 \\ 1 & 0 & 0 & 0 \\ 3/5 & -1/4 & 111/163 & 1 \end{bmatrix} \quad U = \begin{bmatrix} 5 & -5 & -8 & -9 \\ 0 & -4 & 3 & -6 \\ 0 & 0 & 163/20 & 247/10 \\ 0 & 0 & 0 & -3410/163 \end{bmatrix}.$$

Iteration 4: $\bar{z}_n = [10{,}079/557, 1907/124, 5813/245, 2, 3480/181]^T$

1. $u = [0, 0, 0, -163/3410]^T$, $v = [-2/55, 111/3410, 9/3410, -163/3410]^T$,

$$w = N^T v = [-389/682, 251/1705, -1399/3410, -1918/1705, -107/110]^T.$$

2. $J = \{2\}$.
3. $\beta = \min[\bar{z}_2/w_2] = 14{,}312/137, \ q = 2$.
4. $\bar{z}_N = \bar{z}_N - \beta w = [12{,}429/160, 0, 7391/111, 6693/56, 5438/45]^T$,

$$\bar{f} = \bar{f} + \beta = 5164/39,$$
$$\bar{z} = [12{,}429/160, 0, 0, 7391/11, 0, 6693/56, 5438/45, 0]^T.$$

5. $u = [3, 8, -42/5, -502/163]^T$, $\bar{a}_2 = [-695/144, -1135/341, -765/518, 251/1705]^T$.
6. $\max[-695/144, -1135/341, -765/518] = -765/518 \leq 0, \ p = 3, j_p = 3$.
7.(1) Reselect row index $\max[(695/144)0, (1135/341)0, (765/518)0] = 0$.
(2) Optimality achieved. Calculate the basics of the primal optimal solution below:

$$\bar{x}_2 = 1/\bar{a}_{4,2} = 1705/251, \ \bar{x}_8 = -(1705/251)(-695/144) = 2590/79.$$
$$\bar{x}_5 = -(1705/251)(-1135/341) = 5675/251, \ \bar{x}_3 = -(1705/251)(-765/518) = 3461/345.$$

The primal and dual optimal solutions:

$$\bar{x} = [0, 1705/251, 3461/345, 0, 5675/251, 0, 0, 2590/79]^T,$$
$$\bar{z} = [12{,}429/160, 0, 0, 7391/11, 0, 6693/56, 5438/45, 0]^T, \bar{f} = 5164/39.$$

7.7 Generalized Dual Reduced Face Algorithm

This section is devoted to the generalization of the dual reduced face method. The derived method is of particular interest because it can get itself started directly without any phase I procedure.

To solve the variable-bounded problem (3.1), we handle its dual problem (1.37), which can be rewritten as

$$\begin{array}{ll} \max & g = b^{\mathrm{T}} y + u_{\Gamma}^{\mathrm{T}} z_{\Gamma} + l_{\Pi}^{\mathrm{T}} z_{\Pi} \\ \text{s.t.} & A^{\mathrm{T}} y + z = c. \end{array} \tag{7.23}$$

where $\Gamma = \{j \mid z_j < 0,\ j = 1, \ldots, n\}$, $\quad \Pi = \{j \mid z_j > 0,\ j = 1, \ldots, n\}$.

Let $[A, b] = (B, N)$ be a partition with basis B, including the right-hand side b as its rth column, i.e.,

$$B = [a_{j_1}, \ldots, a_{j_{r-1}}, b, a_{j_{r+1}}, \ldots, a_{j_m}].$$

Assume that B is partitioned as

$$B = (B', b)$$

so that

$$B^{-1}[N, B', b] = [\bar{N}, I', e_r],$$

where $I' = [e_1, \ldots, e_{r-1}, e_{r+1}, e_m]$.

It is not difficult to get a dual solution $\bar{z}$ such that

$$A^T y + \bar{z} = c.$$

with $\bar{z}_{B'} = 0$ (see below).

Accordingly, the dual reduced face tableau can be expressed as

$$\begin{array}{cc|c} x_N & x_{B'} & rhs \\ \hline \bar{N} & I' & e_r \\ \hline \bar{z}_N^{\mathrm{T}} & & = \bar{y}_r \end{array} \tag{7.24}$$

with the rth row, termed *dual objective row*, giving the search direction.

Denote the rth row by w^T. But only the following part will be used:

$$w_N^T = e_r^T \bar{N}.$$

Assume for some $q \in N$ it holds that

$$w_q \neq 0, \qquad \bar{z}_q = 0.$$

Introduce the following column and row index sets:

$$N' = N\backslash\{q\}, \qquad R' = \{1, \ldots, m\}\backslash\{r\}. \tag{7.25}$$

To determine a primal solution $\bar{x}$, set $\bar{x}_{N'}$ by

$$\bar{x}_j = \begin{cases} l_j, & \text{if } \bar{z}_j \geq 0 \\ u_j, & \text{if } \bar{z}_j < 0 \end{cases} \qquad j \in N'. \tag{7.26}$$

Then $\bar{x}_q$ has to be determined to satisfy the rth equation, i.e., $w_N^T \bar{x}_N = 1$. Therefore, we set

$$\bar{x}_q = (1 - w_{N'}\bar{x}_{N'})/w_q. \tag{7.27}$$

The basics can be calculated by

$$\bar{x}_{B'} = -N_{R'}\bar{x}_N. \tag{7.28}$$

Then, it is verified that $\bar{x} = (\bar{x}_N, \bar{x}_{B'}))$ is a primal solution, satisfying the system $Ax = b$.

7.7.1 Row Pivoting and Optimality Condition

What to do next is to check whether or not optimality is achieved, in conjunction with the pivot row selection.

Introduce the following deviations:

$$\rho_j = \begin{cases} l_j - \bar{x}_j & \text{if } l_j > \bar{x}_j \\ u_j - \bar{x}_j & \text{if } u_j < \bar{x}_j \\ 0 & \text{otherwise} \end{cases} \qquad j \in B' \cup \{q\}, \tag{7.29}$$

where $B' = \{j_1, \ldots, j_{r-1}, j_{r+1}, \ldots, j_m\}$.

Then, the following rule is logical.

Rule 7.7.1 (Dual Row Rule) Select the row index $p \in R'$ such that

$$p \in \arg\max_{i \in R'} |\rho_{j_i}|, \tag{7.30}$$

where ρ_{j_i} is defined by (7.29).

Obviously, $\rho_{j_p} = 0$ implies that all basics are within bounds, i.e.,

$$l_{B'} \leq \bar{x}_{B'} \leq u_{B'}.$$

Using the preceding notation, the following result gives the optimality condition.

Theorem 7.7.1 *Assume that the primal solution $\bar{x}$ is determined by (7.26)–(7.28). If both ρ_{j_p} and ρ_q vanish, then $(\bar{x}, \bar{z})$ are primal and dual optimal solutions, attaining the optimal objective value below:*

$$\bar{f} = \bar{y}_r + \bar{z}^T \bar{x}. \tag{7.31}$$

Proof Note that $\bar{z}_{j_p}$ is zero. It is obvious that the primal solution $\bar{x}$ satisfies the equality constraint and is complementary with the dual solution $\bar{z}$. In addition, $\bar{x}$ satisfies the bound constraint. Therefore, the optimality condition is fulfilled, and hence $(\bar{x}, \bar{z})$ are primal and dual optimal solutions. The equation corresponding to the bottom row of the optimal tableau gives the expression (7.31). □

Assume that the optimality condition is not satisfied.
There are two cases arising:

(i) $\rho_{j_p} \neq 0$.

In this case, make the basis change by bringing j_p from B' to N and q from N to B'.

It might as well again denote the new dual reduced face tableau by (7.24) and the rth row by $w_N^T = e_r^T \bar{N}$. Note that the leaving index j_p now belongs to the new N, so denote it by $q' = j_p$ instead.

(ii) $\rho_{j_p} = 0$.

Clearly, this case implies $\rho_q \neq 0$. No basis change is made, but set $q' = q$.

To make further progress, introduce the index set below:

$$\begin{aligned} J = &\{q' \neq j \in N \mid \text{sign}(\rho_{q'} w_{q'}) w_j < 0, \ \bar{x}_j = l_j\} \\ &\cup \{q' \neq j \in N \mid \text{sign}(\rho_{q'} w_{q'}) w_j > 0, \bar{x}_j = u_j\}. \end{aligned} \tag{7.32}$$

Lemma 7.7.2 *Under the same assumptions of Theorem 7.7.1, if $\rho_{q'}$ given by (7.29) is nonzero and J is empty, no feasible solution exists.*

Proof Introduce the index sets

$$J_1 = \{q' \neq j \in N \mid \bar{x}_j = l_j\}, \qquad J_2 = \{q' \neq j \in N \mid \bar{x}_j = u_j\},$$

from which and $\bar{x}_N$ satisfying the rth equation, it follows that

$$1 = w_N^T \bar{x}_N = w_{q'}\bar{x}_{q'} + \sum_{j\in J_1} w_j l_j + \sum_{j\in J_2} w_j u_j. \tag{7.33}$$

In addition, emptiness of J implies that

$$\text{sign}(\rho_{q'} w_{q'})w_j \geq 0, \quad \forall\, j \in J_1, \qquad \text{sign}(\rho_{q'} w_{q'})w_j \leq 0, \quad \forall\, j \in J_2. \tag{7.34}$$

Assume that $\hat{x}$ is a feasible solution. Since $l_N \leq \hat{x}_N \leq u_N$, it holds that

$$\sum_{j\in J_1} (\text{sign}(\rho_{q'} w_{q'})w_j) l_j \leq \sum_{j\in J_1} (\text{sign}(\rho_{q'} w_{q'})w_j)\hat{x}_j$$

and

$$\sum_{j\in J_2} (\text{sign}(\rho_{q'} w_{q'})w_j) u_j \leq \sum_{j\in J_2} (\text{sign}(\rho_{q'} w_{q'})w_j)\hat{x}_j$$

Assume that $\text{sign}(\rho_{q'} w_{q'}) = +1$.
Then, adding the preceding two inequalities gives

$$\sum_{j\in J_1} w_j l_j + \sum_{j\in J_2} w_j u_j \leq \sum_{q'\neq j\in N} w_j \hat{x}_j, \tag{7.35}$$

If $\rho_{q'} > 0$, then from $w_{q'} > 0$ and $\hat{x}_{q'} \geq l_{q'}$, it follows that

$$w_{q'}\bar{x}_{q'} = w_{q'}(l_{q'} - \rho_{q'}) < w_{q'}\hat{x}_{q'}$$

If $\rho_{q'} < 0$, from $w_{q'} < 0$ and $\hat{x}_{q'} \leq u_{q'}$, it follows that

$$w_{q'}\bar{x}_{q'} = w_{q'}(u_{q'} - \rho_{q'}) < w_{q'}\hat{x}_{q'},$$

from which, (7.33) and (7.35), it follows that

$$1 = w_{q'}\bar{x}_{q'} + \sum_{j\in J_1} w_j l_j + \sum_{j\in J_2} w_j u_j \leq w_{q'}\bar{x}_{q'} + \sum_{q\neq j\in N} w_j \hat{x}_j < w_{q'}\hat{x}_{q'} + \sum_{q\neq j\in N} w_j \hat{x}_j, \tag{7.36}$$

The preceding means that $\hat{x}$ violates the rth equation, as contradicts that it is a feasible solution.

The same thing is derived in the case of $\text{sign}(\rho_{q'} w_{q'}) = -1$. Therefore, there is no feasible solution to the problem. □

Now, assume that $\rho_{q'}$ is nonzero but J is nonempty,
Determine β and q such that

$$\beta = -\bar{z}_q/w_q, \qquad q \in \arg\min_{j\in J} |\bar{z}_j/w_j|. \tag{7.37}$$

and add β times of the rth row to the bottom row, or in other words, update the dual solution and $\bar{y}_r$ by

$$\bar{z}_N = \bar{z}_N + \beta w_N, \qquad \bar{y}_r = \bar{y}_r - \beta. \tag{7.38}$$

It is clear that the new $\bar{z}_q$ vanishes and signs of $\bar{z}_N$ keep unchanged.

Reset the deviation of $\bar{x}_{q'}$ such that

$$\rho_{q'} = \begin{cases} l_{q'} - \bar{x}_{q'} & \text{if } \bar{z}_{q'} > 0, \\ u_{q'} - \bar{x}_{q'} & \text{if } \bar{z}_{q'} < 0. \end{cases} \tag{7.39}$$

It is notable that the preceding equation shares the same sign with its predecessor. Therefore, updating

$$\bar{x}_{q'} = \bar{x}_{q'} + \rho_{q'}, \tag{7.40}$$

brings it on the "right" bound. But $\bar{x}_q$ should be changed accordingly to maintain the rth equation satisfied, i.e.,

$$w_N^T \bar{x}_N = 1. \tag{7.41}$$

Denoting by δ_q the corresonding variation of $\bar{x}_q$, the new value is then

$$\bar{x}_q = \bar{x}_q + \delta_q,$$

in which substituting (7.40) into (7.41) gives

$$(\bar{x}_q + \delta_q)w_q + (\bar{x}_{q'} + \rho_{q'})w_{q'} + \sum_{j\in N, j\neq q,q'} w_j \bar{x}_j = 1.$$

Subtracting (7.41) from the preceding equation leads to

$$\delta_q w_q + \rho_{q'} w_{q'} = 0,$$

which implies that

$$\delta_q = -\rho_{q'}(w_{q'}/w_q). \tag{7.42}$$

Consequently, we have the following update:

$$\bar{x}_q = \bar{x}_q - \rho_{q'}(w_{q'}/w_q). \tag{7.43}$$

The update of basics is derived similarly, i.e.,

$$\hat{x}_{B'} = \bar{x}_{B'} - \delta_q \bar{a}_{R',q} - \rho_{q'} \bar{a}_{R',q'},$$

which along with (7.42) gives

$$\bar{x}_{B'} = \bar{x}_{B'} + \rho_{q'}((w_{q'}/w_q)\bar{a}_{R',q} - \bar{a}_{R',q'}). \tag{7.44}$$

where $\bar{a}_{R',q}/\bar{a}_{R',q'}$ are the q/q'th columns, excluding the rth entries. If $q' = j_p$, the latter yields from performing the elementary eliminations on e_p and hence can be calculated from the previous entering column, say d, i.e.,

$$\bar{a}_{R',q'} = d' = -d/d_p + (1 + 1/d_p)e_p. \tag{7.45}$$

7.7.2 Generalized Dual Reduced Face Algorithm I

Now, we can formulate the dual reduced face algorithm in the tableau form as well as in the revised form.

Let $A = (B, N))$ be a partition with the normal basis B. Assume that the normal simplex tableau is

$$[I, \bar{N} \mid \bar{b}].$$

To obtain the initial dual solution, compute

$$\bar{z}_N = c_N - N^T \bar{y}, \qquad B^T y = c_B, \qquad \bar{z}_B = 0.$$

Select any index j_r with $\bar{b}_r \neq 0$ to leave the basis and enter the right-hand side to the basis, that is, take $\bar{b}_r$ as the pivot to convert the simplex tableau to the dual reduced face tableau with the rth dual objective row and the column index $q = j_r$ with $\bar{z}_q = 0$.

The initial primal solution $\bar{x}$ is computed by (7.26)–(7.28). Subsequently, it is updated in each iteration.

The overall steps are summarized in the following algorithm.

Algorithm 7.7.1 (Generalized Dual Reduced Face Algorithm I: Tableau Form)

Initial: dual reduced face tableau in the form (7.24) with r, q, and $\bar{x}$. This algorithm solves the variable-bounded problem:

1. *Select the row index p by (7.30).*
2. *Stop if both ρ_{j_p} and ρ_q vanish.*
3. *If $\rho_{j_p} = 0$, set $q' = q$ and go to Step 6.*
4. *Set $q' = j_p$, and adjust $B' = B' \backslash \{q'\} \cup \{q\}$, $N = N \backslash \{q\} \cup \{q'\}$.*

5. *Multiply the pth row by* $1/\bar{a}_{p,q}$ *and add* $-\bar{a}_{i,q}$ *times the row to the ith row for* $p \neq i \in R'$.
6. *Stop if set J defined by (7.32) is empty.*
7. *Determine dual stepsize* β *and the entering index q by (7.37).*
8. *If* $\beta \neq 0$, *add* β *times of the rth row to the bottom row.*
9. *Determine* $\rho_{q'}$ *by (7.39) and set* $\bar{x}_{q'} = \bar{x}_{q'} + \rho_{q'}$.
10. *Compute* $\bar{x}_q$ *by (7.43) and* $\bar{x}_{B'}$ *by (7.44)–(7.45).*
11. *Go to Step 1.*

Theorem 7.7.3 *Algorithm 7.7.1 creates a sequence of dual feasible solutions. If it terminates, it does either at*

(i) Step 2, giving a pair of primal and dual optimal solutions, or at
(ii) Step 4, detecting infeasibility of the problem.

Proof The validity comes from Theorems 7.7.1 and 7.7.2. □

Note: A remarkable feature of the algorithm is that there is no need for phase I process. It starts from any dual reduced face tableau to achieve optimality by pursuing primal feasibility while maintaining complementarity.

To get the revised version, let B be the basis corresponding to the current dual reduced face tableau and its LU factorization be available, i.e.,

$$B = LU.$$

The following items are to be computed in conjunction with systems:
(a) The rth dual objective row $w_N^T = e_r^T B^{-1} N$.
Solve the following two triangular systems first:

$$U^T u = e_r, \qquad L^T v = u, \tag{7.46}$$

and compute

$$w_N^T = v^T N, \tag{7.47}$$

(b) The pivot column $\bar{a}_q = B^{-1} a_q$.
Solve the following two triangular systems:

$$Lu = a_q, \qquad U\bar{a}_q = u, \tag{7.48}$$

(c) Updating of the basics $\bar{x}_{B'}$.
The associated formula (7.44) involves two columns. The current pivot column $\bar{a}_q$ is obtained via (b).

For case (i) where the basis change was made, column $\bar{a}_{q'}$ yields from performing the elementary elimination on e_p. This results in the following formula using the previous pivot column, denoted by d:

$$\bar{a}_{q'} = -d/d_p + (1 + 1/d_p)e_p. \tag{7.49}$$

For case (ii) without basis change, set $\bar{a}_{q'} = d$.

To get ready, we perform the following steps first:

preprocessing:

1. Solve the two triangular systems (7.46) and compute w_N^T by (7.47).
2. Determine primal solution $\bar{x}$ by (7.26)–(7.28).
3. Calculate pivot column $d = \bar{a}_{R',q}$.

Then, the following revised form of Algorithm 7.7.1 is applicable.

Algorithm 7.7.2 (Generalized Dual Reduced Face Algorithm I)

Initial: partition N, B, with b as the rth column of B, $B = LU$, $B' = B\backslash\{b\}$, w, $\bar{z}$, and $\bar{x}$. This algorithm solves the variable-bounded problem:

1. *Select the row index p by (7.30).*
2. *Stop if both ρ_{j_p} and ρ_q vanish (optimality achieved).*
3. *If $\rho_{j_p} = 0$, set $q' = q$, $\rho_{q'} = \rho_q$ and go to Step 7.*
4. *Set $q' = j_p$, and adjust sets 4=$B' = B'\backslash\{q'\} \cup \{q\}$, $N = N\backslash\{q\} \cup \{q'\}$.*
5. *Update LU factors.*
6. *Solve the two triangular systems (7.46) and calculate w_N by (7.47).*
7. *Stop if set J defined by (7.32) is empty (infeasibility detected).*
8. *Determine dual stepsize β and the entering index q by (7.37).*
9. *If $\beta \neq 0$, update $\bar{z}_N = \bar{z}_N + \beta w_N$, $\bar{y}_r = \bar{y}_r - \beta$.*
10. *Determine $\rho_{q'}$ by (7.39) and set $\bar{x}_{q'} = \bar{x}_{q'} + \rho_{q'}$.*
11. *Solve the two triangular systems (7.48) for $\bar{a}_q$.*
12. *If $\rho_{j_p} \neq 0$, compute $\bar{a}_{q'}$ by (7.49); else set $d = \bar{a}_{R;,q'}$.*
13. *Update $\bar{x}_q$ by (7.43) and $\bar{x}_{B'}$ by (7.44)–(7.45).*
14. *Go to Step 1.*

Example 7.7.1 Solve the following variable-bounded problem by Algorithm 7.7.2:

$$A = \begin{bmatrix} 4 & -6 & 3 & 8 & 3 & 1 & -5 & 1 \\ 2 & -11 & -4 & 5 & -2 & 0 & & \\ -6 & 3 & & 3 & -6 & -3 & 3 & \\ -2 & 0 & 2 & -5 & 1 & 0 & -1 & 2 \end{bmatrix} \qquad b = \begin{bmatrix} 26 \\ -7 \\ -21 \\ 5 \end{bmatrix}$$

$c = [0, 0, 0, 2, -4, 3, 0, 5]^T$
$u = [14, 7, 10, 6, 9, 12, 11, 5]^T$
$l = [-8, 0, -6, -7, -5, 0, -3, -8^T$

Initial: $B = \{1, 2, 3, 9\}$, $B' = \{1, 2, 3\}$, $N = \{4, 5, 6, 7, 8\}$, $R' = \{1, \ldots, m - 1\} r = m, q = 7 \bar{z} = c = [0, 0, 0, 2, -4, 3, 0, 5]^T$.

Gauss-Jordan elimination converts the system to the dual reduced face tableau below:

$$[\bar{A} \mid e_m] = \left[\begin{array}{cccccccc|c} 1 & & & 725/97 & -130/97 & -223/388 & -335/388 & -121/97 & \\ & 1 & & -154/97 & 43/97 & 19/97 & -18/97 & -4/97 & \\ & & 1 & 1090/97 & -259/97 & -137/97 & -151/97 & -109/97 & \\ & & & -243/97 & 71/97 & 65/194 & 15/194 & 34/97 & 1 \end{array}\right]$$

$$\text{LU factors of } B\text{: } L = \begin{bmatrix} -2/3 & 2/5 & 1 & \\ -1/3 & 1 & & \\ 1 & & & \\ 1/3 & 1/10 & 12/23 & 1 \end{bmatrix} \quad U = \begin{bmatrix} -6 & 3 & & -21 \\ & -10 & -4 & -14 \\ & & 23/5 & 88/5 \\ & & & 97/23 \end{bmatrix}.$$

1. $u = [0, 0, 0, 23/97]^T$, $v = [-12/97, 5/194, -89/582, 23/97]^T$, $w_N = [-243/97, 71/97, 65/194, 15/194, 34/97]^T$,
2. $\bar{x}_N = [-7, 9, 0, -3, -8]$, $\bar{x}_N(4) = \bar{x}_7 = -1314/5$.
   $-N\bar{x}_N = [-1277, 8149/5, 4317/5, -1454/5]^T$, $u = [4317/5, 9588/5, -36{,}711/25, -97/23]^T$,
   $\bar{x}_{B'} = [-345/2, -321/5, -1577/5]^T$, $\bar{x} = [-345/2, -321/5, -1577/5, -7, 9, 0, -1314/5, -8]^T$.
3. .. $u = [3, 7, -29/5, 15/46]^T$, $d = \bar{a}_{R',7} = [-335/388, -18/97, -151/97]^T$.

Call Algorithm 7.7.2.

Iteration 1:

1. $\max |\rho_{B'}| = \max [|-8-(-345/2)|, |0-(-321/5)|, |-6-(-1577/5)|] = 1547/5$, $p = 3, j_p = 3$.
4. $q' = 3$, $N = \{4, 5, 6, 3, 8\}$, $B' = \{1, 2, 7\}$, $\bar{x}_(B')(3) = \bar{x}_7$, $\bar{x}_N(4) = \bar{x}_3$.
5. LU factors of B:

$$L = \begin{bmatrix} -2/3 & 2/5 & 1 & \\ -1/3 & 1 & & \\ 1 & & & \\ 1/3 & 1/10 & 27/58 & 1 \end{bmatrix} \quad U = \begin{bmatrix} 6 & -11 & & -7 \\ & -91/6 & 1 & 121/6 \\ & & 171/91 & 127/91 \\ & & & 377/57 \end{bmatrix}.$$

6. $u = [0, 0, 0, 29/151]^T$, $v = [-27/302, 5/302, -107/906, 29/151]^T$.
   $w_N = [-294/151, 181/302, 40/151, 15/302, 89/302]^T$.
7. $\rho_3 w_N(4) > 0$, $J = \{4, 5\}$.
8. $\theta = \min [|2/(-294/151)|, |-4/(181/302)|] = 151/147$, $q = 4$, $\beta = \theta$.
9. $\bar{z}_N = [2, -4, 3, 0, 5]^T + \beta w_N = [0, -995/294, 481/147, 5/98, 1559/294]^T$,
   $\bar{y}_m = -151/147$.
10. $\bar{z}_N(4) > 0$, $\rho_3 = -6 - \bar{x}_3 = 1547/5$, $\bar{x}_N(4) = \bar{x}_3 = -6$.
11. $u = [3, 6, 38/5, -294/29]^T$, $\bar{a}_{R',4} = [375/302, -442/151, -1090/151]^T$.
12. $\bar{a}_{R',3} = -d/d(3) + (1 + 1/d(3))e_3 = [-335/604, -18/151, -97/151]^T$.
13. $\bar{x}_N(1) = \bar{x}_4 = \bar{x}_4 - \rho_3(w_N(4)/w_N(1)) = -7 - (1547/5)((15/302)/(-294/151)) = 25/28$,

    $\bar{x}_{B'} = [-345/2, -321/5, -1314/5]^T + \rho_3((15/302)/(-294/151))\bar{a}_{R',4} - \bar{a}_{R',3}) = [-599/56, -59/14, -99/14]^T$,

$\bar{x} = [-599/56, -59/14, -6, 25/28, 9, 0, -99/14, -8]^T$. $d = \bar{a}_{R',4}$

Iteration 2:

1. $\max |\rho_{B'}| = \max [|-8-(-599/56)|, |0-(-59/14)|, |-3-(-99/14)|] = 59/14,\ p = 2, j_p = 2$.
4. $q' = 2,\ N = \{2, 5, 6, 3, 8\},\ B' = \{1, 4, 7\}, \bar{x}_{B'}(2) = \bar{x}_4,\ \bar{x}_N(1) = \bar{x}_2$.
5. LU factors of B:

$$L = \begin{bmatrix} -2/3 & 1 & & \\ -1/3 & 3/5 & 1 & \\ 1 & & & \\ 1/3 & -3/5 & -19/44 & 1 \end{bmatrix} \qquad U = \begin{bmatrix} -6 & 3 & 3 & -21 \\ & 10 & -3 & 12 \\ & & 44/5 & -106/5 \\ & & & 221/22 \end{bmatrix}.$$

6. $u = [0, 0, 0, 22/221]^T,\ v = [15/442, 19/442, 5/1326, 22/221]^T,$
   $w_N = [-147/221, 41/442, 5/221, 57/442, 103/442]^T$.
7. $\rho_2 w_N(1) < 0,\ J = \{6, 3, 8\}$.
8. $\theta = \min [|(481/147)/(5/221)|, |(5/98)/(57/442)|, |(1559/294)/(103/442)|] = 1105/2793, q = 3, \beta = -\theta$.
9. $\bar{z}_N = [0, -995/294, 481/147, 5/98, 1559/294]^T + \beta w_N = [5/19, -65/19, 62/19, 0, 99/19]^T, \bar{y}_m = -151/147 + 1105/2793 = -12/19$.
10. $\bar{z}_N(1) > 0,\ \rho_2 = 0 - (-59/14) = 59/14,\ \bar{x}_N(1) = \bar{x}_2 = 0$.
11. $u = [0, 3, -29/5, 57/44]^T,\ \bar{a}_{R',3} = [-535/884, 9/221, -77/221]^T$.
12. $\bar{a}_{R',2} = -d/d(2) + (1 + 1/d(2))e_2 = [375/884, -151/442, -545/221]^T$.
13. $\bar{x}_N(4) = \bar{x}_3 = \bar{x}_3 - \rho_2(w_N(1)/w_N(4)) = -0 - (299/19)((-147/221)/(57/442)) = 299/19$,
    $\bar{x}_{B'} = [-599/56, \mathbf{25/28}, -99/14]^T + \rho_2((-147/221)/(57/442)\bar{a}_{R',3} - \bar{a}_{R',2}) = [51/76, 55/38, 207/19]^T$,
    $\bar{x} = [51/76, 0, 299/19, 55/38, 9, 0, 207/19, -8]^T, d = \bar{a}_{R',3}$.

Iteration 3:

1. $\max |\rho_{B'}| = 0$.
3. $q' = 3,\ \rho_3 = 10 - 299/19 = -109/19$.
7. $\rho_3 w_N(4) = (-109/19)(57/442) < 0, \bar{x}_N = [0, 9, 0, 299/19, -8]^T,\ J = \{6, 8\}$.
8. $\theta = \min [|(62/19)/(5/221)|, |(99/19)/(103/442)|] = 3108/139, q = 8,\ \beta = -\theta$.
9. $\bar{z}_N = [5/19, -65/19, 62/19, 0, 99/19]^T + \beta[-147/221, 41/442, 5/221, 57/442, 103/442]^T = [1559/103, -566/103, 284/103, -297/103, 0]^T,\ \bar{y}_m = -12/19 + 3108/139 = 2238/103$.
10. $\bar{z}_N(4) < 0,\ \rho_3 = -109/19,\ \bar{x}_N(4) = \bar{x}_3 = 10$.
11. $u = [0, 1, -3/5, 103/44]^T,\ \bar{a}_{R',8} = [-517/884, -7/221, 109/221, 103/442]^T$.
12. $\bar{a}_{R',3} = d = [-535/884, 9/221, -77/221]^T$.
13. $\bar{x}_8 = \bar{x}_N(5) = -8 - (-109/19)((57/442)/(103/442)) = -497/103$,

$\bar{x}_{B'} = [51/76, 55/38, 207/19]^T + \rho_3((57/442)/(103/442)\bar{a}_{R',8} - \bar{a}_{R',3}) = [-389/412, 367/206, 755/103]^T$, $d = \bar{a}_{R',8}$.

Iteration 4:

1. $\rho_{B'} = 0, \qquad \rho_8 = 0.$
2. Optimality achieved. Primal and dual optimal solutions:
   $\bar{x} = [-389/412, 0, 10, 367/206, 9, 0, 755/103, -497/103]^T$, $\bar{f} = c^T\bar{x} = -5826/103$; $\bar{z} = [0, 1559/103, -566/103, 284/103, -297/103, 0]^T$, $\bar{g} = \bar{y}_m + \bar{z}^T\bar{x} = -5826/103.$

7.7.3 *Generalized Dual Reduced Face Algorithm II*

The preceding algorithms achieve feasibility of $\bar{x}_{B'}$ as priority, and then feasibility of $\bar{x}_q$. It is not difficult to derive another algorithm, achieving feasibility of $\bar{x}_q$ first. This seems to be advantageous because no basis change is involved before checking feasibility of $\bar{x}_{B'}$ in conjunction with the row pivoting.

The overall steps are put into the following algorithm.

Algorithm 7.7.3 (Generalized Dual Reduced Face Algorithm II: Tableau Form)

Initial: dual reduced face tableau with r, q, and $\bar{x}$. This algorithm solves the variable-bounded problem:

1. *Go to Step 9 if ρ_q defined by (7.29) vanishes.*
2. *Set $q' = q$ and $\rho_{q'} = \rho_q$.*
3. *Stop if J defined by (7.32) is empty.*
4. *Determine β and index q by (7.37).*
5. *If $\beta \neq 0$, add β times of the rth row to the bottom row.*
6. *Determine $\rho_{q'}$ by (7.39) and set $\bar{x}_{q'} = \bar{x}_{q'} + \rho_{q'}$.*
7. *Update $\bar{x}_q$ and $\bar{x}_{B'}$ by (7.43)–(7.44).*
8. *Go to Step 2 if ρ_q is nonzero.*
9. *Select the row index p by (7.30).*
10. *Stop if ρ_{j_p} vanishes.*
11. *Convert $\bar{a}_{p,q}$ to 1, and eliminate nonzeros in the column by elementary transformations.*
12. *Adjust (B', N) by exchanging j_p and q.*
13. *Set $q' = j_p$ and $\rho_{q'} = \rho_{j_p}$.*
14. *Go to Step 3.*

Theorem 7.7.4 *Algorithm 7.7.3 creates a sequence of basic feasible solutions. Under the dual nondegeneracy assumption, it terminates either at*

(i) Step 10, giving a pair of primal and dual optimal solutions, or at
(ii) Step 3, detecting infeasibility of the problem.

Proof The validity comes from Theorems 7.7.1 and 7.7.2. □

Using the same expressions and the preprocessing steps before Algorithm 7.7.2, Algorithm 7.7.3 can be transformed to the following revised version.

The update formula (7.44) can be written as

$$\bar{x}_{B'} = \bar{x}_{B'} + \rho_{q'}((w_{q'}/w_q)d - d'). \tag{7.50}$$

where $d = \bar{a}_{R',q}$ is the current q column and $d' = \bar{a}_{R',q'}$ is the predecessor, excluding the rth entries. As was indicated, if a basis change is involved, the latter can be calculated by

$$d' = -d/d_p + (1 + 1/d_p)e_p, \tag{7.51}$$

where d is the previous one. However, if no basis change is made, $d' =$ is itself the previous, as shown in the following revised version.

Algorithm 7.7.4 (Generalized Dual Reduced Face Algorithm II)

Initial : N, B with b as the rth column of B, $B = LU$. The preprocessing steps before Algorithm 7.7.2 were taken. This algorithm solves the variable-bounded problem:

1. *Go to Step 10 if ρ_q defined by (7.29) vanishes.*
2. *Set $q' = q$ and $\rho_{q'} = \rho_q$.*
3. *Stop if J defined by (7.32) is empty (infeasible problem).*
4. *Determine β and index q by (7.37).*
5. *If $\beta \neq 0$, update $\bar{z}_N$ and $\bar{y}_r$ by (7.38).*
6. *Determine $\rho_{q'}$ by (7.39) and set $\bar{x}_{q'} = \bar{x}_{q'} + \rho_{q'}$.*
7. *Solve systems $Lu = a_q$, $U\bar{a}_q = u$ and set $d = \bar{a}_{R',q}$.*
8. *Update $\bar{x}_q$ and $\bar{x}_{B'}$ by (7.50).*
9. *If ρ_q is nonzero, set $d' = d$ and go to Step 2.*
10. *Select the row index p by (7.30).*
11. *Stop if ρ_{j_p} vanishes (optimality is achieved).*
12. *Adjust $B' = B' \backslash \{j_p\} \cup \{q\}$, $N = N \backslash \{q\} \cup \{j_p\}$.*
13. *Update the factors L, U.*
14. *Solve systems $U^T u = e_r$, $L^T v = u$ and compute $w_N = N^T v$.*
15. *Set $q' = j_p$, $\rho_{q'} = \rho_{j_p}$ and d' by (7.51).*
16. *Go to Step 3.*

Example 7.7.2 Solve the following variable-bounded problem by Algorithm 7.7.4:

$$A = \begin{bmatrix} 8 & 12 & 4 & -6 & 1 & -6 & -3 & -5 & 3 \\ -6 & 6 & 3 & 4 & 7 & 1 & -6 & 3 & \\ -2 & -1 & 0 & 5 & 4 & & -2 & 0 & 2 \\ 9 & 4 & 1 & 0 & -8 & 3 & -5 & 7 & -4 \end{bmatrix} \qquad b = \begin{bmatrix} 8 \\ 12 \\ 6 \\ 7 \end{bmatrix}$$

$u = [6, 10, 5, 6, 12, 8, 4, 7, 15]^T$

$c = [0, 0, 0, 2, -5, -7, 0, 4, 2]^T$

$l = [-3, -8, -4, 0, -8, 0, -1, -4, -7]^T$

Initial: set $A = [A, b]$, $B = \{1, 2, 3, 10\}$, $B' = \{1, 2, 3\}$, $N = \{4, 5, 6, 7, 8, 9\} R' = \{1, \ldots, m-1\}, r = m, q = 7, \bar{z}_N = [2, -5, -7, 0, 4, 2]^T$,

LU factors of B: $L = \begin{bmatrix} 8/9 & 38/39 & 1 & 0 \\ -2/3 & 1 & 0 & 0 \\ -2/9 & -1/78 & -7/12 & 1 \\ 1 & 0 & 0 & 0 \end{bmatrix}$

$$U = \begin{bmatrix} 9 & 4 & 1 & 7 \\ 0 & 26/3 & 11/3 & 50/3 \\ 0 & 0 & -6/13 & -188/13 \\ 0 & 0 & 0 & -2/3 \end{bmatrix}.$$

Preprocessing:

1. $u = [0, 0, 0, -3/2]^T$, $v = [-7/8, 5/6, -3/21]^T$,
   $w_N = [13/12, -217/24, 109/12, -35/8, 111/8, -77/8]^T$.
2. $\bar{x}_N = [0, 12, 8, -1, -4, -7]^T$, $\bar{x}_N(4) = \bar{x}_7 = -599/105$.
   $-N\bar{x}_N = [696/35, -3998/35, -4768/105, 913/21]^T$,
   $u = [913/21, -3836/45, 5594/87, 2/3]^T$, $\bar{x}_{B'} = [86/105, 1322/35, -11{,}338/105]^T$,
   $\bar{x} = [86/105, 1322/35, -11{,}338/105, 0, 12, 8, -599/105, -4, -7]^T$.
3. $u = [-5, -28/3137/13, 35/12]^T$ $d' = [67/8, -41, 457/4, -35/8]^T$.

Call Algorithm 7.7.4.

Iteration 1:

1. $\rho_7 = -1 - (-599/105) = 494/105$.
2. $q' = 7$, $\rho_{q'} = 494/105$.
3. $\rho_7 w_N(4) < 0$, $\bar{x}_N = [0, 12, 8, -599/105, -4, -7]^T$, $J = \{4, 5, 8\}$.
4. $\theta = \min[|\bar{z}_N(1)/w_N(1)|, |\bar{z}_N(2)/w_N(2)|, |\bar{z}_N(5)/w_N(5)|] = 32/111 q = 8$, $\beta = -\bar{z}_N(5)/w_N(5) = -32/111$.
5. $\bar{z}_N = [2, -5, -7, 0, 4, 2]^T + (-32/111)[13/12, -217/24, 109/12, -35/8, 111/8, -77/8]^T = [562/333, -797/333, -3203/333, 140/111, 0, 530/111]^T$, $\bar{y}_m = -\beta = 32/111.8$
6. $\rho_7 = 494/105$, $x_N(4) = -1$.
7. $u = [7, 23/3, -243/13, -37/4]^T$, $d = \bar{a}_{R',8} = [-231/8, 141, -1577/4]^T$,
8. $x_N(5) = \bar{x}_8 = x_N(5) - \rho_7(w_N(4)/w_N(5)) = -838/333$,
   $\bar{x}_{B'} = \bar{x}_{B'} + \rho_7((w_N(4)/w_N(5))[-231/8, 141, -1577/4]^T - [67/8, -41, 457/4]^T) = [472/111, 2386/111, -20{,}192/333]^T$.
   $\bar{x} - [472/111, 2386/111, -20{,}192/3330, 12, 8, -1, -838/333, -7]^T$.
9. $\rho_8 = 0$.
10. $p = 3$, $j_p = 3$.
11. $\rho_3 = l(3) - \bar{x}(3) = 18{,}860/333$.
12. $B = \{1, 2, 8, 10\}$, $N = \{4, 5, 6, 7, 3, 9\}$, $\bar{x}_B(1) = \bar{x}_5$, $\bar{x}_N(2) = \bar{x}_1$.
13. LU factors of B:

$$L = \begin{bmatrix} 8/9 & 38/39 & 1 & \\ -2/3 & 1 & & \\ -2/9 & -1/78 & -43/486 & 1 \\ 1 & & & \end{bmatrix} \quad U = \begin{bmatrix} 9 & 4 & 7 & 7 \\ & 25/3 & 23/3 & 50/3 \\ & & -243/13 & -188/13 \\ & & & 1577/243 \end{bmatrix}.$$

14. $u = [0, 0, 0, 243/1577]^T$, $v = [43/3154, -107/9462, 243/1577, 23/1577]^T$, $w_N = [987/1534, 755/1739, -111/2249, -550/1553, 111/3154, 537/1847]^T$.
15. $q' = 3$, $\rho_{q'} = 18{,}860/333$, $d' = [-231/3154, 564/1577, -4/1577]^T$.

Iteration 2:

3. $\rho_3 w_N(5) > 0$, $\bar{x}_N = [0, 12, 8, -1, -20, 192/333, -7]^T$, $J = \{5, 7\}$.
4. $\theta = \min[|\bar{z}_N(2)/w_N(2)|, |\bar{z}_N(4)/w_N(4)|] = 958/269q = 7$, $\beta = -\bar{z}_N(4)/w_N(4) = 958/269$.
5. $\bar{z}_N = [562/333, -797/333, -3203/333, 140/111, 0, 530/111]^T + (958/269)w_N = [1524/383, -926/1093, -1048/107, 0, 140/1117, 3643/627]^T$, $\bar{y}_m = \bar{y}_m - \beta = -923/282$.
6. $\rho_3 = 18{,}860/333$, $x_N(5) = -4$.
7. $u = [-5, -28/3, 137/13, -1117/486]^T$, $d = \bar{a}_{R',7} = [23/3154, -220/1577, -457/1577]^T$,
8. $x_N(4) = \bar{x}_7 = x_N(4) - \rho_3(w_N(5)/w_N(4)) = 1643/355$,
   $\bar{x}_{B'} = \bar{x}_{B'} + \rho_3((w_N(5)/w_N(4))d - d') = [3327/398, 2262/1117, -707/953]^T$,
   $\bar{x} = [3327/398, 2262/1117, -4, 0, 12, 8, 1643/355, -707/953, -7]^T$.
9. $\rho_7 = u(7) - \bar{x}_(7) = -941/1498 \neq 0$, $d' = d$.
2. $q' = 7, \rho_{q'} = \rho_7$.
3. $\rho_7 w_N(4) = (-941/1498)(-550/1553) > 0$, $\bar{x}_N = [0, 12, 8, 1643/355, -4, -7]^T$, $J = \{5\}$.
4. $\theta = \min[|\bar{z}_N(2)/w_N(2)|] = 281/144$, $q = 5, \beta = -\bar{z}_N(2)/w_N(2) = 281/144$.
5. $\bar{z}_N = [1524/383, -926/1093, -1048/107, 0, 140/1117, 3643/627]^T + (281/144)w_N = [1963/375, 0, -1810/183, -830/1201, 797/4108, 625/98]^T$, $\bar{y}_m = \bar{y}_m - \beta = -2095/401$.
6. $\rho_7 = -941/1498$, $x_N(4) = u(7) = 4$.
7. $u = [-8, 5/3, 253/39, 2054/729]^T$, $d = \bar{a}_{R',5} = [-1128/1663, -182/4731, -1077/1577]^T$,
8. $x_N(2) = \bar{x}_5 = x_N(2) - \rho_7(w_N(4)/w_N(2)) = 5089/443$,
   $\bar{x}_{B'} = \bar{x}_{B'} + \rho_7((w_N(4)/w_N(2))d - d') = [1475/184, 303/158, -1642/1289]^T$,
   $\bar{x} = [1475/184, 303/158, -4, 0, 5089/443, 8, 4, -1642/1289]^T$.
9. $\rho_5 = 0$.
10. $p = 1$, $j_p = 1$.
11. $\rho_1 = u(1) - \bar{x}_1 = -1978/981$.
12. $B = \{5, 2, 8, 10\}$, $N = \{4, 1, 6, 7, 3, 9\}$, $\bar{x}_B(1) = \bar{x}_5$, $\bar{x}_N(2) = \bar{x}_1$.
13. LU factors of B:

$$L = \begin{bmatrix} -1/8 & 1 & & \\ -7/8 & 19/25 & 1 & \\ -1/2 & 2/25 & 383/1226 & 1 \\ 1 & & & \end{bmatrix} \quad U = \begin{bmatrix} -8 & 4 & 7 & 7 \\ & 25/2 & -33/8 & 71/8 \\ & & 613/50 & 569/50 \\ & & & 3209/613 \end{bmatrix}.$$

14. $u = [0, 0, 0, 613/3209]^T$, $v = [193/6418, -383/6418, 613/3209, 151/3209]^T$,
    $w_N = [551/1028, 345/539, -635/6418, -238/681, -75/6418, 632/2225]^T$.
15. $q' = 1$, $\rho_{q'} = -1978/981$, $d' = [-1663/1128, -182/3209, -1028/1021]^T$.

Iteration 3:

3. $\rho_1 w_N(2) < 0$, $\bar{x}_N = [0, 1475/184, 8, 4, -4, -7]^T$, $J = \{4, 6, 7, 9\}$.
4. $\theta = \min[|\bar{z}_N(1)/w_N(1)|, |\bar{z}_N(3)/w_N(3)|, |\bar{z}_N(4)/w_N(4)|, |\bar{z}_N(6)/w_N(6)|] = 2543/1286$,
   $q = 7$, $\beta = -\bar{z}_N(4)/w_N(4) = -2543/1286$.
5. $\bar{z}_N = [1963/375, 0, -1810/183, -830/1201, 797/4108, 625/98]^T + \beta w_N = [4467/1070, -2839/2243, -4896/505, 0, 487/2243, 1832/315]$, $\bar{y}_m = \bar{y}_m - \beta = -539/166$.
6. $\rho_1 = -1978/981$, $x_N(2) = 6$.
7. $u = [-5, -29/8, -381/50, -1041/569]^T$, $d = \bar{a}_{R',7} = [-69/6418, -449/3209, -342/1151]^T$,
8. $x_N(4) = \bar{x}_7 = x_N(4) - \rho_1(w_N(2)/w_N(4)) = 689/2243$,
   $\bar{x}_{B'} = \bar{x}_{B'} + \rho_1((w_N(2)/w_N(4))d - d') = [9933/1172, 2886/2243, -2117/481^T$,
   $\bar{x} = [6, 2886/2243, -4, 0, 9933/1172, 8, 689/2243, -2117/48, -7]^T$.
9. $\rho_7 = 0$.
10. $p = 3$, $j_p = 8$.
11. $\rho_8 = l(8) - \bar{x}_8 = 900/2243$.
12. $B = \{5, 2, 7, 10\}$, $N = \{4, 1, 6, 8, 3, 9\}$, $\bar{x}_B(3) = \bar{x}_7$, $\bar{x}_N(4) = \bar{x}_6$.
13. LU factors of B:

$$L = \begin{bmatrix} -1/8 & 1 & & \\ -7/8 & 19/25 & 1 & \\ -1/2 & 2/25 & 421/762 & 1 \\ 1 & & & \end{bmatrix} \quad U = \begin{bmatrix} -8 & 4 & -5 & 7 \\ & 25/2 & -29/8 & 71/8 \\ & & 381/50 & 569/50 \\ & & & 1907/762 \end{bmatrix}.$$

14. $u = [0, 0, 0, 762/1907]^T$, $v = [259/1907, -421/1907, 762/1907, 45/1907]^T$,
    $w_N = [572/1907, 3479/1907, -1840/1907, -2243/1907, -182/1907, 2121/1907]^T$.
15. $q' = 8$, $\rho_{q'} = 900/2243$, $d' = [-69/1907, -898/1907, -1151/342]^T$.
    Iteration 4:
3. $\rho_8 w_N(4) < 0$, $\bar{x}_N = [0, 6, 8, -2117/481, -4, -7]$, $J = \{4, 6, 9\}$.
4. $\theta = \min[|\bar{z}_N(1)/w_N(1)|, |\bar{z}_N(3)/w_N(3)|, |\bar{z}_N(6)/w_N(6)|] = 1187/227$, $q = 9$, $\beta = -\bar{z}_N(4)/w_N(4) = -1187/227$.

5. $\bar{z}_N = [4467/1070, -2839/2243, -4896/505, 0, 487/2243, 1832/315]$, $+\beta w_N = [1238/475, -3274/303, -2283/491, 1513/246, 217/303, 0]^T$, $\bar{y}_m = \bar{y}_m - \beta = 1217/614$.
6. $\rho_8 = -900/2243,\ x_N(4) = -4$.
7. $u = [-4, 5/2, -27/5, 707/254]^T,\ d = \bar{a}_{R',9} = [78/1907, 186/1907, 2455/1036]^T$,
8. $x_N(6) = \bar{x}_9 = x_N(6) - \rho_8(w_N(4)/w_N(6)) = -2433/370$,
   $\bar{x}_{B'} = [9933/1172, 2886/2243, 689/2243]^T \bar{x}_{B'} + \rho_8((w_N(4)/w_N(6))d - d') = [9933/1172, 2886/2243, 689/2243]^T$,
   $\bar{x} = [6, 1014/707, -4, 0, 2457/290, 8, 461/707, -4, -2433/370]^T$.
9. $\rho_9 = 0$.
10. $\rho_{B'} = 0$. Optimality achieved.

The primal and dual optimal solutions:
$\bar{x} = [6, 1014/707, -4, 0, 2457/290, 8, 461/707, -4, -2433/370]^T \bar{f} = c^T x = -4718/37 \bar{z} = [-3274/303, 0, 217/303, 1238/475, 0, -2283/491, 0, 1513/246, 0]^T \bar{y}_m = 1217/614 \bar{g} = \bar{y}_m + \bar{z}^T \bar{x} = -4718/37$

Chapter 8
Face Method with Cholesky Factorization

In the previous chapters, we presented the face and dual face methods, in which the normal basis still plays a central role. These methods have the advantage of using LU factorization suitable for solving large-scale sparse LP problems.

This chapter derives the face method based on Cholesky factorization. Even involving the orthogonal transformation, which is expensive compared with the elementary elimination, it is worth exploring because of its remarkable advantages. It is basis-free, potentially creates better search directions, and hence reduces required iterations. Moreover, it solves fewer systems in each iteration, and finally, it has excellent stability.

In the subsequent sections, we address themes such as the steepest descent direction, solution update, face contraction, optimality test, face expansion, face algorithm supported by a phase I procedure using the single artificial variable, and related computational results. Before commenting on the face method in the final section, we spend two sections on its variants: the affine face algorithm and the generalized algorithm.

Consider the following reduced LP problem:

$$\begin{array}{ll} \min & x_{n+1} \\ \text{s.t.} & \begin{bmatrix} A & 0 \\ c^T & -1 \end{bmatrix} \begin{bmatrix} x \\ x_{n+1} \end{bmatrix} = \begin{bmatrix} b \\ 0 \end{bmatrix} \qquad x \geq 0. \end{array} \tag{8.1}$$

Redefine the following notation:

$$A := \begin{bmatrix} A & 0 \\ c^T & -1 \end{bmatrix}$$

At the moment, assume that rank $A = m + 1$.

Let $\bar{x}$ be a feasible solution to the problem (8.1) and let (B, N) be a partition of A such that

P.-Q. Pan, *Face Method*, https://doi.org/10.1007/978-3-031-93594-7_8

$$m + 1 \leq |B| \equiv k \leq n + 1, \tag{8.2}$$

and that

$$\bar{x}_j > 0, \quad n + 1 \neq j \in B; \qquad \bar{x}_N = 0. \tag{8.3}$$

In addition, the last column of B is always equal to the last column of A, i.e.,

$$Be_k = -e_{m+1}. \tag{8.4}$$

Definition 8.0.1 The face associated with partition (B, N) is defined as

$$P(B) = \{x \in A \mid Bx_B = b;\ x_j \geq 0,\ n + 1 \neq j \in B; x_N = 0\}. \tag{8.5}$$

A point/solution on the face is called *face point/solution*. Thus, $\bar{x}$ is a face point. For convenience, $\bar{x}$ and $\bar{x}_B$ will not be distinguished: $\bar{x}_B \in P(B)$ is often used instead of $\bar{x} \in P(B)$.

B and N are called *face matrix* and *nonface matrix*, and the associated variables are *face variables* and *nonface variables*. Without confusion, denote the associated column index sets by B and N, respectively, i.e.,

$$B = \{1, \ldots, k - 1, n + 1\}, \quad N = \{1, \ldots, n + 1\} \backslash B.$$

Definition 8.0.2 A face is a level face if the objective value is constant over it. A level face is an *optimal face* if the related objective value equals the optimal value.

8.1 Steepest Descent Direction

As well-known, the search direction used is crucial to the success of any LP solver. In this section, we derive the "best" one.

Consider the following subprogram:

$$\begin{array}{ll} \min & x_{n+1}, \\ \text{s.t.} & Bx_B = b - N\bar{x}_N, \\ & x_j \geq 0, \quad n + 1 \neq j \in B, \end{array} \tag{8.6}$$

which minimizes the objective over the face $P(B)$.

At the moment, assume that B is of full row rank. Introduce the $k \times k$ orthogonal projection matrix, i.e.,

$$P = I - B^{\mathrm{T}}(BB^{\mathrm{T}})^{-1}B. \tag{8.7}$$

Then, the orthogonal projection of the objective gradient, e_k, of the subprogram onto the null space of B is

$$\Delta_B = Pe_k = e_k - B^{\mathrm{T}}(BB^{\mathrm{T}})^{-1}Be_k. \tag{8.8}$$

From the preceding equation and $Be_k = -e_{m+1}$, it follows that

$$\Delta_B = Pe_k = e_k - B^{\mathrm{T}}\bar{y}, \tag{8.9}$$

where $\bar{y}$ is the unique solution to the following system:

$$BB^{\mathrm{T}}y = -e_{m+1}. \tag{8.10}$$

We have the following result.

Proposition 8.1.1 *Vector* Δ_B *satisfies* $B\Delta_B = 0$.

Proof By (8.9), it is easy to verify the validity of the equality. □

Proposition 8.1.2 *The following statements are equivalent*:

$$(i)\ \Delta_B \neq 0; \qquad (ii)\ \Delta_k > 0; \qquad (iii)\ e_k \notin \text{range}(B^{\mathrm{T}}).$$

Proof Premultiply (8.9) by Δ_B^{T}. Then, combining the result and Proposition 8.1.1 gives

$$\Delta_B^{\mathrm{T}}\Delta_B = \Delta_k.$$

Therefore, statements (i) and (ii) are equivalent. If $\Delta_B = 0$, it is known by (8.9) that $e_k = B^{\mathrm{T}}\bar{y}$, i.e., $e_k \in \text{range}(B^{\mathrm{T}})$; hence, (iii) results in (i). Conversely, if $e_k \in \text{range}(B^{\mathrm{T}})$, then there is some vector u such that

$$e_k = B^{\mathrm{T}}u. \tag{8.11}$$

Premultiplying the preceding equation by B together with (8.4) gives

$$-e_{m+1} = BB^{\mathrm{T}}u.$$

Consequently, $u = \bar{y}$ follows from the preceding equation, (8.10), and the full row rank of B. Further, it holds by (8.11) that

$$e_k = B^{\mathrm{T}}\bar{y}.$$

Thus, substituting the preceding equation into (8.9) gives $\Delta_B = 0$. Therefore, (i) and (iii) are equivalent. □

Proposition 8.1.3 *If $\Delta_B \neq 0$, then it holds that*

$$e_k^{\mathrm{T}} \Delta_B = \Delta_k > 0, \tag{8.12}$$

and

$$-e_k^{\mathrm{T}} \Delta_B / \|\Delta_B\| \leq e_k^{\mathrm{T}} v / \|v\|, \quad \forall\, 0 \neq v \in \mathrm{Null}\,(B). \tag{8.13}$$

Proof The expression (8.12) can be derived directly from Proposition 8.1.2.

For P defined by (8.7), it is verified that

$$Pv = v \neq 0, \quad \forall\, 0 \neq v \in \mathrm{Null}(B). \tag{8.14}$$

Further, it follows from the Cauchy inequality that

$$(-Pe_k)^{\mathrm{T}}(Pv) \leq \|Pe_k\| \|Pv\|.$$

Multiplying the preceding equation by $1/\|Pv\|$ gives

$$(-Pe_k)^{\mathrm{T}}(Pv)/\|Pv\| \leq \|Pe_k\|,$$

i.e.,

$$(-Pe_k)^{\mathrm{T}}(Pv)/\|Pv\| \leq (Pe_k)^{\mathrm{T}}(Pe_k)/\|Pe_k\|.$$

Consequently, the preceding equation together with $P^2 = P$, $P^{\mathrm{T}} = P$, (8.14), and $\Delta_B = Pe_k$ gives (8.13). □

According to Proposition 8.1.3, vector $-\Delta_B \neq 0$ forms the most-obtuse-angle in the null space of B with the objective gradient e_k of the subprogram and is hence the steepest descent one. In this sense, taking the vector as a search direction is the "best" choice.

It is obvious that the computation of Δ_B, given by (8.9), mainly lies on the solution of the $(m+1) \times (m+1)$ system (8.10). Since B is of full row rank, the Cholesky factorization exists (Golub and Van Loan, 1989), i.e.,

$$BB^{\mathrm{T}} = LL^{\mathrm{T}}, \tag{8.15}$$

where L is nonsingular lower triangular. Therefore, system (8.10) becomes

$$LL^{\mathrm{T}} y = -e_{m+1}, \tag{8.16}$$

which can be put into the following two triangular systems:

$$Lv = -e_{m+1}, \qquad L^{\mathrm{T}} y = v.$$

The solution of the first system is readily available, i.e., $v = -(1/\nu)e_{m+1}$, where ν is the $(m+1)$th diagonal of L. Consequently, there is only one triangular system left to be solved, i.e.,

$$L^{\mathrm{T}} y = -(1/\nu)e_{m+1}. \tag{8.17}$$

Computing Cholesky Factor L

Beyond all doubt, the Cholesky factor L is the key to forming the search direction.

As is well-known, for $B \in R^{(m+1)\times k}$ having full row rank, there is the unique QR factorization

$$B^T = Q_1 R_1,$$

where $Q_1 \in R^{k\times(m+1)}$ has orthogonal columns and $R_1 \in R^{(m+1)\times(m+1)}$ is upper triangular with positive diagonals. Moreover, $L = R_1^T$ is just the lower triangular Cholesky factor of BB^T. Although there are multiple choices, Givens rotations seem suitable for generating L in sparse computation. There is no need for storing the rotations [10].

The full row rank assumption on B or A is not essential.

In the general case, a slightly modified QR factorization can be performed to find an orthogonal matrix $[Q_1, Q_2] \in \mathcal{R}^{(n+1)\times(n+1)}$ and a permutation $\Pi \in \mathcal{R}^{(m+1)\times(m+1)}$ such that

$$B^{\mathrm{T}}\Pi = [Q_1, Q_2]\begin{bmatrix} R_1 & T \\ 0 & 0 \end{bmatrix},$$

where $R_1 \in \mathcal{R}^{r\times r}$ is nonsingular upper triangular. This implies that rank $B = r$, and the last $m - r + 1$ equations of

$$\Pi^{\mathrm{T}} Bx = \Pi^{\mathrm{T}}(b - N\bar{x}_N)$$

are redundant. If the original notation is still used to denote the problem, resulting from dropping these redundant rows, then $B^{\mathrm{T}} = Q_1 R_1$ and the Cholesky factorization is

$$BB^{\mathrm{T}} = R_1^{\mathrm{T}} Q_1^{\mathrm{T}} Q_1 R_1 = R_1^{\mathrm{T}} R_1, \tag{8.18}$$

which gives the Cholesky factor $L = R^T$. L will be updated subsequently.

From now on, it is assumed that the face matrix $B \in R^{r\times k}$ has full row rank r.

8.2 Updating of Face Solution

Once the descent search direction $-\Delta_B$ is available, the new iterate can be determined by the following line search scheme:

$$\hat{x}_B = \bar{x}_B - \alpha \Delta_B, \tag{8.19}$$

where α is a stepsize to be determined. If the index set

$$J = \{j \in B \mid \Delta_j > 0, j \neq n+1\}$$

is nonempty, determine α and p such that

$$\alpha = \bar{x}_q/\Delta_q = \min_{j \in J} \bar{x}_j/\Delta_j \geq 0, \tag{8.20}$$

which is the maximum possible stepsize for the remaining new iterate within the feasible region.

Note that the stepsize has nothing to do with the kth (objective) component of $\bar{x}_B$.

Lemma 8.2.1 *Assume that $\bar{x}_B \in P(B)$ and $\Delta_B \neq 0$:*

(i) If $J \neq \emptyset$, then $\hat{x}_B \in P(B)$ is a boundary point. The objective value does not increase and strictly decreases if $\bar{x}_B$ is nondegenerate.
(ii) If $J = \emptyset$, the original problem is lower unbound.

Proof

(i) $\Delta_B \neq 0$ and $J \neq \emptyset$ together ensure that (8.20) is well-defined. As $\bar{x}_B \in P(B)$, it is known from Proposition 8.1.1 that $\hat{x}_B \in P(B)$; further, it is clear that $\hat{x}_q = 0$; hence, $\hat{x}$ is a boundary point associated with the new objective value:

$$\hat{x}_{n+1} = \bar{x}_{n+1} - \alpha \Delta_k. \tag{8.21}$$

It is known from (8.20) that $\alpha \geq 0$. Besides, from $\Delta_B \neq 0$ and Proposition 8.1.2, it follows that $\Delta_k > 0$; hence, $\hat{x}_{n+1} \leq \bar{x}_{n+1}$. If $\bar{x}_B$ is nondegenerate, then $\alpha > 0$, leading to $\hat{x}_{n+1} < \bar{x}_{n+1}$.

(ii) When $J = \emptyset$, it is clear that $\hat{x}_B \in P(B)$ holds for any $\alpha > 0$. It is known by (8.21) that

$$\hat{x}_{n+1} \rightarrow -\infty, \qquad \text{as} \qquad \alpha \rightarrow \infty.$$

Therefore, the objective value is lower unbounded over $P(B)$ and, hence, over the feasible region.

□

8.3 Face Contraction

Assume that $\Delta_B \not\geq 0$ and the new iterate $\hat{x}_B$ was determined by (8.19) with $\hat{x}_q = 0$. The associated face matrix $\breve{B} \in \mathcal{R}^{r\times(k-1)}$ results from the old B by dropping the q-indexed column. According to Definition 8.0.1, the face is contracted.

Then, we have the following result.

Proposition 8.3.1 *If B has full row rank r, so is $\breve{B}$.*

Proof It is clear that rank $\breve{B} \leq r$. It is known by (8.20) that $\Delta_q \neq 0$, and by Proposition 8.1.1 that

$$\sum_{j\in B} \Delta_j a_j = B\Delta_B = 0;$$

thus,

$$a_q = -(1/\Delta_q) \sum_{j\in B, j\neq p} \Delta_j a_j.$$

The above equality implies that $a_q \in$ range $\breve{B}$. Therefore, rank $\breve{B} < r$ leads to rank $B < r$, thereby contradicting the assumption. Therefore, rank $\breve{B} = r$. □

Based on the preceding Proposition, there exists the Cholesky factorization:

$$\breve{B}\breve{B}^{\mathrm{T}} = \breve{L}\breve{L}^{\mathrm{T}},$$

where factor $\breve{L}$ can be calculated from its predecessor L by the following procedure due to Professor M. Saunders [37]:

Subalgorithm 8.3.1 (Subalgorithm: Contraction)
Initial: the Cholesky factor L of BB^{T}. This algorithm finds the Cholesky factor $\breve{L}$ of $\breve{B}\breve{B}^{\mathrm{T}}$:

(1) Solve the lower triangular system $Lu = a_q$.
(2) Compute $\gamma = \sqrt{1 - \|u\|^2}$.
(3) Determine Givens rotations $J_r, \ldots, J_1$, where J_i in the $(r+1, i)$-plane eliminates the entry in row i, so that

$$J_1 \ldots J_r \begin{bmatrix} u \\ \gamma \end{bmatrix} = \begin{bmatrix} 0 \\ 1 \end{bmatrix}.$$

(4) Compute $J_1 \ldots J_r \begin{bmatrix} L^{\mathrm{T}} \\ 0 \end{bmatrix} = \begin{bmatrix} \breve{L}^{\mathrm{T}} \\ a_q^{\mathrm{T}} \end{bmatrix}$ *to obtain $\breve{L} \in \mathcal{R}^{r\times r}$.*

(5) return.

Note: According to M. Saunders, the quantity γ in Step 2 is well-defined if $\breve{B}$ has full row rank.

In general, there would be multiple components of $\hat{x}_B$ attaining boundary; that is, besides p, the set below would include other indices:

$$S = \{j \in B \mid \hat{x}_j = 0\}, \tag{8.22}$$

In this case, we move all such ones from B to N, i.e.,

$$\hat{B} = B \backslash S, \qquad \hat{N} = N \cup S. \tag{8.23}$$

and generate the Cholesky factor from scratch by QR factorization. Consequently, the next iteration will be zero-stepsize-free.

In any case, the dimension of the face reduces, at least by one.

8.4 Optimality Test

The search direction is not well-defined by (8.9) if $\Delta_B = 0$. In particular, this is the case whenever $k = r$.

Lemma 8.4.1 *Assume that* Δ_B *is defined by (8.9). If* Δ_B *vanishes, then* $P(B)$ *is a level face and vice versa if* $0 < \bar{x}_B \in P(B)$.

Proof Assume that $\bar{y}$ satisfies (8.10). From $\Delta_B = 0$ and (8.9), it follows that

$$e_k = B^{\mathrm{T}} \bar{y}.$$

Transpose the two sides of the preceding equality, and postmultiply the result by any $x_B \in P(B)$ to obtain

$$e_k^{\mathrm{T}} x_B = \bar{y}^{\mathrm{T}} B x_B,$$

which along with $B x_B = b$ gives

$$f = \bar{y}^{\mathrm{T}} b.$$

It implies that the objective value is constant over $P(B)$; hence, $P(B)$ is a level face.

Now, assume that $P(B)$ is a level face and that $0 < \bar{x}_B \in P(B)$ with $\Delta_B \neq 0$. In case when $\Delta_B \leq 0$, it is known by (ii) of Lemma 8.2.1 that the objective value is lower unbounded over $P(B)$, as contradicts that $P(B)$ is a level face: when $\Delta_B \not\leq 0$, since x_B is nondegenerate, it is known from (i) of Lemma 8.2.1 that the objective value, associated with $\hat{x}_B \in P(B)$ given by (8.19) and (8.20), strictly decreases, as also contradicts that $P(B)$ is a level face. Therefore, Δ_B vanishes. □

Now, consider the dual problem of (8.1), i.e.,

$$\begin{array}{ll} \max & (b^{\mathrm{T}}, 0)^{\mathrm{T}} y \\ \text{s.t.} & \begin{bmatrix} B^{\mathrm{T}} \\ N^{\mathrm{T}} \end{bmatrix} y + \begin{bmatrix} z_B \\ z_N \end{bmatrix} = \begin{bmatrix} e_k \\ 0 \end{bmatrix}, \quad z_B,\ z_N \geq 0,\ z_{n+1} = 0. \end{array} \tag{8.24}$$

Setting y to the solution $\bar{y}$ of the system (8.10), the preceding constraint gives the dual solution below:

$$\bar{z}_B = e_k - B^{\mathrm{T}} \bar{y}, \qquad \bar{z}_N = -N^T \bar{y}. \tag{8.25}$$

Introduce the following index set:

$$T = \{j \in N \mid \bar{z}_j < 0\}. \tag{8.26}$$

Theorem 8.4.2 *Assume that $\bar{x} \in P(B)$ and $(\bar{y}, \bar{z})$ satisfies (8.10) and (8.25). If $\Delta_B = 0$ and set T defined by (8.26) is empty, then $\bar{x}$ and $(\bar{y}, \bar{z})$ are a pair of primal and dual optimal solutions. Moreover, $P(B)$ is an optimal face.*

Proof Note that $\bar{x}$ is a primal feasible solution and $(\bar{y}, \bar{z})$ is a dual solution satisfying the dual equality constraint.

From (8.9) and (8.25), it follows that

$$\bar{z}_B = \Delta_B = 0, \tag{8.27}$$

which and emptiness of T together ensure that $(\bar{y}, \bar{z})$ is not only dual feasible but also exhibit complementarity slackness with $\bar{x}$. Therefore, they are a pair of primal and dual optimal solutions.

By Lemma 8.4.1, $\Delta_B = 0$ implies that $P(B)$ is a level face. It includes the optimal solution $\bar{x}$ and is therefore an optimal face. □

Interestingly, the search direction Δ_B is just the $\bar{z}_B$ itself, which is the active part of the corresponding dual solution. In addition, according to Proposition 8.1.2, Δ_B vanishes if and only if so does $\Delta_k = 0$.

8.5 Face Expansion

Assume $\Delta_B = 0$ and T defined by (8.26) is nonempty. . In this case, optimality cannot be asserted, but the following column index q' can be determined such that

$$q' \in \arg\min_{j \in N} \bar{z}_j.$$

Accordingly, update the face and nonface index sets by

$$\widehat{B} = B \cup \{q'\}, \quad \widehat{N} = N \backslash \{q'\}. \tag{8.28}$$

Consequently, the q'-indexed column is added to the face matrix, and the dimension of the face increases by one.

Denote the new face matrix by $\widehat{B} \in \mathcal{R}^{(m+1)\times(k+1)}$, which is still of full row rank. The corresponding updating of the Cholesky factor is simple (Golub [10]), as stated as follows.

Subalgorithm 8.5.1 (Subalgorithm: Expansion)
Initial: the Cholesky factor L of BB^{T}. This algorithm finds Cholesky factor $\widehat{L}$ of $\widehat{B}\widehat{B}^{\mathrm{T}}$.

Determine Givens rotations $J_1, \ldots, J_r$, where J_i in the $(i, r+1)$-plane eliminates the entry in row $r+1$ such that

$$J_r \ldots J_1 \begin{bmatrix} L^{\mathrm{T}} \\ a_{q'}^{\mathrm{T}} \end{bmatrix} = \begin{bmatrix} \widehat{L}^{\mathrm{T}} \\ 0 \end{bmatrix}, \tag{8.29}$$

which gives $\widehat{L} \in \mathcal{R}^{r\times r}$.

Proposition 8.5.1 *Subalgorithm 8.5.1 is valid.*

Proof The validity is shown by

$$\widehat{L}\widehat{L}^{\mathrm{T}} = LL^{\mathrm{T}} + a_{q'}a_{q'}^{\mathrm{T}} = BB^{\mathrm{T}} + a_{q'}a_{q'}^{\mathrm{T}} = \widehat{B}\widehat{B}^{\mathrm{T}}.$$

□

Generally, the set T defined by (8.26) would include multiple indices. It seems more favorable to move all of them from N to B, i.e.,

$$\hat{B} = B \cup T, \qquad \hat{N} = N \backslash T. \tag{8.30}$$

8.6 Face Algorithm with Cholesky Factorization

In this section, we formulate the face algorithm with Cholesky factorization.

Let $(\bar{x}, \bar{x}_{n+1})$ be an initial face solution. and let (B, N) be a partition satisfying (8.2)–(8.3).

The search direction is calculated in each iteration, and $\bar{x}_B$ is updated with (B, N) adjusted accordingly. Iteration by iteration, the dimension of the face $P(B)$ decreases strictly unless lower unboundedness is declared. Such a series of iterations is termed *face contraction (process)*, which ends whenever Δ_B vanishes and reaches a level face. Then, optimality is checked. If set T defined by (8.26) is empty, then an optimal face is attained. Otherwise, start another round of face contraction with an updated partition.

The overall steps can be put into the following algorithm.

Algorithm 8.6.1 (Face Algorithm with Cholesky Factorization)
Initial: $(B,\ N),\ r \leq k = |B|$, *feasible solution* $\bar{x}$. *This algorithm solves the standard problem:*

1. *Compute the Cholesky factor L of* BB^T *by Givens rotations.*
2. *Solve the upper triangular system* $L^{\mathrm{T}}\bar{y} = -(1/\nu)e_r$, *where* ν *is the* r*th diagonal of L.*
3. *Go to Step 5 if* $k = r$; *else compute* $\Delta_B = e_k - B^{\mathrm{T}}\bar{y}$.
4. *Go to Step 8 if* $\Delta_B \neq 0$.
5. *Compute* $\bar{z}_N = -N^{\mathrm{T}}\bar{y}$.
6. *Stop if set T defined by (8.26) is empty.*
7. *Go to Step 1 with* $B = B \cup T, N = A \backslash B$.
8. *Stop if* $J = \{j \in B \mid \Delta_j > 0,\ j \neq n+1\} = \emptyset$.
9. *Determine stepsize* α *and index* q *such that* $\alpha = \bar{x}_q/\Delta_q = \min_{j \in J} \bar{x}_j/\Delta_j$.
10. *Update:* $\bar{x}_B = \bar{x}_B - \alpha\Delta_B$.
11. *Update* $B = B \backslash S, N = N \cup S, k = k - |S|$, *where S is defined by (8.22).*
12. *Go to Step 1 if* $|S| > 1$; *else, update L by Subalgorithm 8.3.1.*
13. *Go to Step 2.*

Theorem 8.6.1 *Assume that each contraction process involves a nonzero stepsize. Algorithm 8.6.1 terminates either at*

(i) *Step 6, giving an optimal face together with a pair of primal and dual optimal solutions, or at*
(ii) *Step 8, detecting lower unboundedness.*

Proof Note that each face contraction process reaches a level face in a finite number of iterations.

According to Theorem 8.4.2, optimality is achieved if T is empty at Step 5. Otherwise, all indices of T are moved from N to B, and start a new contraction process. Since each contraction process includes a nonzero stepsize at some iteration, the objective value strictly decreases while reaching a level face.

Thus, the untermination of the algorithm implies that there are infinitely many level faces, contradicting the fact that there are a finite number of level faces. Therefore, the algorithm terminates.

The meaning of Step 8 is obvious, according to Lemma 8.2.1. □

8.7 Double-Phase Face Algorithm with Cholesky Factorization

Any phase I procedure can be utilized to provide a feasible solution to get the face algorithm started. However, taking phase I using a single artificial variable seems especially favorable.

Let $h \geq 0$ be any n-dimensional vector. Phase I deals with the following auxiliary problem:

$$\begin{aligned} & \min\ x_{n+1} \\ & \text{s.t.}\quad Ax + a_{n+1}x_{n+1} = b, \qquad x,\ x_{n+1} \geq 0, \end{aligned} \tag{8.31}$$

where

$$a_{n+1} = (b - Ah/\|b - Ah\|.$$

Assume that $b - Ah \neq 0$. It is verified that

$$\bar{x} = h, \qquad \bar{x}_{n+1} = \|b - Ah\|. \tag{8.32}$$

forms a feasible solution to it. Therefore, Algorithm 8.6.1 can get itself started.

A slight modification of it is helpful for phase I purposes. The index $n + 1$ is always a candidate for leaving B, and set J is nonempty. Once the auxiliary objective value $\bar{x}_{n+1}$ reaches 0, the associated $\bar{x}$ is a feasible solution to the original problem. Otherwise, if optimality is achieved with a positive value of $\bar{x}_{n+1}$, there is no feasible solution to the original problem.

Initially, take the following partition:

$$B = \{1, \ldots, n, n + 1\}, \quad N = \emptyset. \tag{8.33}$$

Thus, $k = |B| = n + 1$.

One notable issue is about the computation of the search direction. Compared with phase II, this time, the end column of B is that of A, i.e.,

$$Be_k = a_{n+1}.$$

Therefore, from (8.8), it follows that

$$\Delta_B = Pe_k = e_k - B^{\mathrm{T}}\bar{y}, \tag{8.34}$$

where $\bar{y}$ is the unique solution to the following two triangular systems:

$$Lu = a_{n+1}, \qquad L^T y = u. \tag{8.35}$$

where L is the Cholesky factor of BB^T.

The related overall steps are summarized in the following algorithm.

Subalgorithm 8.7.1 (Face Phase I: Single Artificial Variable)
Initial: (B, N) *by (8.33),* $(\bar{x}, \bar{x}_{n+1})$ *by (8.32), This algorithm finds a feasible solution to the standard problem.*

1. *Compute the Cholesky factor* L *of* BB^T *by Givens rotations, with row rank* r.
2. *Solve the two triangular systems (8.35) for* $\bar{y}$.
3. *Go to Step 5 if* $k = r$*; else, compute* $\Delta_B = e_k - B^{\mathrm{T}}\bar{y}$.
4. *Go to Step 8 if* $\Delta_B \neq 0$.
5. *Compute* $\bar{z}_N = -N^{\mathrm{T}}\bar{y}$.
6. *Stop if set* T *defined by (8.26) is empty (there is no feasibility solution).*
7. *Go to Step 1 with* $B = B \cup T, N = A \backslash B$.
8. *Determine* $J = \{j \in B \mid \Delta_j > 0\} = \emptyset$.
9. *Determine stepsize* α *and index* q *such that* $\alpha = \bar{x}_q/\Delta_q = \min_{j \in J} \bar{x}_j/\Delta_j$.
10. *Update*: $\bar{x}_B = \bar{x}_B - \alpha\Delta_B$.
11. *Stop if* $\bar{x}_{n+1} = 0$ *(feasibility achieved).*
12. *Update* $B = B\backslash S, N = N \cup S, k = k - |S|$*, where* S *is defined by (8.22).*
13. *Go to Step 1 if* $|S| > 1$*; else, update* L *by Subalgorithm 8.3.1.*
14. *Go to Step 2.*

Once phase I finishes with a feasible solution, the reduced problem is formed by adding the $(m+1)$th row. The corresponding value of the objective function is calculated and set to $\bar{x}_{n+1}$. The end partition (B, N) is still helpful, but the Cholesky factor L has to be recalculated.

Example 8.7.1 Solve the following standard problem using the double-phase face algorithm:

$$A = \begin{bmatrix} 4 & 0 & -3 & 4 & 3 & -6 & 0 & 3 \\ -7 & -6 & 0 & -2 & 0 & 2 & -5 & 8 \\ 5 & -2 & -6 & 0 & 5 & -9 & 6 & -4 \\ -2 & 0 & -6 & 8 & 7 & 3 & 0 & 3 \end{bmatrix}, \quad b = \begin{bmatrix} 3 \\ -5 \\ 0 \\ -2 \end{bmatrix}$$

$$c = [3, 0, 9, -2, 0, -7, 2, -3]^T$$

Phase I. Solve the auxiliary problem (8.31) with $h = e$.

$a_9 = [-91/760, 91/304, 91/304, -273/304]^T$.

Initial: $k = 9$, $B = \{1, \ldots, 9\}$, $N = \emptyset$, $\bar{x} = [1, \ldots, 1, 1520/91]^T$. $B = [A \mid a_9]$, $N = \emptyset$.

Call Subalgorithm 8.7.1.

Iteration 1:

1. Cholesky factor $L = \begin{bmatrix} 1969/202 & & & \\ -397/161 & 4922/371 & & \\ 3517/361 & -1326/223 & 2871/298 & \\ 927/167 & 2841/910 & -1332/931 & 106{,}96/941 \end{bmatrix}$

$r = 4$.

2. $\bar{y} = [-10/8073, 123/23{,}671, 47/9633, -58/9217]^{\mathrm{T}}$.
3. $\Delta_B = [115/26{,}451, 1584/38{,}695, -155/12{,}707, 215/3273, 62/2653, 309/6872, -26/7895, 28/51{,}791, 1012/1021]^{\mathrm{T}} \neq 0$.
8. $J = \{1, 2, 4, 5, 6, 8, 9\}$.
9. $\alpha = \min[1/(115/26{,}451), 1/(1584/38{,}695), 1/(215/3273), 1/(62/2653), 1/(309/6872), 1/(28/51{,}791), (1520/91)/(1012/1021)] = 3273/215, \ q = 4$.
10. $\bar{x}_B = [1, 1, 1, 1, 1, 1, 1, 1, 1520/91]^{\mathrm{T}} - (3273/215)\Delta_B$ $= [776/831, 335/889, 779/657, 0, 1235/1917, 218/691, 1571/1496, 241/243, 975/604]^{\mathrm{T}}$.
11. $\bar{x}_9 \neq 0$.
12. $B = \{1, 2, 3, 5, 6, 7, 8, 9\}, N = \{4\}, k = 8$.
13. $L = \begin{bmatrix} 8809/991 & & & \\ -810/449 & 4813/364 & & \\ 641/60 & -4510/713 & 3230/389 & \\ 2788/1121 & 1032/283 & 2834/1291 & 2449/268 \end{bmatrix}$

Iteration 2:

2. $\bar{y} = [-58/5897, 136/12{,}565, 76/6443, -1343/102{,}665]^{\mathrm{T}}$.
3. $\Delta_B = [170/5673, 410/4631, -173/4648, 376/6055, 337/5205, -181/10{,}867, 140/4771, 647/660]^{\mathrm{T}} \neq 0$.
8. $J = \{1, 2, 5, 6, 8, 9\}$.
9. $\alpha = \min[(776/831)/(170/5673), (335/889)/(410/4631), (1235/1917)/(376/6055), (218/691)/(337/5205), (241/243)/(140/4771), (975/604)/(647/660)] = 1757/1067, \ q = 9$.
10. $\bar{x}_B = [776/831, 335/889, 779/657, 1235/1917, 218/691, 1571/1496, 241/243, 975/604]^{\mathrm{T}} - (1757/1067)\Delta_B$ $= [467/528, 326/1411, 1343/1077, 1362/2513, 697/3337, 1820/1689, 1001/1061, 0]^{\mathrm{T}}$.
11. $\bar{x}_9 = 0$. (feasibility achieved)

Phase-II: Call Algorithm 8.6.1.
Initial: $B = \{1, 2, 3, 5, 6, 7, 8, 9\}, k = 8, N = \{4\}$.
$\bar{x}_B = [467/528, 326/1411, 1343/1077, 1362/2513, 697/3337, 1820/1689, 1001/1061]^T$,
Iteration 3:

1. $L = \begin{bmatrix} 1431/161 & & & & \\ -2558/1421 & 2287/173 & & & \\ 5590/523 & -2395/378 & 3595/434 & & \\ 948/383 & 986/269 & 1867/824 & 880/97 & \\ 322/159 & -3075/622 & -889/1482 & -6952/835 & 3227/436 \end{bmatrix}$
2. $\bar{y} = [92/19{,}253, -19/30{,}698, 95/2913, -107/6387, -214/11{,}723]^T$.
3. $\Delta_B = [-169/9140, 71/25{,}291, 907/9286, 116/1339, -48/2627, 120/8663,$

$-26/30{,}663, 484/493]^{\mathrm{T}} \neq 0$.

8. $J = \{2, 3, 5, 7\}$.
9. $\alpha = \min[(326/1411)/(71/25{,}291), (1343/1077)/(907/9286), (1362/2513)/(116/1339), (697/3337)/(120/8663)] = 4323/691, q = 5$.
10. $\bar{x}_B = \bar{x}_B - \alpha\Delta_B = [6810/6809, 548/2567, 1172/1843, 0, 1475/4564, 980/989, 685/722, 2597/464]^T$.
 $\bar{x} = [6810/6809, 548/2567, 1172/1843, 0, 0, 1475/4564, 980/989, 685/722, 2597/464]^T$.
11. $B = \{1, 2, 3, 6, 7, 8, 9\}, N = \{4, 5\}, k = 7$.
12. $$L = \begin{bmatrix} 4108/491 & & & & \\ -1964/1027 & 1360/103 & & & \\ 3557/372 & -4613/719 & 3049/377 & & \\ 491/4108 & 2462/735 & 2057/2264 & 5951/878 & \\ 2515/1169 & -3720/757 & -439/866 & -4144/383 & 1426/543 \end{bmatrix}.$$

Iteration 4:

2. $\bar{y} = [659/27{,}202, 232/17{,}881, 25/1478, -921/3979, -271/1869]^T$.
3. $\Delta_B = [-604/5093, 241/2158, 988/10{,}935, -181/3698, 806/3181, 231/1534, 1598/1869]^{\mathrm{T}} \neq 0$.
8. $J = \{2, 3, 7, 8\}$.
9. $\alpha = \min[(548/2567)/(241/2158), (1172/1843)/(988/10{,}935), (980/989)/(806/3181), (685/722)/(231/1534)] = 281/147, q = 2$,
10. $\bar{x}_B = \bar{x}_B - \alpha\Delta_B = [1179/961, 0, 812/1753, 453/1087, 1315/2596, 840/1271, 2754/695]^T \bar{x} = [1179/961, 0, 812/1753, 0, 0, 453/1087, 1315/2596, 840/1271, 2754/695]^T$.
11. $B = \{1, 3, 6, 7, 8, 9\}, N = \{2, 4, 5\}, k = 6$.
12. $$L = \begin{bmatrix} 4108/491 & & & & \\ -1964/1027 & 29{,}099/2474 & & & \\ 3557/372 & -2105/256 & 3181/538 & & \\ 491/4108 & 2369/630 & 1257/443 & 3488/583 & \\ 2515/1169 & -1164/211 & -2120/699 & -7961/782 & 1240/549 \end{bmatrix}.$$

Iteration 5:

2. $\bar{y} = [0, 1379/24{,}491, 277/4654, -527/1580, -857/4372]^T$.
3. $\Delta_B = [147/8425, 229/1907, 361/6988, 894/2825, 573/2863, 808/1005]^T$
8. $J = \{1, 3, 6, 7, 8\}$.
9. $\alpha = \min[(1179/961)/(147/8425), (812/1753)/(229/1907), (453/1087)/(361/6988), (1315/2596)/(894/2825), (840/1271)/(573/2863)] = 477/298, q = 7$,
10. $\bar{x}_B = \bar{x}_B - \alpha\Delta_B = [1109/925, 965/3561, 774/2317, 0, 1402/4117, 9991/3734]^T$.

11. $B = \{1, 3, 6, 8, 9\}$, $N = \{2, 4, 5, 7\}$, $k = 5$.

12. $$L = \begin{bmatrix} 4108/491 & & & & \\ -1964/1027 & 831/68 & & & \\ 3557/372 & -1867/417 & 3746/527 & & \\ 491/4108 & 1245/344 & 430/1483 & 3407/509 & \\ 2515/1169 & -3849/857 & -1929/2930 & -2018/183 & 615/418 \end{bmatrix}.$$

Iteration 6:

2. $\bar{y} = [309/1996, 213/4151, -61/5195, -3045/4001, -1524/3299]^T$,
3. $k = 5$.
5. $\bar{z}_N = [769/2704, 2152/463, 2200/447, 1675/13{,}391]^T$.
6. $T = \emptyset$, optimality achieved.

The primal and dual optimal solutions are as follows:
$\bar{x} = [1109/925, 0, 965/3561, 0, 0, 774/2317, 0, 1402/4117]^T$, $\bar{f} = 9991/3734$.
$\bar{z} = [0, 769/2704, 0, 2152/463, 2200/447, 0, 1675/13{,}391, 0]^T$.

In computational experiments, the following two codes without exploiting sparsity were tested and compared:

1. FALP: A prototype of the face Algorithm 8.6.1 supported by Subalgorithm 8.7.1 as phase I.
2. RSA: Our working simplex code.

The test set comprises the 26 smallest Netlib standard problems (Appendix C: Table 4: AFIRO-DEGEN2, Pan [36]). Code RSA failed to solve problem DEGEN2 within 10,000 iterations. Therefore, we report the numerical results on the other 25 problems, by listing the overall ratios of RSA to FALP as follows:

Ratio of RSA to FALP

				Total		Phase 1	
Ave. size (25)	m	n	$m+n$	Itn	Time	Itn	Time
RSA/FALP	213	248	461	1.51	4.91	2.88	11.00

Although the computational experiments are preliminary, the results still acknowledge that the face algorithm's prototype outperformed the simplex code significantly, with a time ratio of **4.91**, on the test set at least. As the iteration ratio is only **1.51**, its computational complexity is much lower than the simplex algorithm in a single iteration. The margin of victory is even more prominent with phase I alone; the overall time ratio is up to **11.00**.

We point out that all optimal faces attained by the FALP are not vertices except for ISREAL. It must contribute a remarkable performance. For details, refer to the Appendix A.

8.8 Affine Face Algorithm with Cholesky Factorization

From (8.20), we see that the sizes of components of $\bar{x}_B$ or, precisely, of $\bar{x}_J$ determine the stepsize α. If the latter includes zero components, the stepsize vanishes.

Too small α is obnoxious. It follows from (8.19) that the difference between the new and old objective values is

$$\hat{x}_{n+1} - \bar{x}_{n+1} = -\alpha \Delta_k \leq 0,$$

where " $\leq$" is from (8.12). Therefore, the decrement in objective value and the stepsize are directly proportional, so the smaller the stepsize, the smaller the decrement.

To avoid too small stepsize, we use the affine technique to modify the search direction of the face algorithm to point to the interior of the feasible region while maintaining descent. This technique can be used at the beginning of a contraction process.

Assume $\bar{x}_B > 0$. Denote by $\bar{X}_B$ the $k \times k$ diagonal matrix, whose diagonals are components of the vector $\bar{x}_B$, except for the (objective) kth diagonal 1, i.e.,

$$\bar{X}_B = \text{diag}(\bar{x}_1, \ldots, \bar{x}_{k-1}, 1). \tag{8.36}$$

Consider affine transformation:

$$x_B = \bar{X}_B x'_B, \tag{8.37}$$

which is invertible and determines a 1-to-1 correspondence between $\bar{x}_B$ and

$$\bar{x}'_B = (\underbrace{1, \ldots, 1}_{k-1}, \bar{x}_{n+1})^{\mathrm{T}}.$$

Even if some of the first $k-1$ components of $\bar{x}_B$ are close to zero, the corresponding components of $\bar{x}'_B$ are balanced, all equal to one, so that the line search along any direction in $\bar{x}'_B$-space will not lead to too small stepsizes.

To form a search direction, consider the following subprogram:

$$\begin{array}{ll} \min & x_{n+1}, \\ \text{s.t.} & (B\bar{X}_B)x'_B = b, \\ & x'_j \geq 0, \quad n+1 \neq j \in B \end{array}$$

The projection matrix onto the null space of $B\bar{X}_B$ is then

$$P = I - \bar{X}_B B^{\mathrm{T}} (B\bar{X}_B^2 B^{\mathrm{T}})^{-1} B\bar{X}_B. \tag{8.38}$$

Therefore, the orthogonal projection of the objective gradient onto the null space is Pe_k, corresponding to the direction in x_B-space, i.e.,

$$\Delta_B = \bar{X}_B P e_k,$$

which together with (8.38) and

$$\bar{X}_B e_k = e_k, \qquad B e_k = -e_r,$$

gives

$$\Delta_B = e_k - \bar{X}_B^2 B^{\mathrm{T}} \bar{y}, \qquad (B\bar{X}_B^2 B^{\mathrm{T}})\bar{y} = -e_r. \tag{8.39}$$

If the Cholesky factorization $B\bar{X}_B^2 B^{\mathrm{T}} = LL^{\mathrm{T}}$ is available, $\bar{y}$ can be obtained by solving an upper triangular system only, i.e.,

$$L^{\mathrm{T}}\bar{y} = -(1/\nu)e_r,$$

where ν is the rth diagonal of L.

If $\Delta_B \neq 0$, it is verified that its kth (objective) component is positive while satisfying $B\Delta_B = e_k - \bar{X}_B^2 B^{\mathrm{T}}\bar{y} = 0$. Therefore, $-\Delta_B$ is eligible to be the search direction. Of course, as we know, it is more than that.

It is also noted that $\Delta_B = e_k - \bar{X}_B^2 B^{\mathrm{T}}\bar{y} = 0$ implies $\bar{z}_B = e_k - B^{\mathrm{T}}\bar{y} = 0$, and optimality is hence achieved if $\bar{z}_N \geq 0$.

When some component of $\bar{x}_B$ is close to zero or even vanishes, numerical difficulties would arise, as the invertibility of the transformation (8.37) becomes questionable. A trick to overcome it is the perturbation below:

Let δ be a given small positive number. Introduce

$$\hat{x}_j = \begin{cases} \bar{x}_j, & \text{If } \bar{x}_j \geq \delta, \\ \delta, & \text{If } \bar{x}_j < \delta, \end{cases} \qquad j \in B, \; j \neq n+1. \tag{8.40}$$

Then use the formula (8.36) as follows:

$$\bar{X}_B = \mathrm{diag}(\hat{x}_1, \ldots, \hat{x}_{k-1}, 1). \tag{8.41}$$

The whole procedure differs from the face algorithm just in its search direction, for which we have a preference, as is formulated as follows.

Algorithm 8.8.1 ([Affine Face Algorithm with Cholesky Factorization) *]* *Initial:* (B, N), $k = |B|$, *face solution* $\bar{x}$, *small real* δ. *This algorithm solves the reduced problem (8.1).*

1. *Compute Cholesky factor* L *of* $B\bar{X}_B^2 B^{\mathrm{T}}$.
2. *Solve triangular system* $L^{\mathrm{T}}\bar{y} = -(1/\nu)e_r$, *where* ν *is the* r*th diagonal of* L.

3. *Go to Step 5 if* $k = r$; *else compute* $\Delta_B = e_k - \bar{X}_B^2 B^{\mathrm{T}} \bar{y}$.
4. *Go to Step 8 if* $\Delta_B \neq 0$.
5. *Compute* $\bar{z}_N = -N^{\mathrm{T}} \bar{y}$.
6. *Stop if set* T *defined by (8.26) is empty (optimality achieved).*
7. *Go to Step 1 with* $B = B \cup T$, $N = A \backslash B$.
8. *Stop if* $J = \{j \in B \mid \Delta_j > 0,\ j \neq n+1\} = \emptyset$. *(There is no feasible solution.)*
9. *Determine stepsize* α *and index* q *such that* $\alpha = \bar{x}_q / \Delta_q = \min_{j \in J} \bar{x}_j / \Delta_j$.
10. *Update:* $\bar{x}_B = \bar{x}_B - \alpha \Delta_B$.
11. *Update* $B = B \backslash S$, $N = N \cup S$, $k = k - |S|$, *where* S *is defined by (8.22).*
12. *Go to Step 1 if* $|S| > 1$; *else, update* L *by Subalgorithm 8.3.1.*

Example 8.8.1 Solve the following reduced problem by the affine face algorithm:

$$A = \begin{bmatrix} 1 & 3 & 1 & & -1 & 7 & & \\ -3 & & -5 & 3 & -2 & & 4 & \\ & -2 & 2 & & 1 & -3 & 5 & \\ -5 & 6 & & 8 & & 9 & 6 & \\ 1 & 5 & & -2 & -7 & & 4 & -1 \end{bmatrix}, \quad b = \begin{bmatrix} 63 \\ 21 \\ -13 \\ 129 \\ 0 \end{bmatrix}$$

Feasible solution: $\bar{x} = [1, 2, 0, 4, 0, 8, 3, 15]^T$.

Initial: $B = \{1, 2, 4, 6, 7, 8\}$, $N = \{3, 5\}$, $k = 6$; $\bar{x}_B = [1, 2, 4, 8, 3, 15]^T$, $\delta = 10^{-3}$.

Iteration 1:

1. $L = \begin{bmatrix} 4788/85 & & & & \\ -85/1596 & 4205/244 & & & \\ -170/7 & 2665/257 & 1991/182 & & \\ 6913/95 & 5243/146 & -4369/430 & 19{,}008/5287 & \\ 1946/1797 & 468/179 & 4415/347 & 2978/339 & 3066/181 \end{bmatrix}$
2. $\bar{y} = [-200/34{,}487, -51/2089, 42/3511, 67/7868, -16/4591]^T$.
3. $\Delta_B = [-209/9776, 120/3919, -176/5935, -41/4072, 35/5639, 765/3544]^{\mathrm{T}} \neq 0$.
8. $J = \{2, 7\}$.
9. $\alpha = \min[(2)/(120/3919), (3)/(35/5639)] = 3919/60$, $q = 2$.
10. $\bar{x}_B = \bar{x}_B - \alpha \Delta_B = [266/111, 0, 659/111, 961/111, 96/37, 609/676]^T$,
    $\bar{x} = [266/111, 0, 0, 659/111, 0, 961/111, 96/37, 609/676]$.
11. $B = \{1, 4, 6, 7, 8\}$, $N = \{2, 3, 5\}$, $k = 5$.
12. $L - \begin{bmatrix} 6429/106 & & & & \\ -611/2151 & 1026/47 & & & \\ -3841/148 & 3154/541 & 3502/301 & & \\ 1006/13 & 12{,}112/237 & -4693/490 & 1741/378 & \\ 554/5851 & -11{,}233/2027 & 4703/323 & -1561/456 & 609/676 \end{bmatrix}$

Iteration 2:

2. $\bar{y} = [1047/691, 1350/833, 1075/1364, -772/843, -3466/2813]^T$.
3. $k = r = 5$.

5. $\bar{z}_N = [2545/293, 1699/339, -2156/463]^T$.
6. $T = \{5\}$.
7. $B = \{1, 4, 5, 6, 7, 8\}$, $N = \{2, 3\}$, $k = 6$.

1. $L = \begin{bmatrix} 6429/106 & & & & \\ -611/2151 & 1026/47 & & & \\ -3841/148 & 3154/541 & 3502/301 & & \\ 1006/13 & 12{,}112/237 & -4693/490 & 2372/515 & \\ 383/4045 & -11{,}632/2099 & 4703/323 & -736/215 & 1509/1675 \end{bmatrix}$.
2. $\bar{y} = [2547/1681, 833/514, 729/925, -1435/1567, -775/629]^T$.
3. $\Delta_B = [2/126{,}965, 4/291{,}147, 0, -1/342{,}988, 0, 1/56{,}824]^T$.
8. $J = \{1, 4\}$.
9. $\alpha = \min[(266/111)/(2/126{,}965), (659/111)/(4/291{,}147)] = 608{,}517/4$, $q = 1$.
10. $\bar{x}_B = \bar{x}_B - \alpha\Delta_B = [0, 1658/431, 532/751, 1529/168, 2042/751, -6599/3715]^T$,
    $\bar{x} = [0, 0, 0, 1658/431, 532/751, 1529/168, 2042/751, -6599/3715]^T$.
11. $B = \{4, 5, 6, 7, 8\}$, $N = \{1, 2, 3\}$, $k = 5$.
12. $L = \begin{bmatrix} 4651/73 & & & & \\ 83/5269 & 3837/241 & & & \\ -3086/113 & 1878/203 & 1346/135 & & \\ 4341/53 & 3237/97 & -3464/399 & 3325/664 & \\ 138/2503 & 16{,}520/7201 & 17{,}413/1393 & -1541/242 & 6599/3715 \end{bmatrix}$.

Iteration 3:

2. $\bar{y} = [827/1537, 569/659, 131/2822, -721/1789, -1069/3373]^T$.
3. $k = r = 5$.
5. $\bar{z}_N = [1019/2878, 1335/538, 2573/698]^T$.
6. $T = \emptyset$. Optimality is achieved.

The primal and dual optimal solutions are as follows:

$\bar{x} = [0, 0, 0, 1658/431, 532/751, 1529/168, 2042/751, -6599/3715]^T$,
$\bar{z} = [1019/2878, 1335/538, 2573/698, 0, 0, 0, 0, 0]^T$, $\quad \bar{f} = -6599/3715$.

8.9 Generalized Face Algorithm with Cholesky Factorization

In this section, we generalize Algorithm 8.6.1 to solve the bounded-variable reduced problem below:

$$\begin{aligned} &\min\ x_{n+1}, \\ &\text{s.t.}\ \ Ax = b; \qquad l_j \le x_j \le u_j,\ j = 1, \ldots, n. \end{aligned} \tag{8.42}$$

where $l_j,\ u_j$ are given reals and A, b are the same as usual.

It is obvious that the search direction Δ_B, defined by (8.9), is still applicable here. As the bound constraints are present, however, the determination of the stepsize α and row index p should utilize the following formula instead:

$$\alpha \equiv \alpha_p = \min\{\alpha_j \mid j \in B,\ j \neq n+1\}, \tag{8.43}$$

where

$$\alpha_j = \begin{cases} (\bar{x}_j - u_j)/\Delta_j, & \text{if } \Delta_j < 0, \\ (\bar{x}_j - l_j)/\Delta_j, & \text{if } \Delta_j > 0, \\ \infty, & \text{if } \Delta_j = 0, \end{cases} \quad j \in B,\ j \neq n+1. \tag{8.44}$$

Instead of (8.22), accordingly, set S is now determined by

$$S = \{j \in B \mid \bar{x}_j = l_j \text{ or } u_j\}, \tag{8.45}$$

which will be moved from B to N before carrying out the next iteration.

In addition, the optimality should be checked according to whether the following index set is empty:

$$T = \{j \in N \mid \bar{x}_j = l_j,\ \bar{z}_j < 0\} \cup \{j \in N \mid \bar{x}_j = u_j,\ \bar{z}_j > 0\}, \tag{8.46}$$

In theory, the variable-bounded problem rules out the possibility of the lower unboundedness. Practically, however, the problem should better be viewed as lower unbounded if the stepsize exceeds some given large number.

We summarize the overall steps in the following algorithm.

Algorithm 8.9.1 (Generalized Face Algorithm with Cholesky Factorization) *Initial: (B, N), $r \leq k = |B|$, feasible solution $\bar{x}$, large number λ. This algorithm solves the variable-bounded problem:*

1. *Compute the Cholesky factor L of BB^T by Givens rotations.*
2. *Solve upper triangular system $L^{\mathrm{T}}\bar{y} = -(1/\nu)e_r$, where ν is the rth diagonal of L.*
3. *Go to Step 5 if $k = r$; else compute $\Delta_B = e_k - B^{\mathrm{T}}\bar{y}$.*
4. *Go to Step 8 if $\Delta_B \neq 0$.*
5. *Compute $\bar{z}_N = -N^{\mathrm{T}}\bar{y}$.*
6. *Stop if set T defined by (8.46) is empty (optimality achieved).*
7. *Go to Step 1 with $B = B \cup T$, $N = A \backslash B$.*
8. *Determine stepsize α and index q by (8.43).*
9. *Stop if $\alpha > \lambda$ (lower unboundedness declared).*
10. *Update: $\bar{x}_B = \bar{x}_B - \alpha\Delta_B$.*
11. *Update $B = B\backslash S$, $N = N \cup S$, $k = k - |S|$, where S is defined by (8.45).*
12. *Go to Step 1 if $|S| > 1$; else, update L by Subalgorithm 8.3.1.*
13. *Go to Step 2.*

Phase I issue:

The matching phase I can be obtained from the preceding algorithm with small changes. Setting a given $l < h < u$, the objective column of the auxiliary problem is

$$a_{n+1} = (b - Ah/\|b - Ah\|.$$

Starting from

$$\bar{x} = h, \qquad \bar{x}_{n+1} = \|b - Ah\|, \tag{8.47}$$

we solve it to attain $\bar{x}_{n+1} = 0$ for feasibility.

These steps can be put into the following procedure.

Subalgorithm 8.9.1 (Phase I for Generalized Face Algorithm with Cholesky Factorization)

Initial: (B, N), $r \leq k = |B|$, *solution* $\bar{x}$ *(8.47), This algorithm finds a feasible solution to the variable-bounded problem.*

The same as Algorithm 8.9.1 except for Steps 2, 6, 9, and 10 replaced by the following items:

2. *Solve triangular systems* $Lu = a_{n+1}$ *and* $L^T\bar{y} = u$*, successively.*
6. *Stop. (There is no feasible solution.)*
9. *Update:* $\bar{x}_B = \bar{x}_B - \alpha\Delta_B$.
10. *Stop if* $\bar{x}_{n+1} = 0$.

Example 8.9.1 Solve the following variable-bounded problem: :

$$A = \begin{bmatrix} 3 & 1 & -5 & 1 & & 3 & -4 \\ -2 & & 6 & & 1 & -6 & \\ -6 & -3 & 3 & & -2 & 4 & 7 \\ 1 & & -1 & 2 & & -1 & 2 \end{bmatrix} \qquad b = \begin{bmatrix} -17 \\ 14 \\ 16 \\ 4 \end{bmatrix}$$

$c = [-2, 0, 4, -6, 0, 5, 8]$.

$l = [1, -2, 0, 0, 0, -6, -5]$, $\qquad u = [9, 8, 12, 10, 15, 9, 5]$

Phase I. Solve the auxiliary problem with $h = [5, 3, 6, 5, 7.5, 1.5, 0]^T$.

Example 8.9.2 Solve the following variable-bounded problem: :

$$A = \begin{bmatrix} 3 & 1 & -5 & 1 & & 3 & -4 \\ -2 & & 6 & & 1 & -6 & \\ -6 & -3 & 3 & & -2 & 4 & 7 \\ 1 & & -1 & 2 & & -1 & 2 \end{bmatrix} \qquad b = \begin{bmatrix} -17 \\ 14 \\ 16 \\ 4 \end{bmatrix}$$

$c = [-2, 0, 4, -6, 0, 5, 8]$.

$l = [1, -2, 0, 0, 0, -6, -5]$, $\qquad u = [9, 8, 12, 10, 15, 9, 5]$

Phase I. Solve the auxiliary problem with $h = [5, 3, 6, 5, 7.5, 1.5, 0]^T$.

$a_8 = [-957/3266, -296/1395, 66/71, -231/3266]^T$.

Initial: $k = 8$, $B = \{1, \ldots, 8\}$, $N = \emptyset$, $\bar{x} = [5, 3, 6, 5, 15/2, 3/2, 0, 1633/33]^T$.

$B = [A \,|\, a_8]$, $N = \emptyset$, $\bar{x}_B = \bar{x}$, $l(8) = 0$, $k = 8$.
Call Subalgorithm 8.9.1.
Iteration 1:

1. $$L = \begin{bmatrix} 4666/597 & & & \\ -2933/425 & 499/92 & & \\ -4267/638 & -3467/444 & 1581/371 & \\ -902/7199 & -1107/2107 & -157/167 & 1461/466 \end{bmatrix}.$$
2. $\bar{y} = [-343/13{,}271, -81/3851, -261/97{,}973, -92/7483]^T$.
3. $\Delta_B = (101/3178, 96/5377, -53/7230, 719/14{,}256, 64/4075, -283/5626, -436/7249, 759/767)^{\mathrm{T}} \neq 0$.
8. $\alpha = \min[(5-1)/(101/3178), (3-(-2))/(96/5377), (6-12)/(-53/7230), (5-0)/(719/14{,}256), (15/2-(0))/(64/4075)), (3/2-15)/(-283/5626), (0-9)/(-436/7249), (1633/33)/(759/767)] = 7801/156$, $q = 8$.
9. $\bar{x}_B = \bar{x}_B - \alpha\Delta_B = [1968/577, 1494/709, 2362/371, 1628/657, 2847/424, 1301/324, 3907/1299, 0]^T$.
10. Feasible solution achieved:
    $\bar{x} = [1968/577, 1494/709, 2362/371, 1628/657, 2847/424, 1301/324, 3907/1299]^T$,
    $\bar{f} = c^T\bar{x} = 5702/119$.

Phase II. The coefficient matrix of the original reduced problem:

$$A = \begin{bmatrix} 3 & 1 & -5 & 1 & & 3 & -4 & \\ -2 & & 6 & & 1 & -6 & & \\ -6 & -3 & 3 & & -2 & 4 & 7 & \\ 1 & & -1 & 2 & & -1 & 2 & \\ -2 & & 4 & -6 & & 5 & 8 & -1 \end{bmatrix}$$

$B = \{1, \ldots, 8\}$, $N = \emptyset$, $k = 8$,
$\bar{x} = [1968/577, 1494/709, 2362/371, 1628/657, 2847/424, 1301/324, 3907/1299, 5702/119]$.
Call Algorithm 8.9.1.
Iteration 2:

1. $$L = \begin{bmatrix} 1523/195 & & & & \\ -6029/872 & 3496/647 & & & \\ -7610/1143 & -739/95 & 1581/371 & & \\ -195/1523 & -448/839 & -1381/1469 & 849/271 & \\ -527/84 & -3947/470 & -2235/1342 & -2180/493 & 2046/551 \end{bmatrix}.$$
2. $\bar{y} = [-821/2964, -214/1091, -360/7069, -147/1436, -543/7487]^T$.
3. $\Delta_B = [205/2267, 846/6811, 264/1993, 451/9684, 544/5769, 1210/10{,}251, 229/6843, 1087/1172]^T \neq 0$. $l = [1, -2, 0, 0, 0, -6, -5]$, $u = [9, 8, 12, 10, 15, 9, 5]$

8. $\alpha = \min[(1968/577 - 1)/(205/2267), (1494/709 - (-2))/(846/6811), (2362/371 - 0)/(264/1993), (1628/657 - 0)/(451/9684), (2847/424 - 0)/(544/5769), (1301/324 - (-6))/(1210/10{,}251), (3907/1299 - (-5))/(229/6843)] = 2426/91, q = 1.$
10. $\bar{x}_B = \bar{x}_B - \alpha\Delta_B = [1, -979/813, 1531/540, 9949/8047, 1151/274, 734/845, 1007/476, 3293/142]^T.$
11. $B = \{2, \ldots, 8\}, N = \{1\}.$
12. $$L = \begin{bmatrix} 9223/1279 & & & & \\ -4107/617 & 616/115 & & & \\ -1009/214 & -4441/604 & 4591/1403 & & \\ -1369/2468 & -690/1001 & -726/3475 & 6630/2189 & \\ -3387/568 & -3105/364 & -668/771 & -797/168 & 4839/1499 \end{bmatrix}.$$

Iteration 3:

2. $\bar{y} = [-1800/5677, -1384/6285, -44/1257, -271/1803, -601/6263]^T.$
3. $\Delta_B = [299/1410, 557/7482, 1182/28{,}199, 453/3016, 248/2493, 851/18{,}892, 179/198]^T \neq 0.$ $l = [1, -2, 0, 0, 0, -6, -5]$, $u = [9, 8, 12, 10, 15, 9, 5]$
8. $\alpha = \min[(-979/813 - (-2))/(299/1410), (1531/540 - 0)/(557/7482), (9949/8047 - 0)/(1182/28{,}199), (1151/274 - 0)/(453/3016), (734/845 - (-6))/(248/2493), (1007/476 - (-5))/(851/18{,}892)] = 987/263, q = 2.$
10. $\bar{x}_B = \bar{x}_B - \alpha\Delta_B = [-2, 1145/448, 1051/974, 1433/394, 845/1706, 1528/785, 4593/232]^T.$
11. $B = \{3, \ldots, 8\}, N = \{1, 2\}.$
12. $$L = \begin{bmatrix} 4999/700 & & & & \\ -3401/506 & 5317/1008 & & & \\ -2496/575 & -6322/897 & 1278/415 & & \\ -601/1073 & -536/751 & -551/3676 & 5431/1795 & \\ -1704/283 & -1066/121 & -92/1237 & -7550/1557 & 539/251 \end{bmatrix}.$$

Iteration 4:

2. $\bar{y} = [-1149/1805, -563/1283, -294/13{,}271, -432/1243, -211/973]^T.$
3. $\Delta_B = [173/4753, 268/8779, 589/1493, 461/4513, 157/4052, 762/973]^T \neq 0.$

   $l = [1, -2, 0, 0, 0, -6, -5], \qquad u = [9, 8, 12, 10, 15, 9, 5]$
8. $\alpha = \min[(11{,}145/448 - 0)/(173/4753), (1051/974 - 0)/(268/8779), (1433/394 - 0)/(589/1493), (845/1706 - (-6))/(461/4513), (1528/785 - (-5))/(157/4052)] = 2397/260, q = 5.$
10. $\bar{x}_B = \bar{x}_B - \alpha\Delta_B = [3488/1571, 3748/4699, 0, -6154/13{,}785, 89/56, 3899/310]^T.$
11. $B = \{3, 4, 6, 7, 8\}, N = \{1, 2, 5\}.$

12. $L = \begin{bmatrix} 4999/700 & & & & \\ -3401/506 & 1937/374 & & & \\ -2496/575 & -1012/149 & 4113/1369 & & \\ -601/1073 & -1408/1937 & -1357/11{,}066 & 765/253 & \\ -1704/283 & -3607/402 & 6869/22{,}406 & -1307/266 & 1 \end{bmatrix}.$

Iteration 5:

2. $\bar{y} = [-11/4, -643/336, 1/28, -13/8, -1]^T$,
3. $k = 5$.
5. $\bar{z}_N = [179/42, 20/7, 667/336]^T$.
6. $T = \emptyset$, optimality achieved.

The primal and dual optimal solutions are as follows:

$\bar{x} = [1, -2, 3488/1571, 3748/4699, 0, -6154/13{,}785, 89/56, 3899/310]^T$, $\bar{f} = 3899/310$.
$\bar{z} = [179/42, 20/7, 0, 0, 667/336, 0, 0, 0]^T$.

8.10 Remarks on Face Method with Cholesky Factorization

The quality of an initial partition (B, N) is important to the face algorithm. Although initially setting $B = A$ would be a convenient choice, a small B should be favorable, ideally containing the "right" (unknown optimal indices) face columns. Therefore, it might be advisable to determine an initial B, whose cardinality is no more than $2m$, especially for large-scale problems with $n \gg m$, by using some crash procedure based on the most-obtuse-angle heuristic, balanced with sparsity considerations.

The computational results with the face algorithm are excellent. The outcome is unsurprising because it uses the steepest descent search direction and involves less computational effort per iteration, solving fewer linear systems than the simplex algorithm.

Based on the *dual elimination* (Pan [34]), it is possible to design another type of algorithm in the following way. By a series of elementary transformations with row exchanges, the standard dual problem can be converted to the following form:

$$\begin{array}{ll} \min & -g = -\bar{g} + \bar{x}^{\mathrm{T}} z, \\ \text{s.t.} & G_2 z = d_2, \quad z \geq 0, \end{array} \tag{8.48}$$

together with $Uy + G_1 z = d_1$. Then, one can apply face Algorithm 8.6.1 or affine face Algorithm 8.8.1 to solve.

Gill and Murray [9] uses the Cholesky factorization $BB^{\mathrm{T}} = LL^{\mathrm{T}}$, where B is the typical basis, to improve the numerical stability of the simplex method. Based on BB^{T} and hence L being independent of the ordering of the columns of B, Saunders

[37] proposes an approach to keeping L sparse throughout the solution process. It initially carries out row and column permutations to turn the matrix A to the lower triangular as much as possible and then uses the resulting row permutation in subsequent iterations. This approach is suitable in the current context.

An interesting result published by Zhang et al. [39] is worth mentioning. They present a variant based on updating $(BB^{\mathrm{T}})^{-1}$ by the Sherman-Morrison formula to avoid the orthogonal transformations, but it turns out to be unstable (Pan [35]).

Chapter 9
Dual Face Method with Cholesky Factorization

The same idea of the face method with Cholesky factorization can be applied to the dual problem. Interestingly enough, the resulting *dual face method with Cholesky factorization* can be developed symmetrically with the primal version. It appears even much more efficient and attractive than the latter. It is the topic of this chapter.

In the next section, we put forward basic concepts first. Then, in the following sections, we address the steepest ascent direction; update the dual solution, dual face contraction, optimality test, and dual face expansion; and then formulate the dual face algorithm with Cholesky factorization. Then, we offer a double-phase algorithm and introduce a variant of the algorithm. The final section is devoted to The generalization of the dual face algorithm is done with Cholesky factorization.

Consider the following problem:

$$\begin{array}{ll} \min & c^T x \\ \text{s.t.} & \begin{bmatrix} A & -b \\ 0 & 1 \end{bmatrix} \begin{bmatrix} x \\ x_{n+1} \end{bmatrix} = \begin{bmatrix} 0 \\ 1 \end{bmatrix}, \qquad x \geq 0, \end{array} \tag{9.1}$$

where $A \in \mathcal{R}^{m\times n}$, $b \in \mathcal{R}^m$, $m < n$, rank $A = m$.

The preceding problem is equivalent to the standard LP problem (2.1). It is beneficial that the dual problem of (9.1) has a so-called reduced form, i.e.,

$$\begin{array}{ll} \max & y_{m+1} \\ \text{s.t.} & \begin{bmatrix} A^T & 0 \\ -b^T & 1 \end{bmatrix} \begin{bmatrix} y \\ y_{m+1} \end{bmatrix} + \begin{bmatrix} z \\ z_{n+1} \end{bmatrix} = \begin{bmatrix} c \\ 0 \end{bmatrix}, \qquad z \geq 0, \quad z_{n+1} = 0. \end{array} \tag{9.2}$$

Assume that the preceding problem is feasible.

Introduce the following notation:

$$\tilde{A} := \begin{bmatrix} A & -b \\ 0 & 1 \end{bmatrix}, \qquad \tilde{c} = \begin{bmatrix} c \\ 0 \end{bmatrix}. \tag{9.3}$$

P.-Q. Pan, *Face Method*, https://doi.org/10.1007/978-3-031-93594-7_9

Let $\tilde{A} = (B, N)$ be a partition of $\tilde{A}$, where

$$1 \leq |B| = k \leq m + 1, \qquad \text{rank } B = k.$$

In particular, it is regulated that the last index of B is $n + 1$ so that the last column and row of the matrix B are, respectively,

$$Be_k = [-b^T, 1]^T \qquad \text{and} \qquad e_{m+1}^T B = e_k^T. \tag{9.4}$$

Definition 9.0.1 The dual face associated with partition (B, N) is defined as

$$\begin{aligned} D(B) = \{(y, z) \in \mathcal{R}^{m+1} \times \mathcal{R}^{n+1} \mid B^T y = \tilde{c}_B, \\ N^{\mathrm{T}} y + z_N = \tilde{c}_N, \ z_B = 0, \ z_N \geq 0\}, \end{aligned} \tag{9.5}$$

where B is termed dual face matrix (or set).

A point on a dual face is termed *dual face point*. The zero-dimensional face ($k = m + 1$) has a unique dual face point—vertex. A dual face is said to be *level dual face* if the dual objective value is constant over it and *dual optimal* if the constant is equal to the optimal value.

Thereafter, $(y, z_N) \in D(B)$ and $(y, z) \in D_N$ will not be distinguished.

9.1 Steepest Ascent Search Direction

Assume that $(\bar{y}, \bar{z}_N) \in D(B)$. To find a new solution in $D(B)$ with a higher objective value, we maximize the objective function over $D(B)$.

To do so, we bring up the subprogram with feasible region D_N, i.e.,

$$\begin{array}{lll} \max & y_{m+1} & \\ \text{s.t.} & B^{\mathrm{T}} y = \tilde{c}_B, & z_B = 0, \\ & N^{\mathrm{T}} y + z_N = \tilde{c}_N, & z_N \geq 0. \end{array} \tag{9.6}$$

Thus, it is crucial to find some search direction Δy in y-space such that

$$B^T \Delta y = 0.$$

In other words, Δy should be an uphill in the null of B^T.

With this aspect, the $(m + 1) \times (m + 1)$ projection

$$P = I - B(B^{\mathrm{T}} B)^{-1} B^{\mathrm{T}} \tag{9.7}$$

is suitable for projecting the objective gradient e_{m+1} onto the null. This leads to $\Delta y = Pe_{m+1}$ and hence the following search direction:

$$\Delta y = e_{m+1} - B\bar{x}_B, \quad B^{\mathrm{T}} B\bar{x}_B = B^T e_{m+1}, \tag{9.8}$$

$$\Delta z_N = -N^{\mathrm{T}} \Delta y. \tag{9.9}$$

By the second expression of (9.4), the system involved in the preceding formulas has a simpler form, i.e.,

$$B^{\mathrm{T}} B\bar{x}_B = e_k. \tag{9.10}$$

Proposition 9.1.1 *For Δy defined by (9.8), it holds that $B^{\mathrm{T}}\Delta y = 0$.*

Proof By (9.8), the equation is verified. □

Proposition 9.1.2 *The following statements are equivalent:*

$$(i)\ \Delta y \neq 0; \qquad (ii)\ (\Delta y)_{m+1} > 0; \qquad (iii)\ e_{m+1} \notin \mathrm{range}(B).$$

Proof Premultiplying the first expression of (9.8) by Δy^{T}, then from Proposition 9.1.1, it follows that

$$0 \leq \Delta y^{\mathrm{T}} \Delta y = (\Delta y)_{m+1},$$

which means the equivalence of (i) and (ii). If $\Delta y = 0$, then it is known by the fist expression of (9.8) that $e_{m+1} = B\bar{x}_B$; hence, $e_{m+1} \in \mathrm{range}(B)$, so (iii) implies (i). Conversely, if $e_{m+1} \in \mathrm{range}(B)$, then there is some vector u such that

$$e_{m+1} = Bu. \tag{9.11}$$

Premultiply the preceding equality by B^{T}; then it follows from the second expression of (9.4) that

$$e_k = B^{\mathrm{T}} Bu,$$

which together with the second expression of (9.8) and B of the full column rank gives $u = \bar{x}_B$. Therefore, (9.11) implies

$$e_{m+1} = B\bar{x}_B.$$

Substituting the preceding equality to the first expression of (9.8) leads to $\Delta y = 0$. So (i) implies (iii). Further, it can be asserted that (i) and (iii) are equivalent. □

Proposition 9.1.3 *If $\Delta y \neq 0$, then*

$$e_{m+1}^{\mathrm{T}} \Delta y = (\Delta y)_{m+1} > 0, \tag{9.12}$$

$$e_{m+1}^{\mathrm{T}}\Delta y/\|\Delta y\| \geq e_{m+1}^{\mathrm{T}}u/\|u\|, \quad \forall\, 0 \neq u \in \mathrm{Null}(B^{\mathrm{T}}). \tag{9.13}$$

Proof Equality (9.12) directly follows from Proposition 9.1.2. For P defined by (9.7), it holds that

$$Pu = u \neq 0, \quad \forall\, 0 \neq u \in \mathrm{Null}(B^{\mathrm{T}}). \tag{9.14}$$

It is known from Cauchy inequality that

$$\|Pe_{m+1}\|\|Pu\| \geq (Pe_{m+1})^{\mathrm{T}}(Pu).$$

Multiplying the preceding inequality by $1/\|Pu\|$ renders

$$\|Pe_{m+1}\| \geq (Pe_{m+1})^{\mathrm{T}}(Pu)/\|Pu\|,$$

i.e.,

$$(Pe_{m+1})^{\mathrm{T}}(Pe_{m+1})/\|Pe_{m+1}\| \geq (Pe_{m+1})^{\mathrm{T}}(Pu)/\|Pu\|,$$

which along with $P^2 = P,\ P^{\mathrm{T}} = P$, (9.14) and $\Delta y = Pe_{m+1}$ leads to (9.13). □

The preceding proposition implies that $\Delta y \neq 0$ forms the most acute angle with objective gradient e_{m+1} of the subprogram, in the null of B^{T}. That is why we call it *the steepest ascent*. Thus, the "best" choice is taking it as a search direction in y-space.

9.2 Computation of Search Direction

As for computing the search direction $(\Delta y, \Delta z_N)$, defined by (9.8)–(9.9), the key is the unique solution to the $k \times k$ system (9.10).

Assume that the QR factorization of B is

$$B = QR_1 = Q\begin{bmatrix} L^{\mathrm{T}} \\ 0 \end{bmatrix}, \tag{9.15}$$

where $Q \in \mathcal{R}^{(m+1)\times(m+1)}$ is orthogonal and $L \in\in \mathcal{R}^{k\times k}$ is nonsingular lower triangular. Then, it holds that

$$B^{\mathrm{T}}B = LL^{\mathrm{T}}.$$

Therefore, L is just the Cholesky factor of $B^{\mathrm{T}}B$. Consequently, the system (9.10) becomes

$$LL^{\mathrm{T}}x_B = e_k,$$

whose unique solution can be obtained by solving the following two triangular systems:

$$Lu = e_k, \qquad L^{\mathrm{T}}x_B = u.$$

Obviously, the solution to the first system is $u = (1/\mu)e_k$, where μ is the kth diagonal of L. Therefore, only the following lower triangular system needs to be solved:

$$L^{\mathrm{T}}x_B = (1/\mu)e_k. \tag{9.16}$$

9.3 Updating of Dual Face Solution

Assume that $(\bar{y}, \bar{z}_N) \in D_N$ and that $(\Delta y, \Delta z_N)$ is available with $\Delta y \neq 0$.

Introduce the following index set:

$$J = \{j \in N \mid \Delta z_j < 0\}. \tag{9.17}$$

If J is nonempty, the new dual solution can be determined by the following line search scheme:

$$\hat{y} = \bar{y} + \beta\Delta y, \qquad \hat{z}_N = \bar{z}_N + \beta\Delta z_N, \tag{9.18}$$

where β is the maximum stepsize, maintaining $\bar{z}_N \geq 0$, defined by

$$\beta = -\bar{z}_q/\Delta z_q = \min_{j\in J} -\bar{z}_j/\Delta z_j. \tag{9.19}$$

If some component of $\bar{z}_J$ vanishes, so is the stepsize β. Consequently, the so-called "new" dual solution is the same as the old. This case is said to be degeneracy.

Lemma 9.3.1 *Assume that $(\bar{y}, \bar{z}_N) \in D_N$ and that $(\Delta y, \Delta z_N)$ is available with $\Delta y \neq 0$.*

(i) If J is nonempty, then $\hat{z}_N$ is a boundary point belonging to D_N; the objective value does not decrease and strictly increases if $(\bar{y}, \bar{z}_N)$ is nondegenerate.
(ii) If set J is empty, the dual problem (9.2) is upper unbounded, or there is no feasible solution to the primal problem (9.1).

Proof

(i) In this case, the new point $(\hat{y}, \hat{z}_N)$ is well-defined. By $(\bar{y}, \bar{z}_N) \in D_N$, (9.19) and Proposition 9.1.1, it is known that $(\hat{y}, \hat{z}_N) \in D_N$. In addition, it is clear that

$\hat{z}_q = 0$; hence, the new iterate is on the boundary, associated with the objective value below:

$$\hat{y}_{m+1} = \bar{y}_{m+1} + \beta(\Delta y)_{m+1}. \tag{9.20}$$

Further, from $\beta \geq 0$ and Proposition 9.1.2, it is known that $(\Delta y)_{m+1} > 0$; hence, $\hat{y}_{m+1} \geq \bar{y}_{m+1}$. If $(\bar{y}, \bar{z}_N)$ is nondegenerate, then $\beta > 0$; hence, $\hat{y}_{m+1} > \bar{y}_{m+1}$, which indicates that the objective value increases strictly.

(ii) In this case, it is obvious that $(\hat{y}, \hat{z}_N) \in D_N \ \forall \beta > 0$. It is then known by (9.20) and Proposition 9.1.2 that the objective value tends to ∞, as so does the stepsize β.

□

9.4 Dual Face Contraction

Assume that $N^{\mathrm{T}}\Delta y \not\leq 0$, in which case $\hat{z}_q = 0$. We execute the following tactic:

$$\breve{B} = B \cup \{q\}, \qquad \breve{N} = N \backslash \{q\}.$$

According to Definition 9.10.1, the dimension of the dual face is reduced by one. We have the following result.

Proposition 9.4.1 *If* Rank $B = k$, *then* Rank $\breve{B} = k + 1$.

Proof It is clear that $k \leq$ Rank $\breve{B} \leq k + 1$. Assume Rank $\breve{B} \neq k + 1$. Then, there is a vector $u \neq 0$ such that $a_q = Bu$. Thus, it is known by Proposition 9.1.1 that $B^{\mathrm{T}}\Delta y = 0$; hence,

$$a_q^{\mathrm{T}}\Delta y = u^{\mathrm{T}}B^{\mathrm{T}}\Delta y = 0.$$

Further, it follows from (9.19) that $a_q^{\mathrm{T}}\Delta y > 0$, which is a contradiction. Therefore, Rank $\breve{B} = k + 1$. □

According to the preceding proposition, there exists the Cholesky factorization:

$$\breve{B}^{\mathrm{T}}\breve{B} = \breve{L}\breve{L}^{\mathrm{T}}.$$

If a_q is inserted as the second last column of $\breve{B}$, the Cholesky factor $\breve{L}$ can be computed by the following procedure.

Subalgorithm 9.4.1 *Assume that the QR factorization of B is given by (9.15).*

1. *Compute* $v = Q^{\mathrm{T}}a_q$.
2. *Insert* v *before the* $(k + 1)$*th column of* R_1, *yielding matrix* H.

3. *Determine Givens rotations* $J_{m+1}, \ldots, J_{k+1}$, *where* J_i *in the* (k, i)*-plane eliminates the entry in row* i, *such that*

$$J_{k+1} \ldots J_{m+1} H = \begin{bmatrix} \breve{L}^{\mathrm{T}} \\ 0 \end{bmatrix},$$

where $\breve{L} \in \mathcal{R}^{(k+1)\times(k+1)}$ *is lower triangular.*

Proposition 9.4.2 *Subalgorithm 9.4.1 is valid.*

Proof It is obvious that

$$\begin{aligned} J_{k+1} \ldots J_{m+1} Q^{\mathrm{T}} \breve{B} &= J_{k+1} \ldots J_{m+1} Q^{\mathrm{T}} [a_1, \ldots, a_{k-1}, a_q, a_{n+1}] \\ &= J_{k+1} \ldots J_{m+1} H = \begin{bmatrix} \breve{L}^{\mathrm{T}} \\ 0 \end{bmatrix}. \end{aligned}$$

By Proposition 9.4.1, it is known that $\breve{L} \in \mathcal{R}^{(k+1)\times(k+1)}$ is nonsingular. Thus, from the preceding expression, it follows that

$$\breve{B}^{\mathrm{T}} \breve{B} = \breve{B}^{\mathrm{T}} Q J_{m+1}^{\mathrm{T}} \ldots J_{k+1}^{\mathrm{T}} J_{k+1} \ldots J_{m+1} Q^{\mathrm{T}} \breve{B} = (\breve{L}, 0) \begin{bmatrix} \breve{L}^{\mathrm{T}} \\ 0 \end{bmatrix} = \breve{L}\breve{L}^{\mathrm{T}}.$$

□

Step 1 of Subalgorithm 9.4.1 indicates that there is a need for storing Givens rotations in product form.

In general, multiple components of $\hat{z}_N$ would vanish; in other words, the following set includes indices other than q:

$$S = \{j \in N \mid \hat{z}_j = 0\}. \tag{9.21}$$

Then, executing the following tactic seems logical:

$$\hat{N} = N \backslash S, \qquad \hat{B} = B \cup S. \tag{9.22}$$

although computing the associated Cholesky factorization becomes somewhat cumbersome. This approach is attractive because of the ensuing nonzero stepsize in the next iteration.

9.5 Optimality Test

Assume that Δy vanishes and cannot be used as a search direction, e.g., whenever $k = m + 1$.

Lemma 9.5.1 *Assume that* Δy *is defined by (9.8). If* $\Delta y = 0$*, then* D_N *is a level dual face and vice versa if there is* $(\bar{y}, \bar{z}_N) \in D_N$ *such that* $\bar{z}_N > 0$*.*

Proof Assume that $\bar{x}_B$ is the solution to $B^{\mathrm{T}} B x_B = e_k$. From $\Delta y = 0$ and the first expression of (9.8), it follows that

$$e_{m+1} = B\bar{x}_B.$$

Transposing the two sides of the preceding equality and post-multiplying the result by y (for any $(y, z_N) \in D_N$) gives

$$e_{m+1}^{\mathrm{T}} y = \bar{x}_B^{\mathrm{T}} B^{\mathrm{T}} y,$$

which with $B^{\mathrm{T}} y = c_B$ leads to

$$y_{m+1} = \bar{x}_B^{\mathrm{T}} c_B.$$

So, the objective value is constant over D_N. Therefore, D_N is a level face.

Now, assume that D_N is a level face but $\Delta y \neq 0$. If $N^{\mathrm{T}} \Delta y \leq 0$, it is known by Lemma 9.3.1(ii) that the objective value is upper unbounded over D_N, contradicting the fact that it is a level face. If, otherwise, $N^{\mathrm{T}} \Delta y \not\leq 0$, then since $(\bar{y}, \bar{z}_N)$ is nondegenerate, it is known from Lemma 9.3.1(i) that $(\hat{y}, \hat{z}_N)$, defined by (9.18)–(9.19), belongs to D_N, with a strictly increased objective value, contradicting the fact that D_N is a level face. Therefore, it holds that $\Delta y = 0$. □

Theorem 9.5.2 *Assume that* $(\bar{y}, \bar{z}_n)$ *belongs to* D_N *and that* Δy *determined by (9.8) vanishes. If the following index set:*

$$T = \{j \in B \mid \bar{x}_j < 0\} \tag{9.23}$$

is empty, then $\bar{x}$ *and* $(\bar{y}, \bar{z})$ $(\bar{x}_N = 0,\ \bar{z}_B = 0)$ *are a pair of primal and dual optimal solutions, and* D_N *is a dual optimal face to (9.2).*

Proof Note that the emptiness of T implies $\bar{x}_B \geq 0$. From (9.8) and $\Delta y = 0$, it follows that $e_{m+1} = B\bar{x}_B$. Therefore, $\bar{x}$ with $\bar{x}_N = 0$ is a feasible solution to (9.1). On the other hand, it is obvious that $(\bar{y}, \bar{z})$ $(\bar{z}_B = 0)$ is a feasible solution to (9.2). The two solutions exhibit slackness complementarity. Therefore, they are a pair of primal and dual optimal solutions. By Lemma 9.5.1, D_N is a (dual) level face. In fact, it is a (dual) optimal face since including an optimal solution. □

9.6 Dual Face Expansion

Assume that Δy vanishes but set T defined by (9.23) is nonempty, so optimality cannot be declared.

Firstly, we determine column index p such that

$$p = \arg\min_{j \in N} \bar{z}_j, \tag{9.24}$$

where $\bar{z}_p$ is negative. Then, move p from B to N, i.e., execute the following tactic:

$$\widehat{B} = B \backslash \{p\}, \quad \widehat{N} = N \cup \{p\}. \tag{9.25}$$

Consequently, the dual face is expanded, according to Definition 9.10.1.

Assume that $\widehat{B} \in \mathcal{R}^{(m+1)\times(k-1)}$ is the matrix, resulting from dropping the pth column of B. It still has a full column rank, and the following steps can generate the associated Cholesky factor.

Subalgorithm 9.6.1 *Initial: Cholesky factorization $B^{\mathrm{T}}B = LL^{\mathrm{T}}$. This algorithm generates the Cholesky factor $\widehat{L}$ of $\widehat{B}^T\widehat{B}$.*

1. *Drop the pth column from L^{T}, yielding Hessenberg matrix H.*
2. *Determine Givens rotations $J_{p-1}, \ldots, J_{k-1}$, where J_i in $(i, i+1)$-plane eliminates the entry in row $i+1$, such that*

$$J_{k-1} \ldots J_{p-1} H = \begin{bmatrix} \widehat{L}^{\mathrm{T}} \\ 0 \end{bmatrix},$$

where $\widehat{L} \in \mathcal{R}^{(k-1)\times(k-1)}$ is lower triangular.

Proposition 9.6.1 *Subalgorithm 9.6.1 is valid.*

Proof It holds that

$$\begin{aligned} J_{k-1} \ldots J_p Q^{\mathrm{T}} \widehat{B} &= J_{k-1} \ldots J_p Q^{\mathrm{T}} [a_1, \ldots, a_{p-1}, a_{p+1}, \ldots, a_{n+1}] \\ &= J_{k-1} \ldots J_p H = \begin{bmatrix} \widehat{L}^{\mathrm{T}} \\ 0 \end{bmatrix}, \end{aligned}$$

Therefore, it follows that

$$\widehat{B}^{\mathrm{T}} \widehat{B} = \widehat{B}^{\mathrm{T}} Q J_p^{\mathrm{T}} \ldots J_{k-1}^{\mathrm{T}} J_{k-1} \ldots J_p Q^{\mathrm{T}} \widehat{B} = (\widehat{L}, 0) \begin{bmatrix} \widehat{L}^{\mathrm{T}} \\ 0 \end{bmatrix} = \widehat{L}\widehat{L}^{\mathrm{T}}.$$

□

If set T includes multiple indices, the following tactic to expand the dual face seems more attractive:

$$\widehat{B} = B \backslash T, \qquad \widehat{N} = N \cup T.$$

However, computing the resulting Cholesky factor is somewhat cumbersome. , Besides, the following claim is noteworthy.

Proposition 9.6.2 *Index set T determined by (9.23) does not include index $n+1$.*

Proof The last equation of (9.16) is

$$\mu x_{n+1} = 1/\mu,$$

from which it follows that

$$\bar{x}_{n+1} = 1/\mu^2 > 0.$$

Therefore, T does not include $n + 1$. □

The preceding proposition ensures that index $n + 1$ remains in B throughout the solution process.

9.7 Dual Face Algorithm with Cholesky Factorization

In this section, we formulate the dual face algorithm with Cholesky factorization.

Each iteration calculates the search direction, and $\bar{z}_N$ is updated with (B, N) adjusted accordingly. Iteration by iteration, the dimension of face $D(B)$ decreases strictly unless the upper unboundedness of the dual problem is detected. Such a series of iterations is referred to as *dual face contraction process*, which ends whenever Δ_y vanishes, or $|B|$ reaches $m + 1$; this implies that a level dual face is attained. If set T defined by (9.23) is empty, then it is an optimal face. Otherwise, we start another round of dual face contraction with an updated partition.

Initially, it seems preferable to take a small B, although we also favor partition (B, N) that

$$\bar{z}_B = 0, \qquad \bar{z}_N > 0. \tag{9.26}$$

The overall steps can be summarized into the following algorithm.

Algorithm 9.7.1 (Dual Face Algorithm with Cholesky Factorization)
Initial: $B,\ N,\ 1 \le k \equiv |B| \le m + 1$; *dual feasible solution* $(\bar{y}, \bar{z})$ *and Cholesky factor* L *of* $B^{\mathrm{T}}B$. *This algorithm solves the standard problem.*

1. *Solve* $L^{\mathrm{T}}\bar{x}_B = -(1/\mu)e_k$ *for* $\bar{x}_B$, *where* μ *is the* $(m + 1)$*th diagonal of* L.
2. *If* $|B| < m + 1$, *compute* $\Delta y = e_{m+1} - B\bar{x}_B$.
3. *Go to Step 11 if* $|B| = m + 1$ *or* $\Delta y = 0$.
4. *Compute* $\Delta z_N = -N^T \Delta y$.
5. *Stop if* J *defined by (9.17) is empty.*
6. *Determine index* q *and stepsize* β *such that*

$$\beta = -\bar{z}_q/\Delta z_q = \min_{j \in J} -\bar{z}_j/\Delta z_j.$$

7. *If* $\beta \neq 0$, *update:*$\bar{y} = \bar{y} + \beta\Delta y,\ \bar{z}_N = \bar{z}_N + \beta\Delta z_N$.

8. *Update* (B, N) *by taking* q *from* N *and inserting it before the last index of* B.
9. *Call Subalgorithm 9.4.1 to update* L.
10. *Go to Step 1.*
11. *Stop if* $\bar{x}_B \geq 0$.
12. *Determine* $p \in \arg\min_{j \in B} \ \bar{x}_j$.
13. *Update:*$B = B \backslash \{p\},\ N = N \cup \{p\}$.
14. *Call Subalgorithm 9.6.1 to update* L.
15. *Go to Step 1.*

Theorem 9.7.1 *Assume that each contraction process involves a nonzero stepsize. Algorithm 9.10.1 terminates either at*

(i) *Step 4, detecting upper unboundedness of (9.2), or at*
(ii) *Step 11, generating a (dual) optimal face with a pair of primal and dual optimal solutions.*

Proof It is evident that each contraction process reaches a level dual face in a finite number of iterations.

According to Theorem 9.5.2, optimality is achieved if T is empty at Step 11. In the other case, set T is moved from B to N, and start another contraction process. As each contraction process includes a nonzero stepsize at some iteration, the dual objective value decreases strictly while reaching a level dual face.

Thus, that the algorithm does not terminate implies that there are infinitely many level dual faces, contradicting the fact that the level dual faces are finite. Therefore, the algorithm terminates.

The meaning of Step 8 is obvious, according to Lemma 9.3.1. □

9.8 Double-Phase Dual Face Algorithm

To provide a dual feasible solution to get Algorithm 9.10.1 started, any dual-phase I approaches are applicable. In particular, the single-artificial-variable approach is recommended. For phase I purpose, in this section, an auxiliary dual problem is formed and solved by modifying the dual face algorithm with Cholesky factorization.

Given any $(\bar{y}, \bar{z})$ with $\bar{z} \geq 0$. If $A^{\mathrm{T}}\bar{y} + \bar{z} = c$, then $(\bar{y}, \bar{z})$ is already a dual feasible solution. Now, assume it is not the case.

Define

$$h = (\bar{z} - c + A^{\mathrm{T}}\bar{y})/\|\bar{z} - c + A^{\mathrm{T}}\bar{y}\|. \tag{9.27}$$

Introducing the "objective" variable y_{m+1}, we construct the following auxiliary problem:

$$\begin{array}{ll} \max & y_{m+1} \\ \text{s.t.} & A^T y + y_{m+1}h + z = c, \quad z \geq 0, \quad y_{m+1} \leq 0. \end{array} \tag{9.28}$$

It is evident that the preceding upper bounded problem has a feasible solution $(\bar{y}, \bar{y}_{m+1}, \bar{z})$, where

$$\bar{y}_{m+1} = -\|\bar{z} - c + A^{\mathrm{T}}\bar{y}\|.$$

Hence, there is an optimal solution. If its optimal value is equal to 0, then the corresponding part (y, z) of the optimal solution is feasible to the original dual problem; otherwise, the dual problem is infeasible.

In order to limit the possibility of zero stepsizes, it is inadvisable for $\bar{z}$ to have many zero components. For this reason, we recommend to set $\bar{y} = 0$ and $\bar{z} = e$, so that the expression (9.38) becomes

$$h = (e - c)/\|e - c\|. \tag{9.29}$$

Consequently, the following feasible solution to the associated auxiliary problem presents:

$$\bar{y} = 0, \quad \bar{y}_{m+1} = -\|e - c\|, \quad \bar{z} = e. \tag{9.30}$$

Now, resetting

$$A^{\mathrm{T}} := [A^{\mathrm{T}}, h] \qquad \text{and} \qquad y^T := [y^T, y_{m+1}] \tag{9.31}$$

converts the auxiliary problem into a special form, the same as (9.2), except for the so-called *objective* column h. Therefore, $\bar{x}_B$ is now determined by solving the following two triangular systems, successively:

$$Lu = h, \qquad L^T\bar{x}_B = u. \tag{9.32}$$

Then, initiating with $B = \emptyset$, the following modification of Algorithm 9.10.1 is a proper solver.

Subalgorithm 9.8.1 (Dual Face Phase I)
Initial: $B = \emptyset$, $N = A$, $\bar{y} = \|e - c\|e_{m+1}$, $\bar{z}_N = e$. *This algorithm finds a feasible solution to the standard problem.*

1. *Set* $\Delta y = e_{m+1}$, *and go to Step 5.*
2. *Solve the two triangular systems (9.32).*
3. *Compute* $\Delta y = e_{m+1} - B\bar{x}_B$.
4. *Go to Step 11 if* $\Delta y = 0$.
5. *Compute* $\Delta z_N = -N^T\Delta y$.
6. *Set* $\beta = \infty$ *if* $J = \{j \in N \mid \Delta z_j < 0\}$ *is empty; else*

determine q and β such that $\beta = -\bar{z}_q/\Delta z_q = \min_{j\in J}\ -\bar{z}_j/\Delta z_j$

7. *Redetermine* $\beta = \min\{-\bar{y}_{m+1}/\Delta y_{m+1},\ \beta\}$.
8. *Update:* $\bar{y} = \bar{y} + \beta\Delta y,\ \bar{z}_N = \bar{z}_N + \beta\Delta z_N$.
9. *Stop if* $\bar{y}_{m+1} = 0$ *(dual feasibility achieved).*
10. *Update* (B, N) *by bringing q from N to B, and go to Step 2.*
11. *Update L by Subalgorithm 9.4.1.*
12. *Stop if* $\bar{x}_B \ge 0$ *(dual infeasible problem).*
13. *Determine* $p \in \arg\min_{j\in B}\ \bar{x}_j$.
14. *Update* (B, N) *by bringing p from B to N.*
15. *Update L by Subalgorithm 9.6.1.*
16. *Go to Step 2.*

Example 9.8.1 Solve the following standard problem using the double-phase face algorithm:

$$A = \begin{bmatrix} 2 & 1 & 1 & 1 & 0 \\ 1 & -1 & -1 & 0 & -1 \\ 2 & 0 & -1 & 0 & 1 \end{bmatrix}, \qquad b = \begin{bmatrix} 2 \\ -1 \\ 0 \end{bmatrix}$$

$$c = [4, -1, 2, 0, 0]^T$$

Phase I

Calculate the following vector:

$$h = (e - c)/\|e - c\| = [-3, 2, -1, 1, 1]^T/4.$$

Form the auxiliary problem (9.28), and reset (A^T, y^T) by (9.31).

Call Subalgorithm 9.8.1.

Initial: $B = \emptyset$, $N = \{1, 2, 3, 4, 5\}$, $\bar{y} = [0, 0, 0, -4]^{\mathrm{T}}$, $\bar{z} = [1, 1, 1, 1, 1]^{\mathrm{T}}$.

Iteration 1:

1. $\Delta y = [0, 0, 0, 1]^{\mathrm{T}}$.
5. $\Delta z_N = [3/4, -1/2, 1/4, -1/4, -1/4]^{\mathrm{T}}$.
6. $J = \{2, 4, 5\}$, $\beta = \min\{-1/(-1/2), -1/(-1/4), -1/(-1/4)\} = 2, q = 2$.
7. $\beta = \min\{4/1, 2\} = 2$.
8. $\bar{y} = [0, 0, 0, -4]^{\mathrm{T}} + 2[0, 0, 0, 1]^{\mathrm{T}} = [0, 0, 0, -2]^{\mathrm{T}}$.
   $\bar{z}_N = [1, 1, 1, 1, 1]^{\mathrm{T}} + 2[3/4, -1/2, 1/4, -1/4, -1/4]^{\mathrm{T}}$
   $= [5/2, 0, 3/2, 1/2, 1/2]^{\mathrm{T}}$.
9. $\bar{y}_4 = -2 \neq 0$.
10. $B = \{2\}$, $N = \{1, 3, 4, 5\}$.
11. $L = [3/2]$.

Iteration 2:

2. $\bar{x}_B = [2/9]$.
3. $\Delta y = [0, 0, 0, 1]^{\mathrm{T}} - [1, -1, 0, 1/2]^{\mathrm{T}}(2/9) = [-2/9, 2/9, 0, 8/9]^{\mathrm{T}}$.
5. $\Delta z_N = [8/9, 2/3, 0, 0]^{\mathrm{T}}$.
6. $J = \emptyset$, $\beta = \infty$.
7. $\beta = -(-2)/(8/9) = 9/4$.

8. $\bar{y} = [0, 0, 0, -2]^{\mathrm{T}} + (9/4)[-2/9, 2/9, 0, 8/9]^{\mathrm{T}} = [-1/2, 1/2, 0, 0]^{\mathrm{T}}$.
$$\bar{z}_N = [5/2, 3/2, 1/2, 1/2]^{\mathrm{T}} + (9/4)[8/9, 2/3, 0, 0]^{\mathrm{T}} = [9/2, 3, 1/2, 1/2]^{\mathrm{T}}.$$
9. $\bar{y}_4 = 0$. (dual feasibility is achieved)
Convert the dual problem of form (9.2) with setting (9.3).

Phase II: Call Algorithm 9.10.1.
Initial: $B = \{2, 6\},\ N = \{1, 3, 4, 5\}\ \bar{y} = [-1/2, 1/2, 0, -3/2]^{\mathrm{T}}$,
$\bar{z}_N = [9/2, 3, 1/2, 1/2]^{\mathrm{T}}$,
$$L = \begin{bmatrix} \sqrt{2} & 0 \\ -(3\sqrt{2})/2 & \sqrt{3/2} \end{bmatrix}.$$
Iteration 3:

1. $\bar{x}_B = [1, 2/3]^{\mathrm{T}}$.
2. $$\Delta y = \begin{bmatrix} 0 \\ 0 \\ 0 \\ 1 \end{bmatrix} - \begin{bmatrix} 1 & -2 \\ -1 & 1 \\ 0 & 0 \\ 0 & 1 \end{bmatrix} \begin{bmatrix} 1 \\ 2/3 \end{bmatrix} = \begin{bmatrix} 1/3 \\ 1/3 \\ 0 \\ 1/3 \end{bmatrix}.$$
3. $\Delta y \neq 0$.
4. $$\Delta z_N = -\begin{bmatrix} 2 & 1 & 2 & 0 \\ 1 & -1 & -1 & 0 \\ 1 & 0 & 0 & 0 \\ 0 & -1 & 1 & 0 \end{bmatrix} \begin{bmatrix} 1/3 \\ 1/3 \\ 0 \\ 1/3 \end{bmatrix} = \begin{bmatrix} -1 \\ 0 \\ -1/3 \\ 1/3 \end{bmatrix}, \qquad J = \{1, 4\}.$$
5. $J = \{1, 4\}$.
6. $\beta = \min\{-(9/2)/-1, -(1/2)/(-1/3)\} = 3/2,\ q = 4$.
7. $\bar{y} = [-1/2, 1/2, 0, -3/2]^{\mathrm{T}} + (3/2)[1/3, 1/3, 0, 1/3]^{\mathrm{T}} = [0, 1, 0, -1]^{\mathrm{T}}$,
$$\bar{z}_N = [9/2, 3, 1/2, 1/2]^{\mathrm{T}} + (3/2)[-1, 0, -1/3, 1/3]^{T} = [3, 3, 0, 1]^{T}.$$
8. $B = \{2, 4, 6\},\quad N\{1, 3, 5\}$.
9. $$L = \begin{bmatrix} \sqrt{2} & 0 & 0 \\ \sqrt{2}/2 & \sqrt{2}/2 & 0 \\ -3/\sqrt{2} & -\sqrt{2}/2 & 1 \end{bmatrix}.$$

Iteration 4:

1. $\bar{x}_B = [1, 1, 1]^{\mathrm{T}}$.
2. $$\Delta y = \begin{bmatrix} 0 \\ 0 \\ 0 \\ 1 \end{bmatrix} - \begin{bmatrix} 1 & 1 & -2 \\ -1 & 0 & 1 \\ 0 & 0 & 0 \\ 0 & 0 & 1 \end{bmatrix} \begin{bmatrix} 1 \\ 1 \\ 1 \end{bmatrix} = \begin{bmatrix} 0 \\ 0 \\ 0 \\ 0 \end{bmatrix}.$$

11. $\bar{x}_B \geq 0$. The optimal solution and optimal value to (9.2)
$$\bar{y} = [0, 1, 0, -1]^{\mathrm{T}}, \quad \bar{z} = [3, 0, 3, 0, 1, 0]^{\mathrm{T}},$$

The related dual optimal face:

$$y_1 = 0, \quad y_2 = 1, \quad -3 \le y_3 \le 1, \quad y_4 = -1.$$

Consequently, the optimal solution and optimal value to the original problem are

$$\bar{x} = [0, 1, 0, 1, 0]^{\mathrm{T}}, \quad \bar{f} = -1,$$

and the dual optimal solution is

$$\bar{y} = [0, 1, 0]^{\mathrm{T}}, \quad \bar{z} = [3, 0, 3, 0, 1]^{\mathrm{T}}.$$

We conducted related computational experiments in the same environment as for Code FALP.

The following two dense codes were tested and compared:

1. DFA: A prototype of dual face Algorithm 9.10.1, supported by Subalgorithm 9.8.1 as phase I.
2. RSA: Our working simplex code.

The test set is the same 26 smallest Netlib problems. The ratios of RSA to DFA are listed below (excluding DEGEN2, with which Code RSA failed to solve):

Ratio of RSA to DFA

				Total		Phase 1	
Ave. size (25)	m	n	$m+n$	Itn	Time	Itn	Time
RSA/DFA	213	248	461	1.19	10.04	2.43	36.70

Code DFA outperformed RSA with great superiority: the overall time ratio is up to **10.04**. It performed even better with phase I alone; the related time ratio is as high as **36.79**! So, it appears that the dual phase I achieves its goal very quickly.

A comparison between the two face codes reveals that DFA is superior to FALP; the overall time ratio of FALP to DFA is **2.04**. It is so partially because DFA involves smaller systems: solving a triangular $k \times k$ system each iteration with a dynamically varying $k \le m$. Actually, k did not reach m for 14 of 26 (**54%**) test problems. For more details, see Appendix A.

9.9 Variant: Updating $(B^T B)^{-1}$

This section presents a variant of the dual face algorithm with Cholesky factorization. It bypasses orthogonal transformation even does linear systems.

Without a doubt, the key to the preceding algorithm is to solve the system

$$B^T B x_B = e_k,$$

related to the computation of the search direction. Using the Cholesky factorization

$$B^T B = LL^T,$$

the algorithm copes with two triangular systems, i.e.,

$$Lu = e_k, \qquad L^T x_B = u.$$

Obviously, the solution to the first system is $u = (1/\mu)e_k$, where μ is the kth diagonal of L. Consequently, only the second triangular system needs to be solved.

The computation of Cholesky factorization involves orthogonal transformations, which would not be very amenable for solving large-scale sparse problems. Therefore, it is natural to think about avoiding Cholesky factorization. Fortunately, it is possible to update $(B^{\mathrm{T}}B)^{-1}$ directly instead.

The derived variant is so simple that it does not solve linear systems. It might be the first algorithm of such type ever.

To begin with, we put forward the following result.

Lemma 9.9.1 *Let B be $(m+1) \times k$ matrix with $1 \leq k < m+1$ and* Rank $B = k$. *For a_q, determined by (9.19), it holds that*

$$a_q^{\mathrm{T}} a_q - a_q^{\mathrm{T}} B(B^{\mathrm{T}}B)^{-1}B^{\mathrm{T}} a_q \neq 0. \tag{9.33}$$

Proof Since B is of full column rank k, $(B^{\mathrm{T}}B)^{-1}$ is well-defined. Denote by $\breve{B}$ the matrix yielding from B by adding column a_q. According to Proposition 9.4.1, Rank $\breve{B} = k+1$, as $a_q \notin$ range B implies. Therefore, it holds that

$$a_q - B(B^{\mathrm{T}}B)^{-1}B^{\mathrm{T}} a_q \neq 0,$$

or equivalently,

$$(I - B(B^{\mathrm{T}}B)^{-1}B^{\mathrm{T}})a_q \neq 0,$$

which means that the orthogonal projection of a_q onto the orthogonal complement space of range B does not vanish. Therefore, it holds that

$$a_q^{\mathrm{T}} a_q - a_q^{\mathrm{T}} B(B^{\mathrm{T}}B)^{-1}B^{\mathrm{T}} a_q = a_q^{\mathrm{T}}(I - B(B^{\mathrm{T}}B)^{-1}B^{\mathrm{T}})a_q \neq 0,$$

Moreover, the claim is hence valid. □

Based on Lemma 9.9.1, the following items are well-defined:

$$v = B^{\mathrm{T}} a_q, \qquad w = (B^{\mathrm{T}}B)^{-1}v, \qquad \mu = (a_q^{\mathrm{T}} a_q - v^T w)^{-1} \tag{9.34}$$

Using the preceding notation, the following claim is valid.

Theorem 9.9.2 *Let B be $(m+1) \times k$ matrix with $1 \le k < m+1$ and* Rank $B = k$. *If $\breve{B} = (B \vdots a_q)$, where $a_q \notin$* range B, *then it holds that*

$$(\breve{B}^{\mathrm{T}} \breve{B})^{-1} = \begin{bmatrix} (B^{\mathrm{T}} B)^{-1} + \mu w w^T & -\mu w \\ -\mu w^T & \mu \end{bmatrix}. \tag{9.35}$$

Proof Note that w and μ are well-defined by (9.33). In addition, it holds that

$$\breve{B}^{\mathrm{T}} \breve{B} = \begin{bmatrix} B^{\mathrm{T}} \\ a_q^{\mathrm{T}} \end{bmatrix} \begin{bmatrix} B \ a_q \end{bmatrix} = \begin{bmatrix} B^{\mathrm{T}} B & B^{\mathrm{T}} a_q \\ a_q^{\mathrm{T}} B & a_q^{\mathrm{T}} a_q \end{bmatrix}.$$

It is straightforward to verify that the product of the right-hand sides of the preceding equation and (9.35) is equal to the unit matrix. □

According to the algorithm, column a_q is inserted before the last column of B. Denote the resulting matrix by $\hat{B}$, which can be obtained from $\breve{B}$ by exchanging its last two columns. If P^{T} is the according column permutation, it holds that

$$(\hat{B}^{\mathrm{T}} \hat{B})^{-1} = ((\breve{B} P^{\mathrm{T}})^{\mathrm{T}} (\breve{B} P^{\mathrm{T}}))^{-1} = P (\breve{B}^{\mathrm{T}} \breve{B})^{-1} P^{\mathrm{T}}.$$

Therefore, the wanted inverse can be obtained by exchanging the last two columns and rows of $(\breve{B}^{\mathrm{T}} \breve{B})^{-1}$, which is referred to as *post-permutation*.

On the other hand, to update the inverse after dropping a column, we have to make some permutations in advance. Assume that $\hat{B}$ is the current dual face matrix and a_p is selected to drop from it. Firstly, we move the column to the end, resulting in a matrix, say $\breve{B}$. If Q^T denotes the according column permutation, it holds that

$$(\breve{B}^{\mathrm{T}} \breve{B})^{-1} = ((\hat{B} Q^{\mathrm{T}})^{\mathrm{T}} (\hat{B} Q^{\mathrm{T}}))^{-1} = Q (\widehat{B}^{\mathrm{T}} \widehat{B})^{-1} Q^{\mathrm{T}},$$

which is termed *pre-permutation*.

Then, the following claim is valid.

Theorem 9.9.3 *Let $\breve{B} \in \mathcal{R}^{(m+1)\times k}$, $1 < k \le m+1$ be a dual face matrix. Assume that the according inverse is available and partitioned as*

$$(\breve{B}^{\mathrm{T}} \breve{B})^{-1} = \begin{bmatrix} D & h \\ h^T & \gamma \end{bmatrix}. \tag{9.36}$$

Then, for matrix B resulting from $\breve{B}$ by dropping its last column, it holds that

$$(B^{\mathrm{T}} B)^{-1} = D - (1/\gamma) h h^{\mathrm{T}}. \tag{9.37}$$

Proof Clearly, adding a_p to the end of matrix B yields $\breve{B}$. According to Theorem 9.9.3, the inverse $(\breve{B}^{\mathrm{T}}\breve{B})^{-1}$ is of the form (9.35) with a_q replaced by a_p. Since the inverse is unique, comparing (9.35) and (9.36) gives

$$\gamma = \mu, \qquad h = -\mu w, \qquad D = (B^{\mathrm{T}}B)^{-1} + \mu w w^T,$$

from which (9.37) follows. □

It seems simple to update the inverse in an expanding iteration by dropping a column: only $(k-1)^2/2 + k - 1$ flops are involved.

In practice, an integer vector is enough to record the positions of columns to avoid moving them all around. It is noted that the inverse is updated as a whole in each iteration, although only its last column is used:

$$\bar{x}_B = (B^T B)^{-1} e_k.$$

In order to limit rounding-off error accumulation, in addition, the solution process should be restarted from scratch periodically.

The overall steps can be put into the following algorithm.

Algorithm 9.9.1 (Dual Face Algorithm: Updating $(B^{\mathrm{T}}B)^{-1}$)
Initial: $B,\ N,\ 1 \le k \le m+1,\ (B^{\mathrm{T}}B)^{-1}$, *dual feasible solution* $(\bar{y}, \bar{z}_N)$. *This algorithm solves the standard problem.*

1. *Take the last column of* $(B^{\mathrm{T}}B)^{-1}$ *as* $\bar{x}_B$.
2. *If* $|B| < m+1$, *compute* $\Delta y = e_{m+1} - B\bar{x}_B$.
3. *Go to Step 11 if* $|B| = m+1$ *or* $\Delta y = 0$.
4. *Compute* $\Delta z_N = -N^T \Delta y$.
5. *Stop if J defined by (9.17) is empty (infeasible problem).*
6. *Determine index q and stepsize* β *such that* $\beta = -\bar{z}_q/\Delta z_q = \min_{j\in J} -\bar{z}_j/\Delta z_j$.
7. *If* $\beta \neq 0$, *update:*$\bar{y} = \bar{y} + \beta\Delta y,\ \bar{z}_N = \bar{z}_N + \beta \Delta z_N$.
8. *Update* $(B^{\mathrm{T}}B)^{-1}$ *by (9.34) and (9.35) with the post-permutation.*
9. *Bring index q from N and insert it before the last index of B.*
10. *Go to Step 1.*
11. *Stop if* $\bar{x}_B \ge 0$ *(optimality achieved).*
12. *Determine* $p \in \arg\min_{j\in B} \bar{x}_j$.
13. *Update* $(B^{\mathrm{T}}B)^{-1}$ *by (9.37) with the pre-permutation.*
14. *Adjust* $B = B\backslash\{p\},\ N = N \cup \{p\}$.
15. *Go to Step 1.*

Example 9.9.1 Solve the following standard problem by Algorithm 9.9.1 :

$$A = \begin{bmatrix} 3 & 0 & -7 & 5 & -4 & -6 & -5 \\ 2 & 1/3 & 0 & -2/3 & 0 & 2 & -2 \\ -5 & 11 & 4 & -3 & 3/2 & 0 & 1 \end{bmatrix}, \qquad b = \begin{bmatrix} -18 \\ 0 \\ 9 \end{bmatrix}$$
$c = [0, 5/2, 1, 7/3, 0, 3, 2]^T \ge 0$

Preparation: form $\tilde{A}$ by (9.3).
Take partition $\tilde{A} = (B, N)$, where $B = \{8\}$, $N = \{1, \ldots, 7\}$.
Dual feasible solution: $\bar{y} = (0, 0, 0, 0)^{\mathrm{T}}$, $\bar{z} = [0, 5/2, 1, 7/3, 0, 3, 2]^T$,
$B = a_8 = [18, 0, -9, 1]^{\mathrm{T}}$.

Iteration 1:

1. $\bar{x}_B = (B^{\mathrm{T}} B)^{-1} = [1/406]$.
2. $\Delta y = e_4 - B\bar{x}_B = [-9/203, 0, 9/406, 405/406]^{\mathrm{T}} \neq 0$.
4. $\Delta z_N = -N^{\mathrm{T}} \Delta y = [99/406, -99/406, -81/203, 117/406, -171/812, -54/203, -99/406]^{\mathrm{T}}$.
5. $J = \{2, 3, 5, 6, 7\} \neq \emptyset$.
6. $\beta = 0/(-171/812), q = 5, a_5 = [-4, 0, 3/2, 0]^{\mathrm{T}}$.
8. $v = B^T a_5 = [-171/2], w = \bar{x}_B = (B^{\mathrm{T}} B)^{-1} v = [1/406][-171/2] = [-171/812]$,
   $\mu = 1/(a_5^T a_5 - v^T w) = 1624/397$.
   $$(\breve{B}^{\mathrm{T}} \breve{B})^{-1} = \begin{bmatrix} 73/397 & 342/397 \\ 342/397 & 1624/397 \end{bmatrix}, \qquad (B^{\mathrm{T}} B)^{-1} = \begin{bmatrix} 1624/397 & 342/397 \\ 342/397 & 73/397 \end{bmatrix}.$$
9. $B = \{5, 8\}$, $N = \{1, 2, 3, 4, 6, 7\}$.

Iteration 2:

1. $\bar{x}_B = [342/397, 73/397]^{\mathrm{T}}$.
2. $\Delta y = e_4 - B\bar{x}_B = [54/397, 0, 144/397, 324/397^{\mathrm{T}} \neq 0$.
4. $\Delta z_N = -N^{\Delta} y = [558/397, -1584/397, -198/397, 162/397, 324/397, 126/397]^{\mathrm{T}}$.
5. $J = \{2, 3\} \neq \emptyset$.
6. $\beta = [(5/2)/(-1584/397), 1/(-198/397)] = 943/1505, q = 2, a_2 = [0, 1/3, 11, 0]^{\mathrm{T}}$.
7. $\bar{y} = \bar{y} + \beta \Delta y = [15/176, 0, 5/22, 45/88]^T$,
   $\bar{z}_N = \bar{z}_N + \beta \Delta z_N = [155/176, 0, 11/16, 1367/528, 309/88, 387/176]^T$,
   $\bar{z} = [155/176, 0, 11/16, 1367/528, 0, 309/88, 387/176]^T$.
8. $v = B^T a_2 = [33/2, -99]^T$, $w = \bar{x}_B = (B^{\mathrm{T}} B)^{-1} v = [-7062/397, -1584/397]^T$,
   $\mu = 1/(a_2^T a_2 - v^T w) = 115/2256$.
   $$(\breve{B}^{\mathrm{T}} \breve{B})^{-1} = \begin{bmatrix} 8432/417 & 2065/461 & 603/665 \\ 2065/461 & 646/649 & 1189/5846 \\ 603/665 & 1189/5846 & 115/2256 \end{bmatrix},$$
   $$(B^{\mathrm{T}} B)^{-1} = \begin{bmatrix} 8432/417 & 603/665 & 2065/461 \\ 603/665 & 115/2256 & 1189/5846 \\ 2065/461 & 1189/5846 & 646/649 \end{bmatrix}.$$
9. $B = \{5, 2, 8\}$, $N = \{1, 3, 4, 6, 7\}$.

Iteration 3:

1. $\bar{x}_B = [2065/461, 1189/5846, 646/649]^{\mathrm{T}}$.
2. $\Delta y = [54/70{,}093, -1189/17{,}538, 37/18{,}010, 107/23{,}148]\mathrm{T} \neq 0$.

4. $\Delta z_N = -N^{\mathrm{T}}\Delta y = [2036/14{,}183, -198/70{,}093, -192/4477, 538/3837, -310/2317]^{\mathrm{T}}$.
5. $J = \{3, 4, 7\}$.
6. $\beta = \min[-(11/16)/(-198/70{,}093), -(1367/528)/(-192/4477),$
   $-(387/176)/(-310/2317)] = 14{,}101/858,\ q = 7.\ a_7 = [-5, -2, 1, 0]^{\mathrm{T}}$.
7. $\bar{y} = \bar{y} + \beta\Delta y = [51/521, -1161/1042, 136/521, 306/521]^{\mathrm{T}}$,
   $\bar{z}_N = \bar{z}_N + \beta\Delta z_N = [1688/521, 334/521, 1562/829, 3030/521, 0]^T$,
   $\bar{z} = [1688/521, 0, 334/521, 1562/829, 0, 3030/521, 0]^T$.
8. $v = B^T a_7 = [43/2, 31/3, -99]^T,\ w = \bar{x}_B = (B^{\mathrm{T}}B)^{-1}v =$
   $[513/785, -198/1751, -310/2317]^T,\ \mu = 1/(a_2^T a_2 - v^T w) = 259/1003$.

$$(\breve{B}^{\mathrm{T}}\breve{B})^{-1} = \begin{bmatrix} 2765/136 & 901/1015 & 2322/521 & -877/5197 \\ 901/1015 & 125/2303 & 108/521 & 81/2774 \\ 2322/521 & 108/521 & 1 & 18/521 \\ -877/5197 & 81/2774 & 18/521 & 259/1003 \end{bmatrix}$$

$$(\bar{B}^{\mathrm{T}}\bar{B})^{-1} = \begin{bmatrix} 2765/136 & 901/1015 & -877/5197 & 2322/521 \\ 901/1015 & 125/2303 & 81/2774 & 108/521 \\ -877/5197 & 81/2774 & 259/1003 & 18/521 \\ 2322/521 & 108/521 & 18/521 & 1 \end{bmatrix}.$$

9. $B = \{5, 2, 7, 8\},\ N = \{1, 3, 4, 6\}$.

Iteration 4:

1. $\bar{x}_B = [2322/521, 108/521, 18/521, 1]^{\mathrm{T}}$.
2. $|B| = 4$.
11. $\bar{x}_B \geq 0$. Optimality is achieved.

Primal and dual optimal solutions:

$\bar{x} = [0, 108/521, 0, 0, 2322/521, 0, 18/521]^{\mathrm{T}}, \qquad \bar{f} = c_B^{\mathrm{T}}\bar{x}_B = 306/521,$
$\bar{z} = [1688/521, 0, 334/521, 1562/829, 0, 3030/521, 0]^T$.

9.10 Generalized Dual Face Algorithm with Cholesky Factorization

In this section, we generalize the dual face algorithm with Cholesky factorization.

The variable-bounded LP problem (3.1) can be transformed to the following form, equivalently:

$$\begin{aligned} &\min\ c^T x \\ &\text{s.t.}\ \begin{bmatrix} A & -b \\ 0 & 1 \end{bmatrix}\begin{bmatrix} x \\ x_{n+1} \end{bmatrix} = \begin{bmatrix} 0 \\ 1 \end{bmatrix}, \qquad l \leq x \leq u, \end{aligned} \tag{9.38}$$

where $A \in \mathcal{R}^{m\times n}$, $b \in \mathcal{R}^m$, $m < n$, rank $A = m$.

Then, the dual problem of (9.38) can be written as (p.121, Pan [36])

$$\begin{array}{ll} \max & y_{m+1} - u^T v + l^T w \\ \text{s.t.} & \begin{bmatrix} A^T & 0 \\ -b^T & 1 \end{bmatrix} \begin{bmatrix} y \\ y_{m+1} \end{bmatrix} + \begin{bmatrix} z \\ z_{n+1} \end{bmatrix} = \begin{bmatrix} c \\ 0 \end{bmatrix}, \quad z_{n+1} = 0, \end{array} \tag{9.39}$$

where

$$\begin{bmatrix} z \\ z_{n+1} \end{bmatrix} = - \begin{bmatrix} v \\ v_{n+1} \end{bmatrix} + \begin{bmatrix} w \\ w_{n+1} \end{bmatrix}, \qquad v, w \geq 0.$$

Introduce the following notation:

$$\tilde{A} := \begin{bmatrix} A & -b \\ 0 & 1 \end{bmatrix}. \tag{9.40}$$

Let $\tilde{A} = (B, N)$ be a partition such that

$$1 \leq |B| \equiv k \leq m+1, \qquad \text{Rank}\, B = k.$$

Definition 9.10.1 The dual face associated with partition (B, N) is defined as

$$D(B) = \{(y, z) \in \mathcal{R}^{m+1} \times \mathcal{R}^{n+1} \mid B^T y = c_B,\ N^{\mathrm{T}} y + z_N = c_N,\ z_B = 0\}. \tag{9.41}$$

A dual face is *level dual face* if the dual objective value is constant over it and *dual optimal* if the constant is maximized. A point (solution) on a dual face is termed *dual face point (solution).*

Assume that $(\bar{y}, \bar{z})$ belongs to $D(B)$. Let $\bar{x}$ be solution such that

$$\bar{x}_j = \begin{cases} l_j & \text{if } \bar{z}_j \geq 0, \\ u_j & \text{if } \bar{z}_j < 0, \end{cases} \qquad j \in N, \tag{9.42}$$

which exhibits zero duality gaps associated with indices of N. If $\bar{x}$ satisfies the equality constraint of (9.38), the dual and primal solutions are termed a *matching pair.*

Now, we regulate that the last column of B has a special form such that

$$B = \begin{bmatrix} B' & -b + N\bar{x}_N \\ 0 & 1 \end{bmatrix}, \tag{9.43}$$

where $B' \in\in \mathcal{R}^{m\times(k-1)}$.

Then, the desired search direction can be derived the same as before (Sect. 9.1), that is,

$$\Delta y = e_{m+1} - B\bar{x}_B, \qquad \Delta z_N = -N^{\mathrm{T}}\Delta y, \tag{9.44}$$

where $\bar{x}_B$ is the solution to the following upper triangle system:

$$L^{\mathrm{T}} x_B = -(1/\mu)e_k,$$

where L is the Cholesky factor of $B^T B$.

Still, we determined the new dual solution by the following line search scheme:

$$\hat{y} = \bar{y} + \beta\Delta y, \qquad \hat{z}_N = \bar{z}_N + \beta\Delta z_N, \tag{9.45}$$

where β is the stepsize to be maximized to maintain 0 duality gaps associated with N.

Introduce the following index set:

$$J = \{j \in N \mid \bar{x}_j = l_j,\ \Delta z_j < 0\} \cup \{j \in N \mid \bar{x}_j = u_j,\ \Delta z_j > 0\} \tag{9.46}$$

It can be shown similarly that if J is empty, the dual problem is upper unbounded, or there is no feasible solution to the primal problem; otherwise, the stepsize is determined such that

$$\beta = -\bar{z}_q/\Delta z_q = \min_{j\in J} -\bar{z}_j/\Delta z_j. \tag{9.47}$$

It can be shown that the new dual solution, corresponding to a higher objective value, reaches a new dual face, if β is nonzero. A series of such iterations are carried out continuously until Δy vanishes.

Lemma 9.10.1 *Assume that $(\bar{y}, \bar{z})$ is a dual face solution and $\bar{x}$ satisfies (9.42). If Δy vanishes, they are a matching pair.*

Proof From $\Delta y = 0$ and the first expression of (9.44), it follows that

$$B\bar{x}_B = e_{m+1},$$

which with (9.43) leads to

$$B'\bar{x}_{B'} + (-b + N\bar{x}_N)\bar{x}_{n+1} = 0, \qquad 0\bar{x}_{B'} + \bar{x}_{n+1} = 1.$$

The preceding two expressions exhibit that

$$B'\bar{x}_{B'} + N\bar{x}_N = b, \qquad \bar{x}_{n+1} = 1 \tag{9.48}$$

Therefore, $(\bar{y}, \bar{z})$ and $(\bar{x}_{B'}, \bar{x}_N)$ is a matching pair. □

Define the following index set:

$$T = \{j \in B \mid x_j < l_j\} \cup \{j \in B \mid x_j > u_j\}. \tag{9.49}$$

The following result gives an optimality criterion.

Theorem 9.10.2 *Assume that $(\bar{y}, \bar{z})$ is a dual face solution and $\bar{x}$ satisfies (9.42). If set T defined by (9.49) is empty, then they are dual and primal optimal solutions.*

Proof According to Lemma 9.10.1, $(\bar{y}, \bar{z})$ and $(\bar{x}_{B'}, \bar{x}_N)$ is a matching pair.

In addition, nonemptiness of T implies that components of $(\bar{x}_{B'}, \bar{x}_N)$ are within bounds. Therefore, the pair are primal and dual feasible solutions. Thus, it is only needed to show complementarity.

For all $j = 1, \ldots, n+1$, set

$$v = \begin{cases} -\bar{z}_j & \text{if } \bar{z}_j \leq 0 \\ 0 & \text{if } \bar{z}_j > 0 \end{cases} \qquad w_j = \begin{cases} 0 & \text{if } \bar{z}_j \leq 0 \\ \bar{z}_j & \text{if } \bar{z}_j > 0 \end{cases} \tag{9.50}$$

Then, complementarity condition (1.40) is verified. Therefore, they are a pair of primal and dual optimal solutions. □

Corollary 9.10.3 *Under the same assumptions, the dual optimal solution reaches a dual optimal level face.*

Proof It is obvious that the direct proposition of Lemma 9.5.1 is valid. □

As for initial preparation, set

$$B = \{n+1\}, N = \{1, \ldots, n\}, y = 0, z_N = c, \tag{9.51}$$

form the following set:

$$S = \{j \in N \mid z_j = 0\}, \tag{9.52}$$

and adjust

$$B = B \cup S, \qquad N = N \backslash S. \tag{9.53}$$

Calculate Cholesky factorization $LL^T = B^T B$. Then, set $\bar{x}_N$ by (9.42).

Favorably, there is no need for the phase I process.

The overall steps are summarized in the following model.

Algorithm 9.10.1 (Generalized Dual Face Algorithm with Cholesky Factorization)
Initial: $B, N, k, \bar{y}, \bar{z}_N, \bar{x}_N$ by (9.53)–(9.42) and L. This algorithm solves the variable-bounded problem.

1. *Solve triangular system $L^{\mathrm{T}}\bar{x}_B = -(1/\mu)e_k$, where μ is the $(m+1)$th diagonal of L.*
2. *Go to Step 5 if $|B| = m+1$.*

3. *Compute* $(\Delta y, \Delta z_N)$ *by (9.44);*
4. *Go to Step 8 if* $\Delta y \neq 0$.
5. *Stop if T defined by (9.49) is empty (optimality achieved).*
6. *Adjust* $B = B\backslash T,\ N = N \cup T$*;*
7. *Update L, and go to Step 1.*
8. *Stop if J defined by (9.46) is empty (infeasible problem).*
9. *Determine index q and stepsize* β *by (9.47).*
10. *If* $\beta \neq 0$*, update:*$\bar{y} = \bar{y} + \beta \Delta y,\ \bar{z}_N = \bar{z}_N + \beta \Delta z_N$.
11. *Adjust* $B = N \cup \{q\},\ N = N\backslash\{q\}$.
12. *Update L, and go to Step 1.*

Example 9.10.1 Solve the following variable-bounded problem using the preceding algorithm:

$$A = \begin{bmatrix} 3 & 1 & -5 & 1 & 0 & 3 & -4 \\ -2 & 0 & 6 & 0 & 1 & -6 & 0 \\ -6 & -3 & 3 & 0 & -2 & 4 & 7 \\ 1 & 0 & -1 & 2 & 0 & -1 & 2 \end{bmatrix}, \qquad b = \begin{bmatrix} -17 \\ 14 \\ 16 \\ 4 \end{bmatrix}$$

$$c = [-2, 0, 4, -6, 0, 5, 8]^T$$
$$l = [1, -2, 0, 0, 0, -6, -5]^T; \qquad u = [9, 8, 12, 10, 15, 9, 5]^T$$

Preparation: form $\tilde{A}$ by (9.40).
Take partition $\tilde{A} = (B, N)$, where $B = \{8\},\ N = \{1, \ldots, 7\}$
Set $\bar{y} = [0, 0, 0, 0, 0]^T$, $\bar{z}_N = [-2, 0, 4, -6, 0, 5, 8]^T$.
$S\{2, 5\}$, reset $B = B \cup S,\ N = N\backslash S$.
Initial input: $B = \{2, 5, 8\}$, $N = \{1, 3, 4, 6, 7\}$, $\bar{x}_N = [9, -2, 10, -6, -5, 1]^T$,

$$L = \begin{bmatrix} 721/228 & & \\ 721/380 & 846/715 & \\ 6304/45 & -1275/397 & 10{,}335/422 \end{bmatrix}$$

Iteration 1:

1. $\bar{x}_B = [-220/2873, 38/8397, 14/8397]^T$
2. $|B| \neq 5$.
3. $\Delta y = e_5 - B\bar{x}_B = [-47/2799, -94/8397, -47/8397, -98/2799, 599/600]^T$,
 $\Delta z_N = [247/8397, -98/2799, 27/311, -247/8397, 353/8397]^T$.
4. $\Delta y \neq 0$.
8. $J = \{1, 3, 4, 6\}$.
9. $\beta = \min[-(-2)/(247/8397), -(4)/(-98/2799), -(-6)/(27/311), -(5)/(-247/8397)] = 8431/124,\ q = 1$.
10. $\bar{y} = \bar{y} + \beta \Delta y = [-282/247, -188/247, -94/247, -588/247, 9503/140]^T$,
 $\bar{z}_N = \bar{z}_N + \beta \Delta z_N = [0, 400/247, -24/247, 3, 2682/247]^T$,
 $\bar{z} = [0, 0, 400/247, -24/247, 0, 3, 2682/247, 0]^T$.
11. $B = \{1, 2, 5, 8\}$, $N = \{3, 4, 6, 7\}$.

12. $B = \begin{bmatrix} 1393/197 & & & \\ 591/199 & 1322/1217 & & \\ 1393/985 & 1319/796 & 239/474 & \\ 5642/79 & 8563/222 & 3864/275 & 2761/157 \end{bmatrix}$

Iteration 2:

1. $\bar{x}_B = [-112/4639, 105/4639, -418/4639, 15/4639]^T$.
2. $|B| \neq 5$.
3. $\Delta y = e_5 - B\bar{x}_B = [-204/4639, -136/4639, -68/4639, -68/4639, 1233/1237]^T$, $\Delta z_N = [-68/4639, 340/4639, 0, -204/4639]^T$.
4. $\Delta y \neq 0$.
8. $J = \{3, 4, 7\}$.
9. $\beta = \min[-(400/247)/(-68/4639), -(-24/247)/(340/4639), -(2682/247)/(-204/4639)] = 757/571, q = 4$.
10. $\bar{y} = \bar{y} + \beta\Delta y = [-6/5, -4/5, -2/5, -12/5, 346/5]^T$,
    $\bar{z}_N = \bar{z}_N + \beta\Delta z_N = [8/5, 0, 3, 54/5]^T$,
    $\bar{z} = [0, 0, 8/5, 0, 0, 3, 54/5, 0]^T$.
11. $B = \{1, 2, 4, 5, 8\}, N = \{3, 6, 7\}$.

12. $L = \begin{bmatrix} 1393/197 & & & & \\ 591/199 & 1322/1217 & & & \\ 985/1393 & -481/475 & 658/353 & & \\ 1393/985 & 1319/796 & 439/1207 & 623/1784 & \\ 4826/75 & 3068/63 & 965/264 & -2643/901 & 1 \end{bmatrix}$

Iteration 3:

1. $\bar{x}_B = [76/5, -61, -18/5, 42/5, 1]^T$, $\bar{x} = [76/5, -61, 0, -18/5, 42/5, -6, -5, 1]^T$.
2. $|B| = 5$.
5. $T = \{1, 2, 4\} \neq \emptyset$.
6. Adjust $B = \{5, 8\}$, $N = \{1, 2, 3, 4, 6, 7\}$;
7. $L = \begin{bmatrix} 2889/1292 & \\ 13{,}081/117 & 9410/139 \end{bmatrix}$

1. $\bar{x}_B = [-50/4583, 1/4583]^T$.
2. $|B| \neq 5$.
3. $\Delta y = e_5 - B\bar{x}_B = [-44/4583, 46/4583, 23/4583, -1/4583, 4582/4583]^T$, $\Delta z_N = [363/4583, 113/4583, -247/2000, 46/4583, 315/4583, -335/4583]^T$.
4. $\Delta y \neq 0$.
8. $J = \{1, 3, 7\}$.
9. $\beta = 0, q = 1$.
10. $\bar{y} = \bar{y} + \beta\Delta y = [-282/247, -188/247, -94/247, -588/247, 9503/140]^T$,
    $\bar{z}_N = \bar{z}_N + \beta\Delta z_N = [0, 400/247, -24/247, 3, 2682/247]^T$

11. $B = \{1, 5, 8\}$, $N = \{2, 3, 4, 6, 7\}$.
12. $L = \begin{bmatrix} 1393/197 & & \\ 1393/985 & 1351/780 & \\ 5023/86 & 4424/99 & 6159/446 \end{bmatrix}$

Iteration 4:

1. $\bar{x}_B = [-31/1907, -258/1907, 10/1907]^T$.
2. $|B| \neq 5$.
3. $\Delta y = e_5 - B\bar{x}_B = [-77/1907, -24/1907, -12/1907, 111/1907, 569/572]^T$,
   $\Delta z_N = [41/1907, -94/1907, 45/1907, 246/1907, -446/1907]^T$.
4. $\Delta y \neq 0$.
8. $J = \{3, 4, 7\}$.
9. $\beta = 0, q = 4$.
11. $B = \{1, 4, 5, 8\}$, $N = \{2, 3, 6, 7\}$.
12. $L = \begin{bmatrix} 1393/197 & & & \\ 985/1393 & 2378/1121 & & \\ 1393/985 & -1121/2378 & 5/3 & \\ 5023/86 & -7922/417 & 616/15 & 3304/279 \end{bmatrix}$

Iteration 5:

1. $\bar{x}_B = [-46/1753, 87/3506, -308/1753, 25/3506]^T$.
2. $|B| \neq 5$.
3. $\Delta y = e_5 - B\bar{x}_B = [-118/1753, -59/1753, -59/3506, 59/1753, 557/561]^T$
   $\Delta z_N = [59/3506, -177/3506, 177/1753, -387/1769]^T$.
4. $\Delta y \neq 0$.
8. $J = \{3, 7\}$.
9. $\beta = \min[-(8/5)/(-177/3506), -(54/5)/(-387/1769)] = 6909/218, q = 3$.
10. $\bar{y} = \bar{y} + \beta\Delta y = [-10/3, -28/15, -14/15, -4/3, 302/3]^T$,
    $\bar{z}_N = \bar{z}_N + \beta\Delta z_N = [8/15, 0, 31/5, 58/15]^T$.
    $\bar{z} = [0, 8/15, 0, 0, 0, 31/5, 58/15, 0]^T$.
11. $B = \{1, 3, 4, 5, 8\}$, $N = \{2, 6, 7\}$.
12. $L = \begin{bmatrix} 1393/197 & & & & \\ -605/93 & 648/121 & & & \\ 985/1393 & -121/270 & 2795/1348 & & \\ 1393/985 & 950/553 & -199/1793 & 371/1942 & \\ 5023/86 & 25{,}242/593 & -2385/233 & 2678/163 & 1 \end{bmatrix}$

Iteration 6:

1. $\bar{x}_B = [27, 59/3, 1/3, -86, 1]^T$.
2. $|B| = 5$.
5. $T = \{1, 3, 5\} \neq \emptyset$.
6. Adjust $B = \{4, 8\}$, $N = \{1, 2, 3, 5, 6, 7\}$;

7. $L = \begin{bmatrix} 2889/1292 & \\ -2889/170 & 13{,}561/117 \end{bmatrix}$

1. $\bar{x}_B = [38/67{,}171, 5/67{,}171]^T$.
2. $|B| \neq 5$.
3. $\Delta y = e_5 - B\bar{x}_B = [42/67{,}171, -47/8308, 149/23{,}008, -21/67{,}171, 13{,}433/13{,}434]^T$ $\Delta z_N = [152/5851, 49/2606, 99/5713, 19/1021, -167/2692, -199/4715]^T$.
4. $\Delta y \neq 0$.
8. $J = \{1, 3, 6, 7\}$.
9. $\beta = 0, q = 1$.
11. $B = \{1, 4, 8\}, N = \{2, 3, 5, 6, 7\}$.
12. $L = \begin{bmatrix} 1393/197 & & \\ 985/1393 & 2378/1121 & \\ -7922/403 & -1759/54 & 6318/61 \end{bmatrix}$

Once phase I finishes with a feasible solution, the reduced problem is formed by adding the $(m+1)$th row. The corresponding value of the objective function is calculated and set to $\bar{x}_{n+1}$. The end partition (B, N) is still functional, but the Cholesky factor L has to be recalculated.

Example 9.10.2 Solve the following standard problem using the double-phase face algorithm:

$$A = \begin{bmatrix} 4 & 0 & -3 & 4 & 3 & -6 & 0 & 3 \\ -7 & -6 & 0 & -2 & 0 & 2 & -5 & 8 \\ 5 & -2 & -6 & 0 & 5 & -9 & 6 & -4 \\ -2 & 0 & -6 & 8 & 7 & 3 & 0 & 3 \end{bmatrix}, \quad b = \begin{bmatrix} 3 \\ -5 \\ 0 \\ -2 \end{bmatrix}$$
$$c = [3, 0, 9, -2, 0, -7, 2, -3]^T$$

Phase I. Solve the auxiliary problem (8.31) with $h = e$.
$a_9 = [-91/760, 91/304, 91/304, -273/304]^T$.
Initial: $k = 9$, $B = \{1, \ldots, 9\}$, $N = \emptyset$, $\bar{x} = [1, \ldots, 1, 1520/91]^T$. $B = [A \mid a_9]$, $N = \emptyset$.
Call Subalgorithm 8.7.1.
Iteration 1:

1. Cholesky factor $L = \begin{bmatrix} 1969/202 & & & \\ -397/161 & 4922/371 & & \\ 3517/361 & -1326/223 & 2871/298 & \\ 927/167 & 2841/910 & -1332/931 & 10{,}696/941 \end{bmatrix}$

   $r = 4$.
2. $\bar{y} = [-10/8073, 123/23{,}671, 47/9633, -58/9217]^{\mathrm{T}}$.
3. $\Delta_B = [115/26{,}451, 1584/38{,}695, -155/12{,}707, 215/3273, 62/2653, 309/6872, -26/7895, 28/51{,}791, 1012/1021]^{\mathrm{T}} \neq 0$.
8. $J = \{1, 2, 4, 5, 6, 8, 9\}$.
9. $\alpha = \min[1/(115/26{,}451), 1/(1584/38{,}695), 1/(215/3273), 1/(62/2653),$

$1/(309/6872), 1/(28/51{,}791), (1520/91)/(1012/1021)] = 3273/215,\ q = 4.$

10. $\bar{x}_B = [1, 1, 1, 1, 1, 1, 1, 1, 1520/91]^{\mathrm{T}} - (3273/215)\Delta_B = [776/831, 335/889, 779/657, 0, 1235/1917, 218/691, 1571/1496, 241/243, 975/604]^{\mathrm{T}}$.
11. $\bar{x}_9 \neq 0$.
12. $B = \{1, 2, 3, 5, 6, 7, 8, 9\}, N = \{4\}, k = 8$.
13. $$L = \begin{bmatrix} 8809/991 & & & \\ -810/449 & 4813/364 & & \\ 641/60 & -4510/713 & 3230/389 & \\ 2788/1121 & 1032/283 & 2834/1291 & 2449/268 \end{bmatrix}$$

Iteration 2:

2. $\bar{y} = [-58/5897, 136/12{,}565, 76/6443, -1343/102{,}665]^{\mathrm{T}}$.
3. $\Delta_B = [170/5673, 410/4631, -173/4648, 376/6055, 337/5205, -181/10{,}867, 140/4771, 647/660]^{\mathrm{T}} \neq 0$.
8. $J = \{1, 2, 5, 6, 8, 9\}$.
9. $\alpha = \min[(776/831)/(170/5673), (335/889)/(410/4631), (1235/1917)/(376/6055), (218/691)/(337/5205), (241/243)/(140/4771), (975/604)/(647/660)] = 1757/1067,\ q = 9$.
10. $\bar{x}_B = [776/831, 335/889, 779/657, 1235/1917, 218/691, 1571/1496, 241/243, 975/604]^{\mathrm{T}} - (1757/1067)\Delta_B = [467/528, 326/1411, 1343/1077, 1362/2513, 697/3337, 1820/1689, 1001/1061, 0]^{\mathrm{T}}$.
11. $\bar{x}_9 = 0$. (feasibility achieved)

    Phase II: Call Algorithm 8.6.1.

    Initial: $B = \{1, 2, 3, 5, 6, 7, 8, 9\}, k = 8, N = \{4\}$.

    $\bar{x}_B = [467/528, 326/1411, 1343/1077, 1362/2513, 697/3337, 1820/1689, 1001/1061]^T$,

Iteration 3:

1. $$L = \begin{bmatrix} 1431/161 & & & & \\ -2558/1421 & 2287/173 & & & \\ 5590/523 & -2395/378 & 3595/434 & & \\ 948/383 & 986/269 & 1867/824 & 880/97 & \\ 322/159 & -3075/622 & -889/1482 & -6952/835 & 3227/436 \end{bmatrix}$$
2. $\bar{y} = [92/19{,}253, -19/30{,}698, 95/2913, -107/6387, -214/11{,}723]^T$.
3. $\Delta_B = [-169/9140, 71/25{,}291, 907/9286, 116/1339, -48/2627, 120/8663, -26/30{,}663, 484/493]^{\mathrm{T}} \neq 0$.
8. $J = \{2, 3, 5, 7\}$.
9. $\alpha = \min[(326/1411)/(71/25{,}291), (1343/1077)/(907/9286), (1362/2513)/(116/1339), (697/3337)/(120/8663)] = 4323/691,\ q = 5$.
10. $\bar{x}_B = \bar{x}_B - \alpha\Delta_B = [6810/6809, 548/2567, 1172/1843, 0, 1475/4564, 980/989, 685/722, 2597/464]^T$.

    $\bar{x} = [6810/6809, 548/2567, 1172/1843, 0, 0, 1475/4564, 980/989, 685/722, 2597/464]^T$.

11. $B = \{1, 2, 3, 6, 7, 8, 9\}$, $N = \{4, 5\}$, $k = 7$.

12. $$L = \begin{bmatrix} 4108/491 & & & & \\ -1964/1027 & 1360/103 & & & \\ 3557/372 & -4613/719 & 3049/377 & & \\ 491/4108 & 2462/735 & 2057/2264 & 5951/878 & \\ 2515/1169 & -3720/757 & -439/866 & -4144/383 & 1426/543 \end{bmatrix}.$$

Iteration 4:

2. $\bar{y} = [659/27{,}202, 232/17{,}881, 25/1478, -921/3979, -271/1869]^T$.
3. $\Delta_B = [-604/5093, 241/2158, 988/10{,}935, -181/3698, 806/3181, 231/1534, 1598/1869]^{\mathrm{T}} \neq 0$.
8. $J = \{2, 3, 7, 8\}$.
9. $\alpha = \min[(548/2567)/(241/2158), (1172/1843)/(988/10{,}935), (980/989)/(806/3181), (685/722)/(231/1534)] = 281/147,\ q = 2$,
10. $\bar{x}_B = \bar{x}_B - \alpha\Delta_B = [1179/961, 0, 812/1753, 453/1087, 1315/2596, 840/1271, 2754/695]^T$.$\bar{x} = [1179/961, 0, 812/1753, 0, 0, 453/1087, 1315/2596, 840/1271, 2754/695]^T$.
11. $B = \{1, 3, 6, 7, 8, 9\}$, $N = \{2, 4, 5\}$, $k = 6$.

12. $$L = \begin{bmatrix} 4108/491 & & & & \\ -1964/1027 & 29{,}099/2474 & & & \\ 3557/372 & -2105/256 & 3181/538 & & \\ 491/4108 & 2369/630 & 1257/443 & 3488/583 & \\ 2515/1169 & -1164/211 & -2120/699 & -7961/782 & 1240/549 \end{bmatrix}.$$

Iteration 5:

2. $\bar{y} = [0, 1379/24{,}491, 277/4654, -527/1580, -857/4372]^T$.
3. $\Delta_B = [147/8425, 229/1907, 361/6988, 894/2825, 573/2863, 808/1005]^T$.
8. $J = \{1, 3, 6, 7, 8\}$.
9. $\alpha = \min[(1179/961)/(147/8425), (812/1753)/(229/1907), (453/1087)/(361/6988), (1315/2596)/(894/2825), (840/1271)/(573/2863)] = 477/298, q = 7$.
10. $\bar{x}_B = \bar{x}_B - \alpha\Delta_B = [1109/925, 965/3561, 774/2317, 0, 1402/4117, 9991/3734]^T$.
11. $B = \{1, 3, 6, 8, 9\}$, $N = \{2, 4, 5, 7\}$, $k = 5$.

12. $$L = \begin{bmatrix} 4108/491 & & & & \\ -1964/1027 & 831/68 & & & \\ 3557/372 & -1867/417 & 3746/527 & & \\ 491/4108 & 1245/344 & 430/1483 & 3407/509 & \\ 2515/1169 & -3849/857 & -1929/2930 & -2018/183 & 615/418 \end{bmatrix}.$$

Iteration 6:

2. $\bar{y} = [309/1996, 213/4151, -61/5195, -3045/4001, -1524/3299]^T$,
3. $k = 5$.

5. $\bar{z}_N = [769/2704, 2152/463, 2200/447, 1675/13{,}391]^T$.
6. $T = \emptyset$, optimality achieved.

The primal and dual optimal solutions are as follows:

$\bar{x} = [1109/925, 0, 965/3561, 0, 0, 774/2317, 0, 1402/4117]^T$, $\bar{f} = 9991/3734$.
$\bar{z} = [0, 769/2704, 0, 2152/463, 2200/447, 0, 1675/13{,}391, 0]^T$.

Appendix A
Empirical Evaluation of Face and Dual Face Methods with Cholesky Factorization

The following three codes in FORTRAN 77 were tested and compared (without exploiting sparsity):

1. RSA: Our working code based on the double-phase simplex algorithm using Harris' two-pass rule.
2. FALP: The face Algorithm 8.6.1, supported by Subalgorithm 8.7.1, with the components of the initial solution all ones.
3. DFA: The dual face Algorithm 9.10.1, supported by Subalgorithm 9.8.1.

We used 10^{-6} as primal and dual feasibility tolerance and 10^{-4} as equality relative tolerance. In FALP, we used $\|\Delta_B\|_\infty < 10^{-8}$ instead of $\Delta_B = 0$, while in DFA, we use $\|\Delta y\|_\infty < 10^{-8}$ instead of $\Delta y = 0$.

Compiled using Visual FORTRAN 5.0, the codes were run under a Windows XP system Home Edition Version 2002 on an IBM PC with an Intel(R) Pentium(R) processor, 1.00 GB of 1.86 GHz memory, and about 16 digits of precision. All reported CPU times were measured in seconds with utility routine CPU_TIME.

The test set comprised 26 Netlib standard problems, i.e., the first 26 problems in the order of increasing m+n (Table C.4, Appendix C, Pan [36]).

Test statistics obtained with RSA, FALP, and DFA are listed in Tables A.1, A.2, and A.3, respectively. Total iterations and time required for solving each problem are displayed in columns labeled Itn and Time, respectively, and those associated with phase I alone in columns labeled Itn1 and Time1. Percentages of total degenerate iterations are listed in columns labeled % Deg; those associated with phase I are in the column labeled % Deg1.

In particular, we display the dimensions of the optimal faces achieved, in the last column labeled k-m of Table A.2, whereas dimensions of dual optimal faces in the column labeled m-k in Table A.3.

As the classic RSA failed to solve DEGEN2 within 10,000 iterations, we list the associated results separately at the bottom of the tables.

P.-Q. Pan, *Face Method*, https://doi.org/10.1007/978-3-031-93594-7

Table A.1 Statistics for code RSA

	Total			Phase 1		
Problem	Itn	Time	%Deg	Itn1	Time1	%Deg1
AFIRO	29	0.03	69.0	28	0.03	71.4
SC50B	57	0.06	80.7	50	0.06	90.0
SC50A	55	0.09	74.5	51	0.09	78.4
ADLITTLE	134	0.13	12.7	69	0.13	24.6
BLEND	115	0.17	40.0	90	0.17	51.1
SHARE2B	196	0.28	57.1	149	0.27	67.8
SC105	123	0.36	73.2	110	0.36	77.3
STOCFOR1	174	0.50	68.4	150	0.48	79.3
SCAGR7	181	0.67	34.3	146	0.64	42.5
ISRAEL	507	1.41	0.8	188	0.97	1.1
SHARE1B	309	1.67	7.1	190	1.58	11.6
SC205	255	2.23	66.7	211	2.16	80.1
BEACONFD	213	2.61	51.2	159	2.53	57.9
LOTFI	354	3.13	16.4	190	2.91	22.6
BRANDY	356	3.86	33.7	238	3.63	49.2
E226	615	5.50	23.7	394	4.94	32.7
AGG	639	13.64	6.4	571	12.89	7.0
SCORPION	405	16.64	61.0	380	16.47	64.5
BANDM	666	20.42	21.8	467	19.39	30.0
SCTAP1	481	23.11	36.6	349	22.44	44.4
SCFXM1	587	26.98	35.1	429	26.03	45.5
AGG2	824	39.47	5.5	629	36.70	5.6
AGG3	837	52.14	5.5	627	49.17	6.1
SCSD1	186	52.31	81.7	80	52.23	97.5
SCAGR25	922	64.52	34.1	596	60.42	44.5
Total(25)	9220	331.93	39.9	6541	316.69	47.3
DEGEN2	10,000	177.83	48.7	1990	87.84	45.7

(a) A comparison between FALP and RSA
Table A.4 gives iteration and time ratios of RSA to FALP. The second bottom line shows that the overall iteration ratio with the 25 problems is **1.51** and the overall time ratio is **4.91**! The time ratio for SCSD1 is exceptionally high (**83.03**), so the bottom line lists the overall ratio with the rest of the 24 problems.

(b) A comparison between DFA and RSA
Table A.5 gives the iteration and time ratio of RSA to DFA. The second bottom line indicates that the overall iteration ratio with the 25 test problems is **1.19** compared with the overall time ratio as high as 10.04!

Table A.2 Statistics for FALP

Problem	Total Itn	Time	%Deg	Phase 1 Itn1	Time1	%Deg1	k-m
AFIRO	26	0.02	0.0	7	0.00	0.0	9
SC50B	30	0.02	0.0	1	0.00	0.0	20
SC50A	32	0.02	0.0	1	0.00	0.0	20
ADLITTLE	175	0.13	0.0	40	0.03	0.0	15
BLEND	49	0.08	0.0	11	0.03	0.0	44
SHARE2B	85	0.06	0.0	55	0.03	0.0	14
SC105	65	0.06	0.0	1	0.02	0.0	45
STOCFOR1	66	0.09	0.0	23	0.03	0.0	63
SCAGR7	134	0.23	0.0	91	0.13	0.0	84
ISRAEL	294	0.67	0.3	79	0.16	0.0	0
SHARE1B	243	0.34	0.0	140	0.16	0.0	89
SC205	139	0.53	0.0	1	0.08	0.0	91
BEACONFD	135	0.44	3.0	60	0.16	5.0	144
LOTFI	353	0.73	0.6	198	0.30	0.0	95
BRANDY	246	0.91	1.2	76	0.23	0.0	142
E226	486	1.81	0.2	192	0.53	0.0	37
AGG	160	4.45	0.0	94	1.81	0.0	49
SCORPION	119	2.36	3.4	23	0.61	17.4	249
BANDM	410	5.02	0.0	167	1.55	0.0	305
SCTAP1	449	3.55	2.9	112	1.02	0.0	120
SCFXM1	528	6.23	0.2	383	3.78	0.0	191
AGG2	282	7.97	0.0	67	2.14	0.0	76
AGG3	290	8.25	0.0	86	2.50	0.0	76
SCSD1	736	0.63	9.8	1	0.03	0.0	77
SCAGR25	556	23.06	0.0	361	13.47	0.0	300
Total(25)	6088	67.66	0.9	2270	28.80	0.9	94
DEGEN2	452	11.13	0.0	190	3.94	0.0	219

Considering that the time ratio is the only indication of efficiency, we conclude that the primal and dual face codes outperformed RSA unambiguously with the test set.

Such an outcome would be astonishing but not surprising:

On the one hand, FALP solves a single triangular $(m + 1) \times (m + 1)$-system in each iteration, compared with RSA solving four triangular $m \times m$-systems.

On the other hand, the last column of Table A.2 shows that all optimal faces achieved by FALP are not vertices except for ISREAL. The average dimensions of the optimal faces are about 94. The average dimension to average row ratio is up to

$$94/213 = 0.44.$$

Achieving an optimal face with a higher dimension is easier in general.

Table A.3 Statistics for code DFA

	Total			Phase 1			
Problem	Itn	Time	%Deg	Itn1	Time1	%Deg1	m-k
AFIRO	30	0.00	20.0	22	0.00	0.0	3
SC50B	50	0.00	0.0	48	0.00	0.0	2
SC50A	53	0.00	0.0	47	0.00	0.0	1
ADLITTLE	164	0.03	15.2	29	0.02	0.0	0
BLEND	104	0.03	2.9	75	0.02	0.0	0
SHARE2B	142	0.03	11.3	69	0.02	1.4	0
SC105	110	0.03	0.0	102	0.02	0.0	1
STOCFOR1	152	0.08	0.0	127	0.05	0.0	0
SCAGR7	254	0.23	11.8	110	0.06	10.9	0
ISRAEL	321	0.36	12.8	140	0.11	0.0	0
SHARE1B	245	0.13	4.5	119	0.06	0.0	0
SC205	209	0.27	0.0	202	0.20	0.0	2
BEACONFD	136	0.06	27.9	1	0.00	0.0	50
LOTFI	264	0.27	8.3	137	0.13	0.0	0
BRANDY	556	0.89	36.7	1	0.00	0.0	16
E226	381	0.78	7.3	156	0.23	0.0	9
AGG	542	3.59	9.0	122	0.63	0.8	11
SCORPION	362	1.44	22.1	1	0.00	0.0	21
BANDM	558	2.75	0.7	294	1.22	0.3	0
SCTAP1	366	0.75	79.0	1	0.02	0.0	33
SCFXM1	713	3.30	5.6	216	0.70	0.5	1
AGG2	571	4.97	5.3	161	1.25	0.0	0
AGG3	588	5.14	5.1	151	1.14	0.0	0
SCSD1	78	0.05	35.9	1	0.00	0.0	0
SCAGR25	822	7.89	16.8	360	2.75	13.3	8
Total(25)	7771	33.07	13.5	2692	8.63	1.1	6
DEGEN2	1627	18.09	6.0	434	3.53	0.2	4

As for comparing the two face codes, DFA is superior to FALP by a time ratio of **2.04**. It is partially because DFA solves a smaller triangular $k \times k$ system with $k \leq m$. The last column of Table A.3 indicates that $k < m$ for 14 of the 26 test problems (**54%**).

Table A.4 Ratio of RSA to FALP

				Total		Phase 1	
Problem	m	n	$m+n$	Itn	Time	Itn	Time
AFIRO	28	32	60	1.12	1.50	4.00	300.00
SC50B	51	48	99	1.90	3.00	50.00	600.00
SC50A	51	48	99	1.72	4.50	51.00	900.00
ADLITTLE	57	97	154	0.77	1.00	1.73	4.33
BLEND	75	83	158	2.35	2.13	8.18	5.67
SHARE2B	97	79	176	2.31	4.67	2.71	9.00
SC105	106	103	209	1.89	6.00	110.00	18.00
STOCFOR1	118	111	229	2.64	5.56	6.52	16.00
SCAGR7	130	140	270	1.35	2.91	1.60	4.92
ISRAEL	175	142	317	1.72	2.10	2.38	6.06
SHARE1B	118	225	343	1.27	4.91	1.36	9.88
SC205	206	203	409	1.83	4.21	211.00	27.00
BEACONFD	174	262	436	1.58	5.93	2.65	15.81
LOTFI	154	308	462	1.00	4.29	0.96	9.70
BRANDY	194	249	443	1.45	4.24	3.13	15.78
E226	224	282	506	1.27	3.04	2.05	9.32
AGG	489	163	652	3.99	3.07	6.07	7.12
SCORPION	358	358	716	3.40	7.05	16.52	27.00
BANDM	306	472	778	1.62	4.07	2.80	12.51
SCTAP1	301	480	781	1.07	6.51	3.12	22.00
SCFXM1	331	457	788	1.11	4.33	1.12	6.89
AGG2	517	302	819	2.92	4.95	9.39	17.15
AGG3	517	302	819	2.89	6.32	7.29	19.67
SCSD1	78	760	838	0.25	83.03	80.00	1741.00
SCAGR25	472	500	972	1.66	2.80	1.65	4.49
Average(25)	213	248	461	1.51	4.91	2.88	11.00
Average(24)	218	226	445	1.69	4.17	2.85	9.19

Table A.5 Ratio of RSA to DFA

Problem	m	n	$m+n$	Total		Phase 1	
				Itn	Time	Itn	Time
AFIRO	28	32	60	0.97	–	1.27	–
SC50B	51	48	99	1.14	–	1.04	–
SC50A	51	48	99	1.04	–	1.09	–
ADLITTLE	57	97	154	0.82	4.33	2.38	6.50
BLEND	75	83	158	1.11	5.67	1.20	8.50
SHARE2B	97	79	176	1.38	9.33	2.16	13.50
SC105	106	103	209	1.12	12.00	1.08	18.00
STOCFOR1	118	111	229	1.14	6.25	1.18	9.60
SCAGR7	130	140	270	0.71	2.91	1.33	10.67
ISRAEL	175	142	317	1.58	3.92	1.34	8.82
SHARE1B	118	225	343	1.26	12.85	1.60	26.33
SC205	206	203	409	1.22	8.26	1.04	10.80
BEACONFD	174	262	436	1.57	43.50	159.00	–
LOTFI	154	308	462	1.34	11.59	1.39	22.38
BRANDY	194	249	443	0.64	4.34	238.00	–
E226	224	282	506	1.61	7.05	2.53	21.48
AGG	489	163	652	1.18	3.80	4.68	20.46
SCORPION	358	358	716	1.12	11.56	380.00	–
BANDM	306	472	778	1.19	7.43	1.59	15.89
SCTAP1	301	480	781	1.31	30.81	349.00	1122.00
SCFXM1	331	457	788	0.82	8.18	1.99	37.19
AGG2	517	302	819	1.44	7.94	3.91	29.36
AGG3	517	302	819	1.42	10.14	4.15	43.13
SCSD1	78	760	838	2.38	1046.20	80.00	–
SCAGR25	472	500	972	1.12	8.18	1.66	21.97
Average(25)	213	248	461	1.19	10.04	2.43	36.70
Average(24)	218	226	445	1.17	8.47	2.40	30.64

References

1. Beale E.M.L, Cycling in the dual simplex algorithm, *Naval Research Logistics Quarterly*, **2** (1955), 269–275.
2. Beale EM, An alternative method for linear programming. Proc Camb Philos Soc **50** (1954),513–523
3. Bixby RE, Solving real-world linear problems: a decade and more of progress. Oper Res **150** (2002), 3–15
4. Bland RG, New finite pivoting rules for the simplex method. Math Oper Res **2** (1977), 103–107
5. Dantzig GB, Programming in a linear structure, Comptroller, USAF, Washington, DC, 1948
6. Dantzig GB, *Linear Programming and Extensions*, Princeton University Press, Princeton, New Jersey, 1963.
7. Dantzig GB, Orchard-Hays W, Alternate algorithm for the revised simplex method using product form for the inverse. Notes on linear programming: part V, RM-1268. The RAND Corporation, Santa Monica, 1953
8. Forrest JJH, Goldfarb D, Steepest edge simplex algorithm for linear programming. Math Program **57** (1992), 341–374
9. Gill PE, Murray W, A numeically stable form of the simplex algorithm. *Linear Algebra Appl.*, **7** (1973), 99–138
10. Golub GH, Numerical methods for solving linear least squares problems. *Numer. Math.*, **7** (1965), 206–216.
11. Guerrero-Garcia F, Santos-Palomo A, Phase I cycling under the most-obtuse-angle pivot rule. Eur J Oper Res **167** (2005), 0–27
12. Harris PMJ, Pivot selection methods of the Devex LP code. Math Program **5** (1973), 1–28
13. Hoffman AJ, Cycling in the simplex algorithm. Technical report 2974, National Bureau of Standards, 1953
14. Karmarkar N, A new polynomial time algorithm for linear programming. Combinatorica **4** (1984), 373–395
15. Klee V, Minty GJ, How good is the simplex algorithm? In: Shisha O (ed) Inequalities-III. Academic, New York, 159–175, 1972
16. Koberstein A and Suhl UH, Progress in the dual simplex method for large scale LP problems: Practical dual-phase 1 algorithms, Computational Optimization and Applications **37** No.1 (2007), 49–65
17. Lemke CE, The dual method of solving the linear programming problem. Nav Res Logist **1** (1954), 36–47
18. Murtagh BA, Saunders MA, MINOS 5.5 user's guid. Technical report SOL 83-20R, Department of Engineering Economics Systems & Operations Research, Stanford University, Stanford, 1998

P.-Q. Pan, *Face Method*, https://doi.org/10.1007/978-3-031-93594-7

19. Orchard-Hays W, Advanced linear programming computing techniques. McGraw-Hill, New York, 1971
20. Pan PQ, Practical finite pivoting rules for the simplex method. OR Spektrum **12** (1990), 219–225
21. Pan PQ, Simplex-like method with bisection for linear programming, Optimization, **2** (1991) No.2, 717–743
22. Pan PQ, A variant of the dual pivot rule in linear programming. J Inf Optim Sci **15** (1994), 405–413
23. Pan PQ, The most-obtuse-angle row pivot rule for achieving dual feasibility in linear programming: a computational study. Eur J Oper Res **101** (1997), 164–176
24. Pan PQ, A basis-deficiency-allowing variation of the simplex method, Comput Math Appl, **36** (1998) No. 3, 33–53
25. Pan PQ, A new perturbation simplex algorithm for linear programming. J Comput Math **17** (1999a), 233–242
26. Pan PQ, Primal perturbation simplex algorithms for linear programming. J Comput Math **18** (2000b), 587–596
27. Pan PQ, A dual projective pivot algorithm for linear programming. Comput Optim Appl **29** (2004), 333–344
28. Pan PQ, A revised dual projective pivot algorithm for linear programming. SIAM J Optim **16** (2005), 49–68
29. Pan PQ, A largest-distance pivot rule for the simplex algorithm. Eur J Oper Res **187** (2008a), 393–402
30. Pan PQ, Efficient nested pricing in the simplex algorithm. Oper Res Lett **38** (2008b), 309–313
31. Pan PQ, A primal deficient-basis algorithm for linear programming. Appl Math Comput **198** (2008c), 898–912
32. Pan PQ, A fast simplex algorithm for linear programming. J Comput Math **28** (2010) No.6, 837–847
33. Pan PQ, Linear Programming Computation (in Chinese), Science Press, Beijing, 2012
34. Pan PQ, *Linear Programming Computation*, Springer, Berlin Heidelberg, Germany, 2014
35. Pan PQ, On "On an efficient implementation of the face algorithm for linear programming ", *Operations Research Transactions*, **19** (2015) No.3, 78–84
36. Pan PQ, *Linear Programming Computation*, Second Edition, Springer Nature Singapore, 2023
37. Saunders MA, Large Scale Linear Programming Using the Cholesky Factorization. Technical report STAN-CS-72-152, Stanford University, 1972
38. Shi YY, Zhang LH.and Zhu WX, A review of Linear Programming Computation by Ping-Qi Pan, *European Journal of Operational Research*, **267** (2018) No. 3, 1182–1183
39. Zhang LH, Yang WH and Liao LZ, On an efficient implementation of the face algorithm for linear programming, *Journal of Computational Mathematics*, **31** (2013) No.4, 335–354

www.ingramcontent.com/pod-product-compliance
Lightning Source LLC
Chambersburg PA
CBHW081814090925
32282CB00002B/39
9783031935930